Graphene-Based Electrochemical Sensors for Toxic Chemicals

Edited by

Alagarsamy Pandikumar[1], Perumal Rameshkumar[2]

[1](Leading Editor), Scientist, Materials Electrochemistry Division, CSIR-Central Electrochemical Research Institute, Karaikudi-630003, Tamil Nadu, India

[2] (Associate Editor), Assistant Professor, Department of Chemistry, Kalasalingam Academy of Research and Education, Krishnankoil, 626 126, Tamil Nadu, India

Published by **Materials Research Forum LLC**
Millersville, PA 17551, USA

Published as part of the book series
Materials Research Foundations
Volume 82 (2020)
ISSN 2471-8890 (Print)
ISSN 2471-8904 (Online)

Print ISBN 978-1-64490-094-9
eBook ISBN 978-1-64490-095-6

Distributed worldwide by

Materials Research Forum LLC
105 Springdale Lane
Millersville, PA 17551
USA
https://www.mrforum.com

Manufactured in the United States of America
10 9 8 7 6 5 4 3 2 1

Table of Contents

Preface

This edited book discusses the importance of detecting toxic molecules and the application of graphene and its nanocomposite materials in the detection of a wide variety of toxic chemicals including heavy metals, inorganic anions, phenolic compounds, pesticides, and chemical warfare reagents. In addition to the instrumental simplicity, the electrochemical methods show the improved sensor performance through the synergetic effect of graphene and other electroactive nanomaterial present in the nanocomposite. Hence, gathering detailed information on the electrochemical sensing of toxic molecules using graphene-based nanomaterials is highly essential in order to realize the potential importance of these materials and the electroanalytical techniques to the maximum. The book is divided into several chapters and the chapters are dedicated to various aspects such as the toxicity of chemicals, importance of electrochemical sensors, the development of different types of graphene-based nanomaterials and the electroanalytical techniques for detection. The recent progress in developing electrochemical detection of toxic molecules with the aid of graphene-based nanocomposite materials and the perspectives on future opportunities in sensor research are also discussed.

This book is for early carrier researchers and engineers who are working in both academic institutions and industries. In addition, the book will be helpful for researchers to establish their own research in the area of nanomaterials-based electrochemical sensors. It can also serve as a reference book for bachelor and master level students pursuing courses in nanoscience and nanotechnology and interdisciplinary subjects. The audience will easily understand the principle of electrochemical sensing systems and role of graphene-based nanocomposite materials in the current scenario of electrochemical sensor research and development.

We are very much thankful to all the authors who contributed the chapters to make a valuable book and for the successful completion of the process. The involvement of several authors from different institutions in diverse geographical locations are highly appreciated. We are thankful to the editor, Thomas Wohlbier, Materials Research Forum LLC for accepting our proposal and giving a chance to edit this book.

Dr. Alagarsamy Pandikumar (Leading Editor)
Scientist, Materials Electrochemistry Division, CSIR-Central Electrochemical Research Institute, Karaikudi-630003, Tamil Nadu, India

Dr. Perumal Rameshkumar (Associate Editor)
Assistant Professor, Department of Chemistry, Kalasalingam Academy of Research and Education, Krishnankoil, 626 126, Tamil Nadu, India

Materials Research Forum LLC
https://doi.org/10.21741/9781644900956-1

Chapter 1

Graphene Modified Electrochemical Sensors for Toxic Chemicals

T. Ramya[1,a], L. Vidhya[2,b], S. Vinodha[3,c], D. Anuradha[4,d], S. Sivanesan*[,5,e]

[1]Department of Environmental Sciences, Bharathiar University, Coimbatore, Tamil Nadu India

[2]Department of Chemical Engineering, Sethu Institute of Technology, Pulloor, Kariapatty-626115, India

[3]Department of Chemical Engineering, Sethu Institute of Technology

Pulloor, Kariapatty- 626115, India

[4]Department of Biotechnology, Anna University, Chennai, Tamil Nadu, India

[5]Department of Applied Sciences, Anna University, Chennai, Tamil Nadu, India

[a]rtramya1@gmail.com, [b]vidhuram236@gmail.com, [c]vinodha.harris@gmail.com, [d]anushivan@gmail.com, [e]sivanesh1963@gmail.com

Abstract

Electrochemical sensing is a broad analytical field related to the generation of useful analytical information through electroactive species with a conductive surface or a functional group. Electrochemical sensors have several advantages such as cost effectiveness, high sensitivity and selectivity, applicability to a vast range of chemicals, ease of handling, functionalization and rich nature. Two key units are present in a chemical sensor, one is a receptor and another one is a transducer. An electrochemical sensor generally consists of three electrodes, namely working electrode, reference electrode and counter electrode which are associated with a potentiostat /galvanostat. The working electrode acts as receptor and is a component of the transducer as well. Electrochemical sensing is a predicting analytical area and it is known for its importance in various fields such as energy, pharmaceutical industry, environment and food. Chemical sensor leaves the analytical information about a particular quantity of certain chemical species in the surrounding environment. In this chapter we mainly focus on graphene modified electrochemical sensors for toxic chemical sensing. Also, we have focused on the graphene modified electrochemical sensors for the electrochemical detection of toxic chemicals. Graphene modified electrochemical sensors are selected based on their superior nature as compared with other sensors and also these sensors support sensing of very toxic chemical pollutants.

Keywords

Nitroaromatics, Cadmium, Bisphenol A, Organoposphate, Nanomaterials

Contents

1.   Introduction

Graphene is a one-atom thick individual planar carbon layer and it possesses characteristics of unique and exclusive nature. This resulted in a number of scientific applications that make use of graphene in it. Though the graphene was first discovered in the 1940 theoretically and known to exist since 1960s, a large number of scientists started shifting their research interest towards graphene only after the research conducted by Geim and Novoselov in 2004-5. They leveraged the 'scotch tape method' in graphene production and also uncovered its unique electronic properties. These investigations were crucial in exposing the advantages of graphene which makes it superior in superfluity of areas. As a result in these investigations, various researchers proposed novel methodologies to produce high quality graphene on industrial scale so as to ensure the global need for graphene is met. Electrochemistry is one of the areas which gained much attention for its wide usage of graphene. This research arena makes use of graphene in numerous applications that range from sensors to energy storage and carbon-based molecular electronics. Having been identified as an interfacial technique, electrochemistry is loaded with processes that happen in a solid-liquid surface. This makes it mandatory to understand the surface chemistry in a detailed manner for further applications.

The purest form of carbon, graphene, is very thin in nature and one-atom thick, which is nearly transparent. Though it has low weight, it is remarkable stronger i.e., 100 times stronger than steel and it has the capability to conduct heat and electricity efficiently. The future techniques in electronics, material sciences and nanotechnology are focused on leveraging the myriad and complex physical properties of graphene in combination with bonded carbon atom structures. Graphene can be best utilized as a sensor device, thanks to its minute diameter and specialized structure. Further, it can also act as a semiconductor and integrated circuit components. The characteristics and its unique applications of this 2-D carbon structure paved the way for wide range of opportunities to create the future technology today.

In electrochemical sensor, the electrode strip is provided to measure a compound in a sample. This sensor is also inclusive of electrode support, a reference or counter electrode which is disposed on the support, a working electrode which is kept at a specific distance from the reference or counter electrode on the support, a covering layer which can be deduced as an enclosed space over the reference and working electrodes. This is further inclusive of an aperture to receive a sample into the enclosed space along with a plurality of mesh layers interposed in the enclosed space between the covering layer and the support. There is also a covering layer present with a sample application aperture, which is distanced from the said electrodes whereas the said reference electrode is distanced from the said working electrode at a position which is remote from and on the opposite side of said working electrode from said aperture. There is also an enzyme included in the working electrode which can catalyze a reaction that involves a substrate for the enzyme or a substrate which is catalytically reactive with an enzyme in the presence of a mediator. In this reaction, the latter has the capability to transfer the electrons which are transferred between enzyme-catalyzed reaction and the working electrode to produce a current representative of the enzyme activity and representative of the compound.

2. Overview on graphene

In the recent years, global researchers started focusing on characterization and development of new production methodologies in graphene. The current review of literature focuses on graphene-modified electrodes and its important applications. According to the calculation and experimental analysis, the 2D electrodes cannot be turned as a 3D electrode. The reason behind the thermodynamic stability of structure and size of graphene exist in the number of carbon atoms in it i.e., structure has a total of

6,000 carbon atoms whereas size has 24,000 carbon atoms. Andre Geim and Konstantin Novoselov (Nobel Prize winners in Physics, 2010) prepared the top of a silicon wafer graphene sheet following scotch tape method. Graphene is generally utilized in various industrial applications, thanks to its mechanical, thermal, optical and electronic properties. For instance, the optical transparency is high in graphene due to which it is frequently used in optical electronic materials.

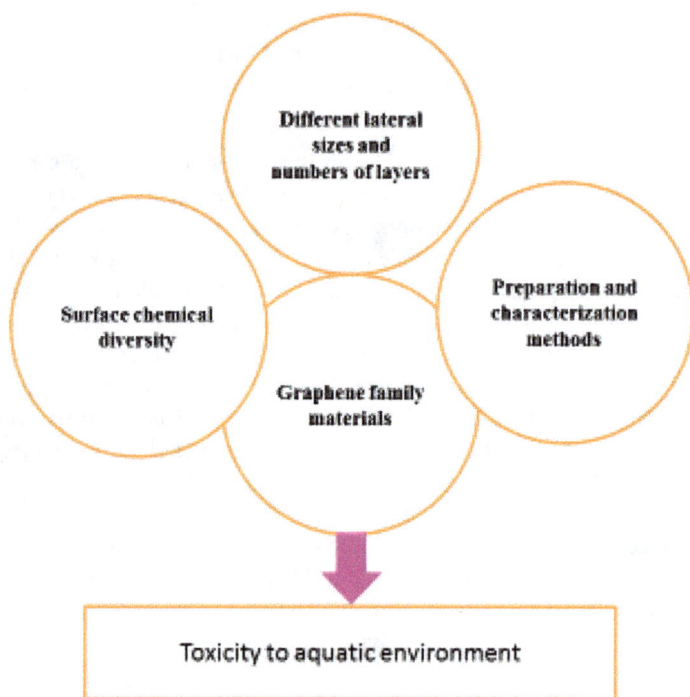

Figure 1 Graphene family materials.

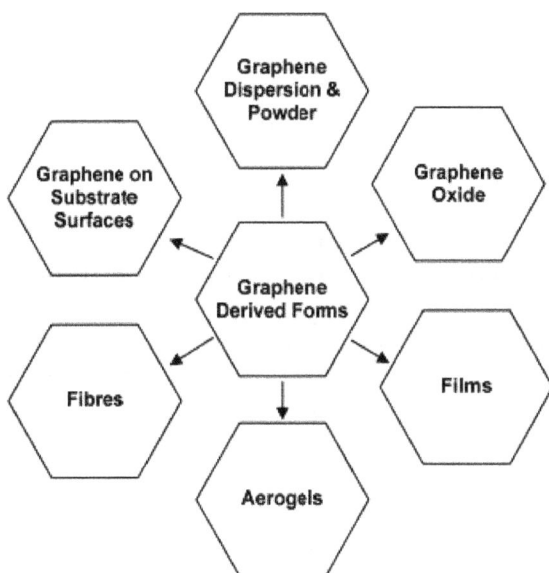

Figure2 Graphene and its derived forms.

3. Application

Graphene is found to be an optimum candidate for trace electrochemical identification of nitroaromatics. A group of researchers indeed conducted an investigation on a wide range of graphene-based materials, for instance, nanosheets hydrogenated graphene, graphenenanoribbons and graphenes manufactured based on hummers, staudenmaier and hofmann syntheses methods. This research is aimed at isolating the most optimum probe material in sensing the nitroaromatics explosives in seawater. Various researchers conducted different studies which altogether inferred the high selectivity and sensitivity of functionalized graphene such as graphene-modified with Ag nanoparticles, Pt-Pd concave nanocubes, ionic liquid, as well as other carbon-based electrochemical sensors, for instance, carbon quantum dots for detecting high-efficient nitro aromatics.

For the mass production of graphene, a novel and highly efficient method was recently proposed i.e., electrochemical exfoliation of graphite under anodic or cathodic conditions. This method allows the electrolytes to be intercalated between graphene layers which eventually leads to exfoliation. It is an established fact that the graphene possesses excellent physicochemical properties, yet this can be impacted by exfoliation conditions such as applied potential and the type of electrolyte used in the procedure.

Currently, there are a number of nanomaterials developed by the researchers in the modification of electrodes that aim to enhance the detection performance of nitrate and nitrite, metals, metal oxide, conductive polymers and carbon nanomaterials. Among these, the mostly widely investigated are carbon nanomaterials. Being a 2-D carbon nanomaterial, graphene consists of carbon atoms in its sp2 hybrid orbitals. Having been developed as a honeycomb hexagonal lattice network, it has the capability to extend indefinitely. This is prevalently utilized to alter the electrodes of electrochemical sensors, thanks to its unique properties such as higher conductivity, stronger mechanical strength, higher electrochemical activity and larger specific surface area. Based on the various molecular structures of graphene, it can be sub-divided into reduced graphene oxide, graphene oxide and graphene (as shown in the figure 1). Further, it can also be sub-divided based on shape, structure and performance too. Due to its limited performance, graphene mostly produces composite nanomaterials with that of the polymers, metals, metal oxides and so forth so as to improve its capability of electrocatalysis.

In food industry, Sulphite (SO_3^{2-}) is mostly utilized as a reducing agent, preservative, bleaching agent and antioxidant. It is also utilized as a food additive in different countries. The level of Sulphite is a critical safety index since if the food has too much Sulphite, it results in dizziness, asthma, headaches, nausea and other allergic reactions. Industrially, nitrite (NO_2^-) is the agent used for coloring and also as a preservative. Nitrite, in *in vivo* conditions, can perform irreversible oxidation of haemoglobin to methaemoglobin resulting in the lack of oxygen-carrying ability. Further, the nitrite has the ability to enter reaction with amines to produce carcinogenic nitrosamine, a potential threat to human beings. Incidentally, SO_3^{2-} and NO_2^- widely coexist in nature that cause environmental pollution and create harm to human beings. It is highly important to determine the levels of SO_3^{2-} and NO_2^-, when it comes to water quality evaluation and food analysis. So, there exists a need for standard methods to determine sulphite and nitrite levels. There are various methods currently deployed to determine the SO_3^{2-} and NO_2^- inclusive of electrochemical method which is quick, simple and incurs less cost. So, this method gained attraction among the researchers.

But the hyper potential of SO_3^{2-} and NO_2^- direct electrooxidation at traditional electrodes is very high. Additionally, the oxidation potentials of two species of bare electrode are close proximity to each other and it eventually led to overlapping in voltammetric responses with rather poor selectivity and reproducibility. So the important research objective, that remains still, is to develop fresh electrode materials for both sensitive as well as selective parallel determination of two species.

A number of pesticides are toxic in nature and their impacts create a lasting impression upon human health. The main reason behind the organophosphate (OP) toxicity is its inhibition of acetyl cholinesterase (AChE) enzyme. Different significant researches have been conducted in electrochemical sensors for OP detection and these investigations proved to be cost-effective, simple and versatile in nature. There is a wide range of important applications found in the areas of agriculture, environment, industry and clinical technology. It is possible to enhance the sensitivity, reproducibility and specific selectivity towards target OP molecules through the application of different coating materials at electrode surface.

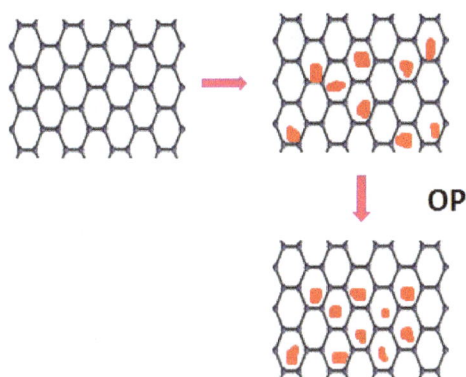

Figure 3 A graphene-ZrO2 nanocomposite: application for detection of organophosphorus agents.

4.   Fundamental electrochemistry of graphene

In literature, it is a common practice to block a graphene upon an electrode surface so as to electrically wire the graphene and conducted an investigation of its electrochemical activity, while the former results in the formation of heterogeneous electrode surface. At the time of utilizing graphene as an electrode surface, the heterogeneous electrode surface is formed [1]. In the research to enhance the fundamental electrode surface which has its own edge plane, basal plane sites with electrochemical activity by itself, of different Butler–Volmer terms, ko and a (see above), when graphene is introduced, this makes the investigation interesting since the latter has its own edge, basal plane sites with own ko and a values.

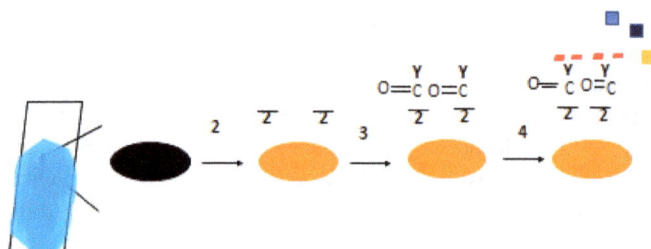

- Step 1: Drop –casting of graphene on carbon screen printed electode
- Step 2: Electro-catalyzed amine (-NH$_2$) functionalization of graphene with 1 aminobenzyamine
- Step 3: Immobilization of anti-parathion antibodies on –NH$_2$ functionalized graphene SPE
- Step 4: Immunosensing of parathion with above sensor

Figure 4 Schematic of the graphene-based, screen-printed immunosensor for parathion.

The metallic electrode modification can deceive at some times, for instance, a combination of gold macro electrode taken in addition to three prominent electrochemical sites, underlying gold and modified graphene along with edge and basal plane sites. In general, the electrochemical activity of gold is comparatively and prevails over graphene. However, if no genuine attempts are made to understand such reactions, the excellent electrochemical activity might get hidden by graphene's name or else it may be misunderstood that graphene exhibited excellent electrochemical activity

When a graphene sheet is kept between the electrode surface along with a duplicate replacing the erect, it can be inferred that the graphene made modifications to the electrode surface as shown in the figure 1. There are edge plane sites, basal plane sites present near in the dispersal area in graphene. This has a close relationship with nano edge bands of thickness which has an approximate length of carbon-carbon bond i.e., B0.142 nm in graphene. This can be taken as a constant one though there are various irregularities found in graphene classes with changes in their field radius. The changes might be due to the increased or decreased size of the graphene flake and upon the fabrication methodology.

(A)

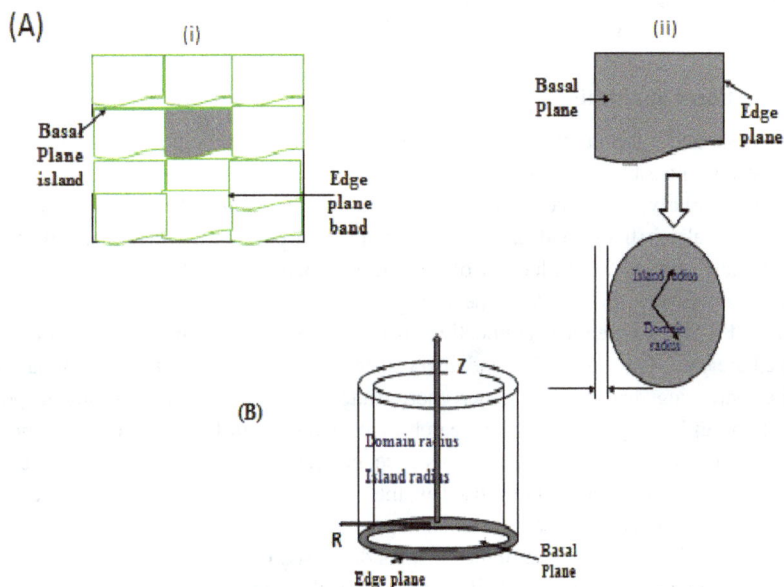

Figure 5 Schematic diagrams showing: (a), (i) the overhead view of a section of the basal plane HOPG surface and (ii) the approximation of each island/band combination as a partially covered circular disc of the same area; (b) the resulting diffusion domain from the approximation in (a) (ii) and the cylindrical coordinate system employed. Reproduced from ref. [45] with permission of The Royal Society of Chemistry. Note that the island radius is termed Rb and the domain radius is R0.

In the example cited earlier, the edge plane performs rapid electron transmission whereas the primary plan has an approximate trifling activity (B109 cms1). Since the voltammetry hasn't been dispatched, it supported the assumption that edge plane side of the graphene is an electroactive site much alike the edge plane nano band. Being a repetitive scenario, this is further investigation in the later stages of research with experimental background.

So, when it comes to modification of HOPG surface using graphene, mostly the basal plane HOPG surface (BPPG) and the ko basal can be avoided so that in alignment with figure 2, only two primary domains such as ko edge (HOPG) and ko edge (graphene) remain. If there is an assumption that the above share similar electrochemical properties to that of the DOS, it can be inferred that the edge plane sites act as the dominant factor in graphene-modified electrode.

5. Pertinent applications

Knowledge and information acquisition about different scientific principles and theories have become possible only due to the fundamental electrochemical studies conducted so far [2]. Similarly, it is the electrochemical applications in industries, medicine and academia make it lively and more important [3]. To be specific, electrochemistry has been widely applied in production of sensors that helps the global workface to detect harmful substances that threaten the very existence of environment and human being. Further, the electrochemical application in energy production and energy storage has opened the gates for wider usage and also gained much importance in the recent days due to the increasing demand of greener energy alternatives without compromising power output and efficiency [4]. Indeed, research investigations must be carried out on how to utilize various electrode materials, for instance, graphene. As discussed above, the carbon graphite forms have been widely used in the production of one-time usable materials, thanks to its cheap cost and easy production processes. Further these are also non-toxic in nature and highly conducive attributes [5]. As discussed in section 2.3 graphene too is in much demand, thanks to its excellent electrochemical properties and unique assortment of physiochemical properties. The potential of the graphene can be unraveled when it is used for sensing and energy-related electrochemical applications.

The following section is a precise overview about the electrode design and also provides an all-inclusive insight and acuity about the development of the electrode utilizing graphene. It further concludes whether such addition is either beneficial or not by making use of few selected electrochemical devices.

6. Sensing applications

Various studies have described how graphene can be executed as a perfect heightened sensor substrate using different layouts of analytes that let the legion bio-molecules, gases and mixed organic and inorganic materials. Few studies interpret helpful insights about the electrochemical detection pertinency of graphene [6,7,8].

7. Direct electro-catalysis

There are earlier mentions about the electrocatalytic behavior of graphene and the associated structures. For instance, identification of dopamine, cadmium, bisphenol A, hydrazine, morphine and vanillin. These pollutants are contrasted only with the specific electrode, not with any other graphite materials [9]. This is touted as the most significant effect, provided most of the analyses conducted so far utilizing graphene, contained multilayers (4-6+ layers) [10]. This is more or less nearby the structural piece of the graphite theoretically. In addition to the above, when the graphene is changed through liquid distribution, this kindles doubt about the graphene's characteristics, reporting and taste on the electrode surface with potential institution of layered/ disconnected graphene structures. This is similar to the findings earlier i.e., graphitic anatomical structure since it can notably be added with the observed electrochemical response and might also be the original beginning point of the observed electro-catalysis instead of graphene.

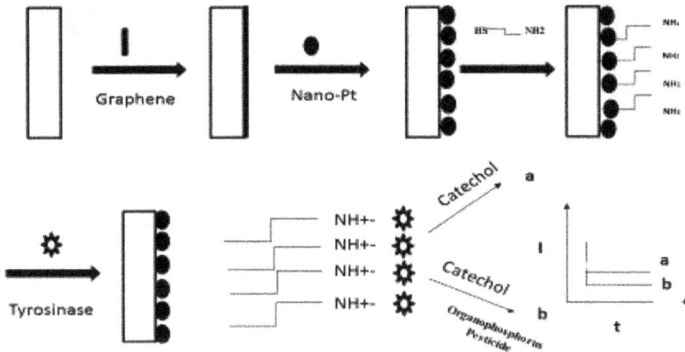

Figure 7: Graphene and tyrosinase immobilized on platinum nanoparticles for sensing organophosphorus pesticides.

While reducing the GO, the graphene is produced as a modified GC electrode which exhibits exemplary electro-catalytic property in sensing kojic acid. To be specific, it showcases an adulterated overpower and magnanimous peak current in comparison with the graphite-modified option and the unadapted GC electrode. These kinds of data are collected only to permit a highly-amended detection capacity at the graphene fictional electrode.

Though there is a description about graphite assures present, the authors mistakenly proceeded with reserve control experiments using GO. So, in this case, the authors did not accepted the role played by oxygenated species that may exist on the reduced GO. This would've been added to the detected electro-catalysis in a significant manner. In this specific instance, the blocked graphene that was synthesized by reducing GO, underwent electro-catalysis and was exacted only when it was equated to the fundamental GC electrode reaction and advances in the discovered voltammetry such as simplification in DEP (lower over-potential oxidation) that points a transmission from permanent to quasi-two-sided redox process along with a drastically-increased peak current [11].

At present, there is no assured investigations available to exact the electro-catalysis given the fact that it would've occurred due to the incomplete reduction of GO or via graphitic impurities that got formed due to production process issues. This can also add to a highly-disordered and porous graphene structure, but it is often observed as an impact of thin-layer effect and the large background current that is projected by the graphene-modified electrode. Further, the reason behind this would be the surface area which is vast, easily accessible and porous in nature. Thus the example cited above promotes the effects which are involved in the control experiment when assessing graphene electrochemistry. The GO is simplified during which the part-functionalized graphene sheets or chemically-reduced GO are produced with the resultant being graphene, a widely applied material in electrochemistry. There exists various structural shortcomings such as edge plane like sites/defects in the graphene synthesized by the above-said method [12,13]. The working radicals also present in this which is proved to significantly determine the electrochemical functions of the graphene that was production in heterogeneous electron carry-over places. This can either benefit or create harm when attempts are made to detect direct analyte [14,15,16,17]. In the presence of numerous investigations which deduce the graphene electro-catalysis with mended predisposition and sensing, it is an established fact that the graphene exhibits excellent electrochemical functioning while the equivalent experiments need to be conducted to counteract the ill-reporting of the good electrocatalysis of graphene [18,19,20,21].

When using a straight graphene-modified electrode, the utilization of pure graphene as a possible electrode material is to be mentioned for the electroanalytical detection of various elements such as p-benzoquinone, DA, UA (uric acid) and APAP [22]. For the purpose of understanding the holistic electroanalytical pertinence of graphene and to assess its wide application in electrochemistry, the study modified the HOPG electrode substances. The latter was proved to have been degraded or exhibited dumb electron carry-over kinetics while the study made use of commercially qualified graphene with no chemical coverage, high amount of low oxygen capacity (o5%) and complementary to the surface-active agent [23].

It is interesting to observe that the graphene-modified electrodes exhibited inhibition in analytics when performed under conditions such as sensibility, one-dimensionality and various catching demarcation lines towards every analyte, considered equivalent to the unadapted fundamental electrode substrates (constructed from graphite) [24]. As cited in the figure 3C and D for DA, there is an increase found in the graphene-modified electrodes when this was equated with graphite options, specifically the lower peak elevations (currents) were found to be occupied [25].

C: Cyclic voltammetric profiles recorded for 50 mM DA in pH 7 phosphate buffer solution (PBS) using unmodified EPPG (solid line) and BPPG (dot-dashed line) electrodes, and 20 ng graphene modified EPPG (dashed line) and BPPG (dotted line) electrodes.

D: Calibration plots towards the detection of DA depicting the peak height as a function of concentration, obtained via cyclic voltammetric measurements performed using unmodified EPPG (squares) and BPPG (circles) electrodes in addition to EPPG (inverted triangles) and BPPG (triangles) electrodes following modification with 20 ng graphene. C and D were obtained at a scan rate of 100 mV s1 (vs. SCE) – reproduced from ref. 79 with permission of The Royal Society of Chemistry.

There seems to be a conflict between 'cut down carrying out' with that of the previous studies since most of the studies infer that when graphene is covered, it significantly enhances the electroanalytical reaction [26,27]. But, to note here, such kind of poor carrying out was apparent in both EPPG (fast electrode kinetics) and BPPG (slow electrode kinetics) graphene modified electrodes. In line with the key possibilities, it is expected to incorporate the pristine graphene-modified graphitic electrodes in the study [28-55]. Graphene, as documented earlier, has a mellow electron compactness when it is concentrated while the unparalleled geometry of the graphene exhibits edge flat like area. It also exhibits reduced number of heterogeneous electron carry-over kinetics, specifically near graphite [56]. So, the reactions are made to occur only based on the proportionate number of usable edge plate sites i.e., electroactive sites in electrode surface. The graphene is ensured to have appropriate number of edges and if excess, those are cut down, or substituted with comparatively cleaner basal plane sites. This effect is otherwise termed as 'blocked up electrode surface and it further stops the electron transfer kinetics.

This results in the exchange of overall changeability / irreversibility of the electrode kinetics relevant to analytical signal, IP. According to the researcher, the significant modifications in the analytic denote the underlying absence of aerated form of clean graphene which is appropriately showcased in HOPG and also beneficial in this case [57]. Further, the literature cites that if future studies are conducted about lotion of graphene-modified graphite electrodes to catch heavy metals, it may shed insights [58]. The graphene tend to inhibit the electroanalytical identification of cadmium (II) ions through anodic cleaning voltammetry. This occurs through the increase in graphene blockage on the electrode surface that results in analytical operation and duplicability when unadapted (graphite) electrode is counterpointed [59]. It is an established fact that the metals undergo nucleation alone at the edge planar sites of graphitic materials due to which one can understand the graphene's nature of small edge-to-radical surface area alike the fundamental graphic electrode. Further, the expected voltammetric response, in the event of bringing graphene, can be understood as the low edge plane capacity of the latter that obviously results in slow electron carry-over kinetics [60]. In addition to the above, previous studies had the same basal structure as their methodology to detect hydrogen peroxide at wetter sorbed/polluted graphene altered electrodes.

The authors conducted controlled experiments using different wetter and graphite-modified electrodes. The study concluded that the graphite-modified electrode exhibited an extraordinary electrochemical reaction, thanks to the higher share of edge plane places in line with graphene. The sensibility values were 32.8 and 49.0 mA mM1 for graphene and graphite-modified electrodes. It is interesting to note that at the time of introducing Nafiont, which is generally used in the allotment of biosensors, into graphene and graphite-modified electrodes, the direction got changed in both the cases. This concluded to be good during earlier stages while it caused damages later, thanks to the accessibility of every material at different edge plane places [61].

Pumera et al. conducted a study which inferred that ace, elite and multi-layered graphene possess no significant advantage alike graphite microparticles when it comes to one-dimensionality, predisposition and electro-analytical sensing of UA. The graphene-modified and graphite-modified electrodes, respectively achieved the predispositions of 4.65 and 5.11 mM1 respectively [62]. The determined values were covered through electrochemical sensing of 2,4,6-trinitrotoluene (TNT) in which a graphite-micro particle altered electrode was used to increase the electro-analytical sensitivity upon single-, few-, and multi-layer grapheneoptions [63]. It was found that the morphology of the graphene makes the pristine graphene inapplicable for electro-analytics. This can be otherwise described as when analytes are present here and there, the rapid electrode kinetics occur and increase the good analytic public presentation. Among the selected ones, the EPPG electrode seems to be systematically displaying the best result due to its favorable preference for the graphite sycamores, resulting in optimal edge plane reporting

[64,65,66,67]. These studies cited above were conducted to check the modification of electrodes with graphene and its impact on electro analysis of direct analytes in the origin. One should be aware that the graphene-based nanomaterials may not agree for all the reactions when the former is synthesized with unlikely methods and with fluctuations in a single method [68].

So, there may be possibilities of misunderstanding whenever characterizing a graphene nanomaterial. Further, there may an impact on the electrochemical reactions when the graphene undergoes ineffective handling and cutting during particular coatings. So, if one is able to forecast the edge plane shortcomings due to the check/control of the graphene preferences, one can also expect the possible amendment in the electroanalytical process [69]. Brownson et al. recently conducted a study in which they compared electroanalytical performance of a CVD-based graphene electrode commercially procured with that of the EPPG and BPPG electrodes that were synthesized from HOPG. This performance had a threshold of sensing the biologically-important analytes, for instance NADH and UA. The study determined the analytical sensitivities with the help of cyclic voltammetry for NADH when using CVD graphene, EPPG and BPPG electrodes. The respective values were 0.26, 0.22 and 0.15 A cm^2 Ml respectively [70]. The study also observed similar responses for EPPG and CVD graphene electrodes, thanks to the similar electronic structures of the electrode surfaces. Further the AFM was used to confirm similar kind of coverages for CVD graphene as well as EPPG. This could've occurred due to the surface of CVD graphene which is highly disordered naturally with parallel and perpendicular-oriented graphene while in some cases graphitic islands. Figure 4 illustrates an AFM image of the CVD graphene surface, also termed to as highly-disordered, polycrystalline and full of defects as it has been developed on nickel substrates. The performance of CVD graphene was higher as expected than the BPPG electrode, thanks for the former's high degree of edge plane sites.

In case of UA, the CVD graphene, EPPG and BPPG electrode registered the analytical sensitivities to be 0.48, 0.61 and 0.33 A cm^2 Ml respectively. The CVD graphene electrode outperformed BPPG electrode. However, it still remained second, only next to EPPG electrode, on the basis of sensitivity of respective surface of UA in the presence of oxygenated species. This further denotes the variants of O/C ratios at both the electrodes [71,72,73]. From the studies discussed so far, the CVD graphene electrode almost performed similar to EPPG in terms of electroanalytical performance. However, there is no value-added advantage to use CVD graphene in analytical perspective. This research work also emphasizes the advantage of edge plane sites (and respective orientation of graphene sheets) as well as oxygenated species (present or absent) when it comes to electroanalysis [74]. This is because these two factors control the progression of analytical sensors. For instance, if the graphene needs to be efficiently made use of and follow various aspects such as control in size, shape and orientation of the graphene sheets with atomic precision in the supporting substrate (edge plane favorable stacking, or the production of nano-ribbons with high edge plane to basal plane surface area ratios) production of nano-ribbons with high edge plane to basal plane surface area ratios), degree of defect density across the basal plane of the graphene sheets (where more defects are beneficial for electrochemical electron transfer mechanisms), the degree of functionalization (where specific oxygenated species are either helpful or damage-creating) and control over the vast array of unique physicochemical properties which must be made use of, in whole (beneficial surface area, optical transparency, flexibility and the possible doping of pristine graphene to name a few).

It has been established that when true graphene is made use of, in the context discussed above (during beneficial rapid electron transfer), the best response is achieved alike the one attained in EPPG electrode governed by Randles–Sevcik equations. In case of any alterations in this result, it would've been due to thin-layer behavior or mass transport characteristic changes or at times, few other contributing factors too, instead of the wrong assumption of graphene to be electro-catalytic.

Conclusions

The current review summarizes the most important developments in graphene electrochemistry. It first introduces the research area and shed insights with specific examples with a short summary of applications. Graphene has a prism of outstanding characteristics thanks to its low defects' rate across its surface. The nano-engineered graphene-based devices need to be introduced without structural defects or impurities so that the electronic structures as well as the desired functionalities can be achieved through electrochemical activity. In addition to the above, this modification should also be extended to graphene oxide in which the carbon-oxygen groups benefit or damage the electrochemical response. Various novel methodologies are devised to synthesize graphene in terms of electrochemistry. So the current research article aims to help the researchers to appropriately characterize their material and scrutinize it in line with other graphene. Actions must be taken to conduct appropriately-controlled experiments using graphite and graphene electrodes. This research arena is attractive and provides scope for future researchers since the graphene can be produced in different ways and can be widely applied.

Reference

[1] D. A. C. Brownson, E. Banks, The electrochemistry of CVD graphene: progress and prospects, Phys. Chem. Chem. Phys. 14 (2012) 8264-8281. https://doi.org/10.1039/c2cp40225d

[2] R. G. Compton, C. E. Banks, Understanding voltammetry, world scientific, Singapore, 2007. https://doi.org/10.1142/6430

[3] E. S. Reich, Nobel document triggers debate. Nature, 468 (2010) 486-486. https://doi.org/10.1038/468486a

[4] D. A. C. Brownson, D. K. Kampouris, C. E. Banks, An overview of graphene in energy production and storage applications, J. Power Sources, 196 (2011) 4873-4885. https://doi.org/10.1016/j.jpowsour.2011.02.022

[5] K. S. Novoselov, D. Jiang, F. Schedin, T. J. Booth, V. V. Khotkevich, S. V. Morozov, A. K. Geim, Two-dimensional atomic crystals, Proc. Natl. Acad. Sci. U. S. A. 102 (2005) 10451-10453. https://doi.org/10.1073/pnas.0502848102

[6] D. A. C. Brownson, C. E. Banks, Graphene electrochemistry: an overview of potential applications, Analyst, 135 (2010) 2768-2778. https://doi.org/10.1039/c0an00590h

[7] Y. Shao, J. Wang, H. Wu, J. Liu, I. A. Aksay,Y. Lin, Graphene based electrochemical sensors and biosensors: A review, Electroanalysis (N. Y.). 22 (2010) 1027-1036. https://doi.org/10.1002/elan.200900571

[8] K. R. Ratinac, W. Yang, J. J. Gooding, P. Thordarson, F. Braet, Graphene and related materials in electrochemical sensing, Electroanalysis (N. Y.). 23 (2011) 803-826. https://doi.org/10.1002/elan.201000545

[9] X. Kang, J. Wang, H. Wu, J. Liu, I. A. Aksay, Y. Lin, A graphene-based electrochemical sensor for sensitive detection of paracetamol, Talanta. 81(2010)754-759. https://doi.org/10.1016/j.talanta.2010.01.009

[10]   Richa Sharma, Joon Hyun Baik, Chrisantha J. Perera, and Michael S. Strano, anomalously large reactivity of single graphene layers and edges towards electron transfer chemistries. NanoLett. 10 (2010) 398-405. https://doi.org/10.1021/nl902741x

[11]   X. Kang, J. Wang, H. Wu, J. Liu, I. A. Aksay, Y. Lin, Graphene-based electrochemical sensor for sensitive detection of paracetamol, Talanta, 81 (2010)754-759. https://doi.org/10.1016/j.talanta.2010.01.009

[12]   S. Park, R. S. Ruoff, Chemical methods for the production of graphenesnat, Nanotechnol. 4 (2009) 217-224. https://doi.org/10.1038/nnano.2009.58

[13]   Y. Shao, J. Wang, H. Wu, J. Liu, I. A. Aksay, Y. Lin, Graphene based electrochemical sensors and biosensors: A review. Electroanalysis (N. Y.), 22 (2010) 1027-1036. https://doi.org/10.1002/elan.200900571

[14]   D. A. C. Brownson, C. E. Banks, Graphene electrochemistry: an overview of potential applications. Analyst, 135(2010) 2768-2778. https://doi.org/10.1039/c0an00590h

[15]   R. L. McCreery, Advanced carbon electrode materials for molecular electrochemistry. Chem. Rev, 108 (2008) 2646-2687. https://doi.org/10.1021/cr068076m

[16]   C. E. Banks, T. J. Davies, G. G. Wildgoose, R. G. Compton, Investigation of modified basal plane pyrolytic graphite electrodes: Definitive evidence for the electrocatalytic properties of the enhs of carbon nanotubes. Chem. Commun, (2005) 829-841.

[17]   X. Ji, C. E. Banks, A. Crossley, R. G. Compton, Oxygenated edge plane sites slow the electron transfer of the ferro-/ferricyanide redox couple at graphite electrodes. Chem. Phys. Chem, 7 (2006) 1337-1344. https://doi.org/10.1002/cphc.200600098

[18]   D. A. C. Brownson, L. J. Munro, D. K. Kampouris, C. E. Banks, Electrochemistry of graphene: not such a beneficial electrode material? RSC Adv. 1 (2011) 978-988. https://doi.org/10.1039/c1ra00393c

[19]   D. A. C. Brownson, J. P. Metters, D. K. Kampouris and C. E. Banks, Surfactant-exfoliated 2D hexagonal boron nitride (2D-hBN): role of surfactant upon the electrochemical reduction of oxygen and capacitance applications. Electroanalysis (N. Y.), 23 (2011)894-899. https://doi.org/10.1039/C6TA09999H

[20]   S. Y. Chee, M. Pumera, Metal-based impurities in graphenes: application for electroanalysis. Analyst, 137 (2012) 2039-2041. https://doi.org/10.1039/c2an00022a

[21]   D. A. C. Brownson, C. W. Foster, C. E. Banks, The electrochemical performance of graphene modified electrodes: An analytical perspective. Analyst, 137 (2012) 1815-1823. https://doi.org/10.1039/c2an16279b

[22]   D. A. C. Brownson, C. W. Foster, C. E. Banks, The electrochemical performance of graphene modified electrodes: An analytical perspective. Analyst, 137 (2012) 1815-1823. https://doi.org/10.1039/c2an16279b

[23]   D. A. C. Brownson, C. W. Foster, C. E. Banks, The electrochemical performance of graphene modified electrodes: An analytical perspective. Analyst, 137 (2012) 1815-1823. https://doi.org/10.1039/c2an16279b

[24] B. Ntsendwana, B. B. Mamba, S. Sampath, O. A. Arotiba, Electrochemical detection of bisphenol A using graphene modified glassy carbon electrode. Int. J. Electrochem. Sci, 7 (2012) 3501-3512.

[25] W.J. Lin, C.S. Liao, J.H. Jhang, Y.C. Tsai, Graphenemodofird basal and edge plane pyrolytic graphite electrodes for electrocatalytic oxidation of hydrogen peroxide and b0nicotinamide adenine dinucleotide. Electrochem. Commun, 11(2009) 2153-2156. https://doi.org/10.1016/j.elecom.2009.09.018

[26] X. Kang, J. Wang, H. Wu, J. Liu, I. A. Aksay, Y. Lin, A graphene-based electrochemical sensor for sensitive detection of paracetamol. Talanta, 81(2010)754-759. https://doi.org/10.1016/j.talanta.2010.01.009

[27] D. A. C. Brownson, L. J. Munro, D. K. Kampouris, C. E. Banks, Electrochemistry of graphene: not such a beneficial electrode material? RSC Adv, 1(2011) 978-988. https://doi.org/10.1039/c1ra00393c

[28] P. Chen, R. L. McCreery, Control of electron transfer kinetics at glassy carbon electrodes by specific surface modification. Anal. Chem., 68 (1996) 3958-3965. https://doi.org/10.1021/ac960492r

[29] A. G. Guell, N. Ebejer, M. E. Snowden, J. V. Macpherson, P. R. Unwin, Structural correlations in heterogeneous electron transfer at minelayer and multilayer graphene electrodes. J. Am. Chem. Soc, 134 (2012) 7258-7261. https://doi.org/10.1021/ja3014902

[30] D. K. Kampouris, C. E. Banks, Exploring the physicoelectrochemical properties of graphene. Chem. Commun, 46 (2010) 8986-8988. https://doi.org/10.1039/c0cc02860f

[31] C. X. Lim, H. Y. Hoh, P. K. Ang, K. P. Loh, Direct voltammetric detection of DNA and pH sensing on epitaxial graphene: an insight into the role of oxygenated defects. Anal. Chem, 82 (2010) 7387-7393. https://doi.org/10.1021/ac101519v

[32] X. Ji, C. E. Banks, A. Crossley, R. G. Compton, Oxygenated edge plane sites slow the electron transfer of the ferro-/ferricyanide redox couple at graphite electrodes. Chem.Phys.Chem, 7 (2006) 1337-1344. https://doi.org/10.1002/cphc.200600098

[33] G. P. Keeley, A. O. Neill, N. Mc Evoy, N. Peltekis, J. N. Coleman, G. S. Duesberg, Electrochemical ascorbic acid sensor based on DMF-exfoliated graphene. J. Mater. Chem., 20 (2010) 7864-7869. https://doi.org/10.1039/c0jm01527j

[34] A. Ambrosi, A. Bonanni, M. Pumera, Electrochemistry of folded graphene edges. Nanoscale, 3 (2011) 2256-2260. https://doi.org/10.1039/c1nr10136f

[35] C. Tan, J. Rodriguez-Lopez, J. J. Parks, N. L. Ritzert, D. C. Ralph, H. D. Abruna, Reactivity of monolayer chemical vapor deposited graphene imperfections studied using scanning electrochemical microscopy. ACS Nano, 6 (2012) 3070-3079. https://doi.org/10.1021/nn204746n

[36] H. L. Poh, F. Sanek, A. Ambrosi, G. Zhao, Z. Sofer, M. Pumera, Graphenes prepared by staudenmaier, hofmann and hummers methods with consequent thermal exfoliation exhibit very different electrochemical properties. Nanoscale, 4 (2012) 3515-3522. https://doi.org/10.1039/c2nr30490b

[37] B. Guo, L. Fang, B. Zhang, J. R. Gong, Co/CoO Nanoparticles assembled on graphene for electrochemical reduction of oxygen. Insciences J, 1 (2011) 80-89. https://doi.org/10.5640/insc.010280

[38] D. A. C. Brownson, J. P. Metters, D. K. Kampouris, C. E. Banks,The electrochemical performance of graphene modified electrodes: An analytical perspective. Electroanalysis (N. Y.), 23 (2011) 894-899. https://doi.org/10.1002/elan.201000708

[39] S. Y. Chee, M. Pumera, Metal-based impurities in graphenes: application for electroanalysis. Analyst, 137 (2012) 2039-2041. https://doi.org/10.1039/c2an00022a

[40] P. M. Hallam, C. E. Banks, A facile approach for quantifying the density of defects (edge plane sites) of carbon nanomaterials and related structures. Phys. Chem. Chem. Phys, 13 (2011) 1210-1213. https://doi.org/10.1039/C0CP01562H

[41] I. Streeter, G. G. Wildgoose, L. Shao and R. G. Compton, Cyclic voltammetry on electode surfaces covered with porous layers an analysis of electron transfer kinetics at single walled carbon nanotube modified electrode. Sens. Actuators. B, 133 (2008) 462-466. https://doi.org/10.1016/j.snb.2008.03.015

[42] S.X. Guo, S.F. Zhao, A. M. Bond and J. Zhang, Simplifying the evaluation of graphene modified electrode performance using rotating disk electrode voltammetry. Langmuir, 28 (2012) 5275-5285. https://doi.org/10.1021/la205013n

[43] D. R. Dreyer, S. Park, C. W. Bielawski, R. S. Rouff, The chemistry of graphene oxide. Chem. Soc. Rev., 39 (2010) 228-240. https://doi.org/10.1039/B917103G

[44] D. A. C. Brownson, A. C. Lacombe, M. Gomez-Mingot, C. E. Banks, Graphene oxide gives rise to unique and intriguing voltammetry. RSC Adv, 2 (2012) 665-668. https://doi.org/10.1039/C1RA00743B

[45] M. Zhou, Y. Wang, Y. Zhai, J. Zhai, W. Ren, F. Wang, S. Dong, Controlled synthesis of large-area and patterned electrochemically reduced graphene oxide films. Chem.Eur. J., 15 (2009) 6116-6120. https://doi.org/10.1002/chem.200900596

[46]   D. A. C. Brownson, C. E. Banks, CVD graphene electrochemistry: the role of
graphitic islands. Phys. Chem. Chem. Phys, 13 (2011)15825-15828.
https://doi.org/10.1039/c1cp21978b

[47]   A. N. Obraztsov, E. A. Obraztsova, A. V. Tyurnina, A. A. Zolotukhin, Chemical
vapor deposition of thin graphite films of nanometer thickness. Carbon, 45 (2007)
2017-2021. https://doi.org/10.1016/j.carbon.2007.05.028

[48]   A. Guermoune, T. Chari, F. Popescu, S. S. Sabri, J. Guillemette, H. S. Skulason,
T. Szkope, M. Siaj, Chemical vapor deposition synthesis of graphene on copper with
methanol, ethanol, and propanol precursors. Carbon, 49 (2011) 4204-4210.
https://doi.org/10.1016/j.carbon.2011.05.054

[49]   D. A. C. Brownson, D. K. Kampouris , C. E. Banks, Electrochemistry of graphene:
not such a beneficial electrode material? J. Power Sources, 196 (2011) 4873-4885.
https://doi.org/10.1016/j.jpowsour.2011.02.022

[50]   K. R. Ratinac, W. Yang, J. J. Gooding, P. Thordarson, F. Braet, Carbon
nanomaterials in biosensors: should you use nanotubes or graphene? Electroanalysis
(N. Y.), 23 (2011) 803-826. doi: 10.1002/anie.200903463.
https://doi.org/10.1002/elan.201000545

[51]   B. Ntsendwana, B. B. Mamba, S. Sampath, O. A. Arotiba, Electrochemical
detection of bisphenol A using graphene-modified glassy carbon electrode. Int. J.
Electrochem. Sci., 7 (2012) 3501-3512.

[52]   W.J. Lin, C.S. Liao, J.H. Jhang, Y.C. Tsai, Graphene modified basal and edge
plane pyrolytic graphite electrodes for electrocatalytic oxidation of hydrogen peroxide
abd b-nicotinamide adenine dinucleotide. Electrochem. Commun, 11 (2009) 2153-
2156. https://doi.org/10.1016/j.elecom.2009.09.018

[53]   Y. Wang, D. Zhang, J. Wu, Determination of kojic acid based on a poly (l-
arginine)-electrochemically reduced graphene oxide modified electrodej. Electroanal.
Chem, 664 (2012) 111-116. https://doi.org/10.1016/j.jelechem.2011.11.004

[54]   X. Kang, J. Wang, H. Wu, J. Liu, I. A. Aksay, Y. Lin, A graphene-based
electrochemical sensor for sensitive detection of paracetamol. Talanta, 81 (2010) 754-
759. https://doi.org/10.1016/j.talanta.2010.01.009

[55]   D. A. C. Brownson, C. W. Foster, C. E. Banks, The electrochemical performance
of graphene modified electrodes: An analytical perspective. Analyst, 137 (2012)
1815-1823. https://doi.org/10.1039/c2an16279b

[56]   D. A. C. Brownson, L. J. Munro, D. K. Kampouris, C. E. Banks, The fabrication, characterisation and electrochemical investigation of screen-printed graphene electrodes. RSC Adv, 1 (2011) 978-988. https://doi.org/10.1039/c1ra00393c

[57]   D. A. C. Brownson, L. J. Munro, D. K. Kampouris, C. E. Banks, The fabrication, characterisation and electrochemical investigation of screen-printed graphene electrodes. RSC Adv, 1 (2011) 978-988. https://doi.org/10.1039/c1ra00393c

[58]   D. A. C. Brownson, A. C. Lacombe, D. K. Kampouris, C. E. Banks, Grapheneelectroanalysis: Inhibitory effects in the stripping voltammetry of cadmium with surfactant free graphene. Analyst, 137(2012) 420-423. https://doi.org/10.1039/C1AN15967D

[59]   D. A. C. Brownson, A. C. Lacombe, D. K. Kampouris, C. E. Banks, Grapheneelectroanalysis: Inhibitory effects in the stripping voltammetry of cadmium with surfactant free graphene. Analyst, 137 (2012) 420-423. https://doi.org/10.1039/C1AN15967D

[60]   D. A. C. Brownson, L. J. Munro, D. K. Kampouris, C. E. Banks, In situ electrochemical characterisation of graphene and various carbon-based electrode materials: an internal standard approach. RSC Adv, 1(2011) 978-988. https://doi.org/10.1039/c5ra03049h.

[61]   D. A. C. Brownson, J. P. Metters, D. K. Kampouris, C. E. Banks, Graphene electrochemistry: surfactants inherent to graphene can dramatically effect electrochemical processes. Electroanalysis (N. Y.), 23 (2011) 894-899. https://doi.org/10.1002/elan.201000708

[62]   M. S. Goh, M. Pumera, Number of grapheneLyers exhibiting an influence on oxidation of DNA bases: analytical parameters. Anal. Chem,82(2010) 8367-8370. https://doi.org/10.1021/ac101996m

[63]   M. S. Goh, M. Pumera, Graphene-based electrochemical sensor for detection of 2,4,6-trinitrotoluene (TNT) in seawater: the comparison of single-, few-, and multilayer graphenenanoribbons and graphite microparticles. Anal. Bioanal. Chem, 399 (2011) 127-131. https://doi.org/10.1007/s00216-010-4338-8

[64]   D. A. C. Brownson, L. J. Munro, D. K. Kampouris, C. E. Banks, Electrochemistry of graphene: not such a beneficial electrode material? RSC Adv, 1(2011) 978-988. https://doi.org/10.1039/c1ra00393c

[65]   D. A. C. Brownson, M. Gomez-Mingot, C. E. Banks, CVD graphene electrochemistry: biologically relevant molecules. Phys. Chem. Chem. Phys, 13 (2011) 20284-20288. https://doi.org/10.1039/c1cp22648g

[66]   D. A. C. Brownson, C. W. Foster, C. E. Banks, The electrochemical performance of graphene modified electrodes: An analytical perspective. Analyst, 137 (2012) 1815-1823. https://doi.org/10.1039/c2an16279b

[67]   D. A. C. Brownson, R. V. Gorbachev, S. J. Haigh, C. E. Banks, CVD graphenevs. highly ordered pyrolytic graphite for use in electroanalytical sensing. Analyst, 137 (2012) 833-839. https://doi.org/10.1039/C2AN16049H

[68]   D. A. C. Brownson, C. E. Banks, The electrochemistry of CVD graphene: progress and prospects. Phys. Chem. Chem. Phys, 14 (2012) 8264-8281. https://doi.org/10.1039/c2cp40225d

[69]   M. Pumera, Graphene-based nanomaterials and their electrochemistry. Chem. Soc. Rev, 39 (2010) 4146-4157. https://doi.org/10.1039/c002690p

[70]   A. N. Obraztsov, E. A. Obraztsova, A. V. Tyurnina, A. A. Zolotukhin, Chemical vapor deposition of thin graphite films of nanometer thickness. Carbon, 45 (2007) 2017-2021. https://doi.org/10.1016/j.carbon.2007.05.028

[71]   J. Premkumar, S. B. Khoo, electrocatalytic oxidations of biological molecules (ascorbic acid and uric acids) at highly oxidized electrodes. J. Electroanal. Chem, 576 (2005) 105-112. https://doi.org/10.1016/j.jelechem.2004.09.030

[72]   D. A. C. Brownson, M. Gomez-Mingot, C. E. Banks, CVD graphene electrochemistry: biologically relevant molecules. Phys. Chem. Chem. Phys, 13 (2011) 20284-20288. https://doi.org/10.1039/c1cp22648g

[73]   D. A. C. Brownson, R. V. Gorbachev, S. J. Haigh, C. E. Banks, Graphene oxide electrochemistry: the electrochemistry of graphene oxide modified electrodes reveals coverage dependent beneficial electrocatalysis. Analyst, 137 (2012) 833-839. https://doi.org/10.1098/rsos.171128.

[74]   D. A. C. Brownson, R. V. Gorbachev, S. J. Haigh, C. E. Banks, In situ electrochemical characterisation of graphene and various carbon-based electrode materials: an internal standard approach. Analyst, 137 (2012) 833-839.

[75]   P. Raghu, B.E.K. Swamy, T. M .Reddy, B. N. Chandresekhar, K. Rrddaiah, Sol gel immobilized biosensor for the detection of organo phosphorous pesticides a voltammetric method. Bioelectrochem, 83(2011) 19- 25. https://doi.org/10.1016/j.bioelechem.2011.08.002

[76]   Y. Zhang, T.F. Kang, Y.W. Wan, S.Y. Chen,Gold nanoparticles-carbon nanotubes modified sensor for electrochemical determination of organophosphate pesticides. Micro chimicaActa, 165(2009) 307-311. https://doi.org/10.1007/s00604-008-0134-y

[77]   P. Reddy Prasad, A.E. Ofamaja, C.N. Reddy, E.B. Naidoo, Square Wave Volta mmetric Detection of Dimethyl vinphos and Naftalofos in Food and Environmental Samples Using RGO/CS modified Glassy Carbon Electrode. Int. J. Electro chem. Sci, 11(2016) 65-79.

[78]   J. Zhang, A. Luo, P. Liu, S. Wei, G. Wang, S. Wei, Detection of organo phosphorus pesticides using potentiometric enzymatic membrane biosensor based on methylcellulose immobilization. Analytical Science, 24 (2011) 511-693.

[79]   R. Sinha, M. Ganesana, S. Andreescu, L. Stanciu, AChE biosensor based on zinc oxide sol-gel for the detection of pesticides. Anal. Chim. Acta, 195 (2010) 661-675. https://doi.org/10.1016/j.aca.2009.12.020

[80]   J.L.Marty, N. Mionetto, T. Noguer, F. Ortega, C. Roux, Enzyme sensors for the detection of pesticides. Biosens. Bio electron, 8 (1993)273-298. https://doi.org/10.1016/0956-5663(93)85007-B

[81]   D. Du, S. Chen, J. Cai, A. Zhang, Electro chemical pesticide sensitivity test using acetyl cholinesterase biosensor based on colloidal gold nanoparticle modified sol-gel interface. Talanta, 74 (2008)766-798. https://doi.org/10.1016/j.talanta.2007.07.014

[82]   R. Sinha, M. Ganesana, S. Andreescu, L. Stanciu, AChE biosensor based on zinc oxide sol-gel for the detection of pesticides. Anal. Chim. Acta, 195 (2010) 661-675. https://doi.org/10.1016/j.aca.2009.12.020

[83]   J.L.Marty, N. Mionetto, T. Noguer, F. Ortega, C. Roux, Enzyme sensors for the detection of pesticides. Biosens. Bio electron, 8 (1993)273-298. https://doi.org/10.1016/0956-5663(93)85007-B

[84]   D. Du, S. Chen, J. Cai, A. Zhang, Electro chemical pesticide sensitivity test using acetyl cholinesterase biosensor based on colloidal gold nanoparticle modified sol-gel interface. Talanta, 74 (2008)766-798. https://doi.org/10.1016/j.talanta.2007.07.014

Chapter 2

Functionalized Graphene Modified Electrochemical Sensors for Toxic Chemicals

Bhawana Singh*, Anurag Kumar Kushwaha*, Siddharth Sankar Singh*, Shyam Sundar

Institute of Medical Sciences, Banaras Hindu University, Varanasi-221005,Uttar Pradesh, India.

*All authors contributed equally

bhavanasonali9@gmail.com

Abstract

Toxic chemicals including pesticides, organic/inorganic compounds etc. are potential threat to the environment as well as the living beings. Indeed, their end products are potential carcinogen and/or mutagens. Use of graphene has been the cutting-edge innovation in nanoengineering due to its unique electrochemical properties. The wide array of chemical modifications of graphene (also termed as functionalization) confers million adjunctive properties to the graphene and significantly improves its performance. Further, graphene-based nanomaterials provide inexpensive, reusable and eco-friendly option overcoming the limitations associated with commercially available electrochemical sensors. Recent development in functionalized graphene-based sensors holds promises for rapid and simultaneous detection of several chemicals. Future efforts aim to further improve the existing electrochemical sensors in terms of their performance for proper management of toxic and hazardous waste and protecting our ecosystem. This chapter aims to provide a comprehensive understanding on the latest researches on development of functionalized graphene-based nanomaterials for designing electrochemical sensors, engineering and fabrication of graphene for detection of different group of toxic chemicals and its role in improving the existing nanocomposite based electrochemical sensors.

Keywords

Nanoengineering, Pesticides, Drugs, Neurotoxins, Hepatotoxins

Contents

1. Introduction

There is plethora of chemicals that humans encounter in their day to day lives that have become an integral part of current lifestyle. Each and every moment we are exposed to different chemicals; many chemicals will contribute significantly to enhancing our quality of living, safety and well- being when correctly utilized [1]. Despite potential health concerns, production and use of chemicals are still increasing globally, particularly in developing countries. This calls upon for stringent management policies for these contaminants; any negligence can undoubtedly cause adverse health consequences. In the household and at workplace there is accidental poisoning of toxic products such as heavy metals, pesticides, solvents, paints, detergents, kerosene, carbon monoxide and drugs. It is estimated that accidental poisonings lead to 193,000 deaths per year, mainly due to preventable chemical exposure. Furthermore, over one-third of ischemic heart disorders can be avoided by limiting or removing toxic contaminants such as indoor air emissions, household air pollution, second-hand smoke and lead (Pb). These toxic contaminantsremain the leading cause of global deaths, injuries and approximately 42 percent of stroke cases, that attributes for the second largest cause for global mortality [2] . One of the most significant challenges of this century is to tackle the problems related to the abundant chemicals that are released in the atmosphere by human activities. In recent years, the progress in nanoengineering has served to replace the expensive and time consuming conventional chemical detection technologies with easy-to-use electrochemical sensors. The development of novel electrochemical sensor-based technology holds hope for the evaluation of contaminants level in human body, fluids and in the atmosphere. Additionally, these sensor platforms have provided pace to development of remediation techniques in order to eliminate such toxic chemicals and protect the ecosystem [3].

2. Toxic chemicals and types

The U.S. Environmental Protection Agency (EPA) defines a toxic chemical as any substance which may be harmful to the environment or hazardous to health if inhaled, ingested or absorbed through the skin. Chemicals are omnipresent compounds with positive as well as negative consequences in environments around the world. It is well known that several chemicals are toxic to human health and the environment (animals and plants). Depending on their mode of action the toxic chemicals can be categorized depending upon their target of action, into following categories:

A) Toxin that affect vital organ in humans

Many chemicals are distributed throughout the body and often only affect particular target organs. Depending on the dosage and route of exposure the target organs affected may be different therefore, depending upon the target organ affected by the toxins, it can be further categorized as neurotoxins, hepatotoxins, nephrotoxins, cardiotoxins, reproductive toxins etc. For example, the major causes of livers diseases are hepatotoxins, which include conventional drugs, herbal medicinal products or chemical products. There are documented evidences for hepatic injury such as elevated hepatic enzymes after unregulated occupational exposure for many of chemicals like aliphatic nitros, aromatic amines, aromatic nitros, chlorinated solvents, halogenated solvents, halowaxes, and nitrosamines. Hepatic injury cannot manifest until a prolonged exposure for toxic chemicals with long biological half-lives such as heavy metals, PCBs and organochlorine insecticides occurs. Nervous system activities may be affected by metals such as lead (Pb) and mercury (Hg). Nerve agents, certain pesticides (e.g.

organophosphate) and some organic solvents, such as hexane, are the most diffused synthetic chemicals that affect the central nervous system. These neurotoxins influence chemical signal transmission between neurons which causes a number of disorders and even death. These nerve agents and organophosphate pesticides act as inhibitors of acetylcholinesterase activity [4]. The adverse effect of substances on renal function can be described as nephrotoxicity. Kidney plays a significant role in mitigating the toxicity of many medications, however, the environmental toxins and natural substances, cadmium, mercury, arsenic, lead, trichloroethylene, bromate, brominated-flame retardants, diglycolic acid, and ethylene glycol are known to affect the kidneys [5]. There are a large number of chemicals manufactured and used around the world, some of which can adversely affect reproductive health. Exposure to environmental toxins like alkylating agents, arsenic, benzene, carbon monoxide, chlorinated hydrocarbons, dioxin, ionizing radiation, organic mercury compounds, polychlorinated biphenyls and metals may affect parents, fetuses and newborns before and after conception. Different chemicals that interact with endocrine may affect sex and reproductive organs of men and women as well [6]. In regeneration and reproduction of newborn or adult specimens, the azo textile Disperse Red 1 caused adverse effects [7]. The risk of adverse pregnancy outcomes such as low birth weight, premature and dead births are air pollution and second-hand smoke. It was calculated that the total risk for mortality was increased by 23% and congenital malformations by 13% for antenatal exposure to second-hand smoke [8].

B) Toxic chemicals in food additives, drugs or pesticides

Food is a most common cause of toxicity exposure to human beings. There is a huge range of contaminants that exist naturally in the food to improve taste, color, stability and texture, develop or increase nutrient values, or to lower their costs and more than 2500 chemicals are applied to the food. An estimated 12,000 substances are also used which enters the food supply unintentionally [9]. Bisphenols and phthalates are most commonly used in plastic packaging, garden hoses, inflatable toys, but also in things like nail polish, hairsprays, lotions and fragrances. These may also imitate body's hormones, impair male genital growth, immune and nervous dysregulation, increased risk of obesity and cardiovascular diseases. Perfluoroalkyl chemicals (PFCs) can lead to babies with a low birthweight, anomalous immune system and thyroid abnormalities.. Perchlorate (found in packaging of dry food) also affects thyroid function and may interfere with early development of the brain. In children with hyperactivity disorder (or ADHD), artificial food colors have been found to worsens the symptoms. Nitrates and nitrites (used for retaining food color in processed food products)" can interfere with the thyroid, as well

Materials Research Foundations **82** (2020) 25-60   https://doi.org/10.21741/9781644900956-2

as oxygen delivering capacity of blood to the body and increases the risk of certain types of cancers [10].

Drug toxicity is the development of severe, drug-related complications. In pharmacology, drugs can become toxic if the dosage becomes too high and the liver or kidneys cannot remove the drug from the blood stream, eventually it accumulates in the body. There is a fine line of difference between the effective and toxic dose;for one person, a therapeutic dosage can be toxic to another [11, 12]. The majority of therapeutic pharmaceutical products generate beneficial reactions by modification of human homeostatic mechanisms; the therapeutic response without direct impact to human metabolic or physiological processes is only conceived of antimicrobial and parasiticidal agents. For example atropine (belladonna), tubocurarine (curare), ergot alkaloids (causing St. Anthony's fire), digoxin (digitalis) and dicoumarol (causing haemorrhagic disease in cattle), are some therapeutic agents having pharmacological properties which are unintentionally or deliberately recognized as a consequence of poisoning [13, 14].

Pesticides are synthesized agents used to attract, seduce, kill or reduce any insect. Depending upon chemical structures, pesticides may be divided into the following groups; 1) Organochlorines (e.g., endosulfan, hexachlorobenzene); 2) Organophosphates (e.g., diazinon, omethoate, glyphosate); 3) Carbamic and thiocarbamide derivatives (e.g., aldicarb, carbofuran, oxamyl, carbaryl); 4) Carboxylic acids and their derivatives (e.g., pentanal, butanamide, butanamide); 5) Urea derivatives (e.g., fenuron, metoxuron, diuron, linuron, monuron); 6) Heterocyclic compounds (e.g., benzimidazole, triazole derivatives); 7) Phenol and nitrophenol derivatives (e.g., dinocap, dinoseb); 8) Hydrocarbons, ketones, aldehydes and their derivatives (e.g., benzene, toluene, cerenox); 9) Fluorine-containing compounds (e.g., cryolite, acetoprole, dichlofluanid); 10) Copper-containing compounds (e.g., champion WP, caocobre, macc 80); 11) Metal-organic compounds (e.g., mancozeb, maneb, zineb, nabam); 12) Synthetic pyrethroids and others (e.g., allethrin, cypermethrin, fluvalinate). Therefore, pesticides are a heterogeneous category amongst the synthetic substances, and human encounters them on a daily basis. The pesticides can distinguish the target and non-target species by varying degrees of toxicity. Due to the cumulative characteristic of many pesticides, these chemicals circulate in ecosystems and can accumulate and migrate across food chains by many living organisms. Pesticides can cause endocrine disruptions and neurological disturbances, influence immune system, reproduction and development [15-18].

C) Natural plant/ animal toxins

All forms of life, including animals, plants and microbes, produce toxins to interfere with and disrupt other organism's physiological processes and, on the other hand, to favor their struggle for existence [19]. Natural toxins may be present inherently in plants. They are usually metabolites produced by plants to defend themselves against various threats such as bacteria, fungi, insects and predators, which may be species specific and give the plant its particular characteristics, e.g. colours and flavors. Common examples of natural toxins in food plants include lectins (in beans such as green beans, red kidney beans and white kidney beans), cyanogenic glycosides (in bitter apricot seed, bamboo shoots, cassava, and flaxseeds), glycoalkaloids (in potatoes), 4'-methoxypyridoxine (in ginkgo seeds), colchicine (in fresh lily), and muscarine (in some wild mushrooms). Animal toxins are a complex mixture of polypeptides, enzymes and chemicals which can cause cellular injury. On numerous occasions, venoms (mixture of toxins) in animals have evolved. Cnidarians (jellyfishes, sea anemones and hydra), molluscs (cone snails), annelids (leeches), arthropods (spiders, scorpions, centipedes, bees and wasps, ants, ticks and horseflies, crustaceans), echinoderms (sea urchins and starfishes), vertebrates (fishes, snakes and lizards, as well as mammals) are the common and well-known venous animals [19].

D) Toxic chemicals as carcinogen/mutagen

The industrial world has exposed humanity to different chemicals, such as drugs, cosmetics, food preservatives, pesticides, industrial additives, chemicals, etc. Many have shown to be carcinogenic and mutagenic in these compounds. For example, the food preservative AF-2, the food fumigant ethylene dibromide, the anti-schistosome drug hycanthone, several hair-dye additives and the industrial compound like benzidine, 2-naphthylamine, 4-aminobiphenyl, bis(chloromethyl) ether, vinyl chloride, arsenic, diethylstilbestrol, phenacetin, melphalan, phenytoin, chloramphenicol, chromateproducing industries, hemitite mining, nickel refining, isopropyl oil, and asbestos; all are potent [20].

3.   Electrochemical sensors: Characteristics features and types

An electrochemical sensor's core components usually involve a working electrode, counter electrode and reference electrode. However, only two electrodes are used in potentiometry and four electrodes are used in ion-transfer voltammetry [21]. Enzymes, antibodies, aptamers and any other components of biorecognition are not used for building chemical sensors [22]. An ideal example of electrochemical sensor is a simple enzyme-free glucose sensor which have strong reproducibility, interference-free and reliably fast determination with low cost. Among non-enzymatic glucose sensors, electrochemical sensors have been considered one of the most common and efficient techniques for the detection of glucose due to their advantages in terms of simplicity, portability, fast response time, high sensitivity and selectivity by specificity of the substrates [23].

Undoubtedly, there is no single technology platform that can satisfy different parameters (sensitivity, specificity, linearity, response time, reproducibility etc.) of diverse range of chemicals and principles of chemical sensing therefore, there has been a substantial need for new inventions with better selectivity and sensitivity. Keeping in mind the aforementioned facts, an ideal chemical sensor should be a cost efficient, portable and full-proof device which responds to a certain target chemical substance (analyte) present in any desired medium with perfect and immediate selectivity to produce a measurable signal output at any analyte concentration.

Current chemical sensors are complex devices which are designed for a particular application, these can be categorized in two groups:

1. Direct-reading, selective sensors (e.g., electrochemical sensors, optical fibers);

2. Sensors that use a preliminary chromatographic or electrophoretic sample separation step followed by sensitive, but not necessarily selective detection.

There are different ways in which these sensors can be used to detect chemical toxins. Given the exhaustive list of known chemicals, molecular sensing relies on accurate determination of analytes depending on their dimension (that can range from liters to picoliters or nanoscale to monomolecular scale), phase (gas, liquid or solid), analysis over a time scale (repetitive or one-time over a period of time). The design of chemical sensors also requires an assessment of the quantitative reliability (precision or accuracy). Thus, there are different chemical sensing technologies which have emerged over a period of time for diverse range of chemicals and their measurement. The different kinds of sensor platforms for toxic chemicals includes: Gas chromatograph-mass spectroscopy; Ion mobility spectrometer; Patch chemical reactions showing color change; Fiberoptic and related methods; Electrochemical sensors; Acoustic-wave chemical sensors; and Sensors for immunochemical assays [24].

4.   Brief history of development of electrochemical sensors

The history of the development of sensor technology has been driven by the advances in materials science and engineering. For example, in the early 1860s, Wilhelm von Siemens developed a temperature sensor based on a copper resistor for the temperature sensitivity of electrical resistance in a variety of materials, which was noted in 1800s. A sensor is defined by the American National Standards Institute (ANSI) standard MC6.1 as "a device which provides a usable output in response to a specific measurand" [25].

An output is defined by the term "electrical quantity," and a measurand is a measured physical quantity, property or condition. The committee acknowledged that any form of energy can be the output of a sensor. An array of early sensors transformed a physical measurand into mechanical energy. Lion (1969) has introduced a principle classification in accordance with the form of energy which sensors received and produced leading to an effect matrix [26]. Later in 1989, Göpel *et al.* proposed physical and chemical transduction principles excluding biological and nuclear effects [27]. Similarly, in 1982, Middlehoek and Noorlag undertook a rigorous classification process in which the input and output energy were only represented as the principle of transduction and no "internal" or compound transduction effects could be achieved [28]. Invention of solid-state electronics has opened up new possibilities for sensor production and control, resulting in sensors that generates almost exclusively an electrical output that finds applications in computational controls, archiving / recording and visual display.

A new class of sensors involving the transduction of energy into an optical form is likely with advancement in optical computing and information processing. In the future, fluid components, as controls and actuating devices, can also be controlled with sensors based on microelectromechanical systems. Thus, it can be postulated that the definition of a "sensor" will continue to evolve. Continuous advancement in the field of material science can boost better control of material properties and behavior; thereby delivering new sensors with improved functionality such as increased accuracy, reduced costs and greater efficiency that have played an important role in everyday life [3].

5.   Electrochemical sensors for toxic chemicals

Electrochemical measurements are based on detection or transport of charges across an electrode. Chemical species, including molecular ions, are referred to as electroactive species if oxidized (losing electrons) or reduced (gaining electrons) can be found on the electrode surface by electrons movement. The electrochemical sensors are mostly separated into three types: potentiometric, conductometric, and amperometric/voltammetric. A detector that measures current upon the contact of electroactive solute with the working electrode is known as an amperometric detector; whereas measurement of current generated as a function of variable potential is known as voltammetry. Amperometric and voltammetric detectors both generate quantitative information of electrode current versus time. Since many analytes are electro-active so, electrochemical methods provide the key alternatives for the trace detection of analytes through atomic and/or mass spectroscope methods. The application of electrochemical methods was limited due to the fact that electrodes with specific electrical and selective properties were required. Recent advances in printing technology and material science are now making it possible to manufacture extremely cost efficient, reliable, and selective electrodes for the detection of different analytes. Such developments may be classified into three electrochemical sensing technologies: A) screen-printed electrodes B) integrated nanomaterials, and C) electrochemical immunoassays. Electrochemical sensors are advantageous over other kinds to recognize and verify dicey compounds, due to their ease, rapid operation, response and detection times. However, in order to make them acceptable for extensive use, sensitivity, selectivity and cost difficulties associated with sensors must be improved. Therefore, semiconductor nano-materials have been used extensively as a redox mediator in chemical sensors and for many other applications. Highly selective electrochemical sensors capable of pM sensitivity, high-throughput and low sample requires low sample (<50uL). The presence of toxic gasses like H_2S, Cl_2, SO_2 and variability of oxygen in the air are identified by electrochemical sensors [29-33].

6.   Nanotechnology for development of electrochemical sensing platforms

The combination of measurements and modeling of dosimetry represents a promising strategy for deciphering the connection between chemical exposure and disease outcome. Quantitative technologies for measuring xenobiotic exposure are needed in order to support the development and implementation of biological monitoring programs. Over the years, the growth of xenobiotics and various hazardous substances in soil and water has generated thousands of hazardous waste sites [34-36]. Many attempts to evaluate hazardous chemicals using chromatographic and spectroscopic techniques have been made, but they suffered from anomalies associated with their impediment and uncertainty [35, 37].

Nanomaterials possess unique electrical, chemical and physical properties (i.e., size, composition, conductivity, magnetism, mechanical strength, light absorbing and emitting properties). Amalgamation of a nanomaterial into other material system is one of the major significant means to adjust the texture and confer beneficial properties to the nanostructures. This has been made feasible with a simple and inexpensive method of creating composite nanoparticles which are then tested for their efficiency with XRD, FESEM, EDS, FTIR and UV- spectroscopy. Synthesized doped nanoparticles were used to develop sensors and to recognize toxic chemicals at micro level [38]. Additionally, the evolution of carbon nanotubes, metal nanoparticles, quantum dots, colloid gold and zirconium oxide nanoparticles specifically designed for creating novel electrochemical sensors revolutionized the environmental and biomedical sector [39-41]. Nanomaterials are meant to be used as signal transducers for analyte detection by two modes: 1) use nanomaterials as electrical signal transducers and 2) nanomaterials using indirect analyte detection as electroactive markers.

Intriguingly, the use of carbon nanotubes and metal nanoparticles has enhanced the electrochemical response, while silicon nanowires and conducting polymer nanowires has application as field effect transistors. Alternatively, some nanoparticles have been used as the electroactive reporters for indirect detection of analyte or signal amplification purposes; the most common of which are metal (e.g., colloid gold) and inorganic nanoparticles (e.g., Fe_2O_3 and Cd-Se quantum dots).

Electrochemical sensing technology in collaboration with nanomaterials (with the use of a variety of biological matrices) has accelerated the detection of chemical exposure/biomarkers of disease. At present, nanoparticles such as colloidal metals (e.g., gold or silver), carbon, inorganic crystals (e.g., quantum dots), and silica have been used as labels, markers, or probes for detection of a wide-range of biomolecules [42-45]. Amongst carbon nanomaterials, graphene and its derivatives are highly demanded nanomaterial, being the building block for fullerenes, nanotubes and 3-D graphite. Due to its layered morphology, unique physio-chemical properties (high surface area to volume ratio, thermal conductivity and high tensile strength), it remains the top priority for multiple scientific application. The limitation of graphene is its zero-band gap and hydrophobicity called upon for chemical modifications to overcome the anomalies. The different modifications in the graphene structure improves its sensitivity, electrocatalytic behavior while reducing fouling to serve the need to be used as electrochemical sensor. The invention of portable electro-chemical sensors based on nanotechnology is expected to meet demands for a selection of low cost, fast, high-performance and ultra-sensible detector for biomonitoring an array of chemical marker.

This chapter provides an insight into the progress made in the field of electrochemical sensing devices after the introduction of graphene (and its modifications). Further, we aim to explore how the modified graphene has revolutionized the electrochemical sensing platforms for detecting different chemical toxins.

7.   Functionalization of graphene for development of sensors

Graphene with its very high 2D electrical conductivity, large surface area and low cost, is an ideal substrate for various applications. Two advantages of graphene are evident when compared with carbon nanotubes (CNTs):

- Graphene has no metallic impurities, as the CNTs. These impurities dominate CNTs electro-chemistry in many situations, which contribute to misleading conclusions.

- Graphene production can be realized using cheap and usable graphite [46]

In addition to its outstanding mechanical properties, graphene and its derivatives can be functionalized with many bioactive molecules, enabling them to be integrated into the various scaffolds that are used in the biomedical sector. The graphene sheet surface functionalization is very similar to the methods used for carbon nanotubes (CNTs) functionalization. Nonetheless, the high application potential, it is important to remember that graphene's zero band gap as well as reaction inertia weakens its competitive strength in the field of semiconductor and sensor manufacturing. This is one the motives why there has been a significant increase in the number of research projects aimed at graphene functionalization including graphene (and its derivatives) reactions with organic and inorganic molecules and chemical alteration of large surface of graphene [47]. The functionalization of graphene sheet is achieved *via.* two strategies- covalent and non-covalent approaches.

(a) Covalent: In general, all of the graphene's covalent functionality is based on exchanging sp^2 orbitals and transferring it to sp^3 hybridized orbitals where it affects the local symmetry and electronic structure [48]. Graphene functionalization by covalent binding may be performed by different types of covalent attachment [49], which are as follows-

1. Covalent binding to pristine graphene functionalities
   i.   Covalent binding to pristine graphene with organic functionality
   ii.   Addition of free radicals to sp^2graphene carbon atoms
   iii. Dienophiles are attached to carbon– carbon bonds
2. Covalent binding of graphene oxide functionalities
   i.   Insertion of chromophores
   ii.   Covalent binding with polymers
   iii. Insertion of other organic molecules
   iv.   Begin with slightly reduced graphene oxide
3. Covalent hydrogen and halogens attachments against graphene derivatives
   i.   Graphene
   ii.   Fluorographene (Graphene fluoride)

(b) Non-covalent: The non-covalent method is of interest in bio-applications where polymer/graphene nanocomposite formation is typically of interest. The interaction between the graphene and polymeric compounds may be caused by the powers of Van der Waals forces, the interaction between π-π and electrostatic interactions and so far, chemical binding. The interaction between the graphene and polymers is often based on the graphene structure integrating Van der Waals forces and other interactions. Additionally, the π-bond polymers, e.g. PVA and PMMA, may have π-π interactions with graphene [50]. These non-covalent interactions give the graphene stability and improve the thermal, electrical and mechanical properties and Graphene nanosheets, similar to polymers, may be applied to bio-ceramics or metallic structures to enhance their bioactivity, mechanical and thermal properties [51].

(c) Intercalation: Organized graphene surfaces and intercalate segments are named "stages," this process is observed widely with the graphite recipients. The "stages" can be broadly categorized- Stage 1 (n=1) suggests the existence of intercalate between all graphene sheets; Stage 2 (n=2) occurs as a byproduct of intercalate between alternate graphene sheets, and so on. Graphite can integrate a large range of intercalate guests to create graphite intercalation compounds (GICs) between planar graphene sheets. Graphite is oxidized within the former case to compensate anionic intercalates and within the latter case graphite is reduced to cationic intercalates. The extraction of valence band electrons into acceptor GICs requires strong oxidants, while the incorporation of electrons into the conductive band to form donor-type GICs requires strong reduction agents [52].

(d) Functionalization with nanoparticles: The immobilization of metallic and other nanoparticles on graphene sheets has become the subject of many researches due to the growing interest in graphene resulting from its exceptional electrical and mechanical properties. Because of its large active surface area per mass unit, a typical pristine graphene sheet can be described as an ideal substratum for the dispersion of nanoparticles compared to the carbon nanotubes, amorphous carbon or graphite which have a lower active surface area as only the outer surface is active. In addition, pure graphene has high conductivity and mechanical strength and is free of metallic or carbon impurities because the preparatory methods are not catalytic, such as in carbon nanotubes, and the starting material for graphite is of the highest purity [53].

Functionalization of graphene with nanoparticles performed by different types which are as follows-

1) Deposition of nanoparticles of precious metals

2) Deposition of metal oxide nanoparticles

3) Deposition of quantum dots

4) Deposition of other nanoparticles

(e) Substitutional Doping: Substitutional graphene doping involves exchange of carbon atoms, nitrogen or boron atoms from the hexagonal graphene honeycomb membrane. The doped graphene sheets display n- or p-attitude, based on the electrophilic nature of the atoms which replaces the carbon atoms. In contrast, the electrical features of graphene may theoretically be modified by regulating the degree of this doping adjustment, thus considerably extending the implementation of graphene in nanoelectronics.The integration of nitrogen atoms is achieved through its three orbitals sp^3; it points to the conjugation of its lone pair of electrons with the graphitic π-system. The N-doped graphene sheets are abundant in electrons, and while that semi-conductive nature of the n-type is anticipated. N-doped graphene sheets are typically created by the exchange of O or C atoms with N during reduction or annealing [54, 55].

8.   Graphene and functionalized graphene based electrochemical sensors

According to IUPAC, chemical sensor is defined as a tool that transforms chemical information into an analytically useful signal, ranging from the concentration of a particular sample component to the total composition analysis. The chemical details can be obtained from a chemical reaction by the analyte or from a physical property of the investigated system [56]. Graphene nanosheets provide an ideal medium for the preparation of usable sensor materials. The combination of metal nanoparticles and graphene nanosheets not only prevents the metal nanoparticles (NPs) aggregation and oxidation but also enhance electrode efficiency [57].

Using graphene sheets in biosensor modification phase increases the biosensor surface area to immobilize large quantities of antibodies, enzymes, ssDNAs probe or cells. So, they amplify the response to electrochemical detection. Graphene has attracted considerable attention as potential matrix for electrochemical biosensors [58]. The electrochemical properties of graphene and its derivatives contribute not only to the development of electron transfers in redox systems but also to the development of new, basic and efficient electrochemical biosensors. The results of the various studies have shown that graphene- and graphene-derived electrodes exhibit excellent electrochemical activity in terms of their effective surface area for the implementation of a large number of active sites and the provision of electron-conductive pathways for the electron transfer of electroactive species [59].

9.   Engineering functionalized graphene based electrochemical sensors for toxic chemicals

a. Detection of heavy metals and oxidizing agents: Strong metal ions like Hg^{2+}, Pb^{2+}, Cd^{2+}, Ag^+ and Cu^{2+} found in water and food supplies have a crucial negative environmental impact and people's health due to the growing agricultural and industrial activity and the inappropriate disposal of metal ions from wastewater and household effluents [60]. Huge efforts have been made to develop analytical techniques for the sensing of heavy metal ions [61], where the most common methods such as atomic absorption spectrometry (AAS), X-ray fluorescence spectrometry (XRF), atomic emission spectrometry (AES), and inductively coupled plasma mass spectrometry (ICP-MS) still lag far behind because of time consuming procedures, expensive instrumentation and complex spectrometry protocols. To overcome these limitations, various methods of electrochemical, optical and colorimetric strategies have been explored extensively to create simple and cost-effective platforms for the realization of sensitive, rapid and selective analysis of heavy metal ions [62]. Impressive progress in the analysis of the two-dimensional graphene and its variants as sensors suggests the unparalleled possibility of detecting even individual molecules and adsorbents due to their extremely high conductivity and sensitivity [63]. Graphene's three key attributes make it desirable for these applications:

(a) a surface area of $2600 \ m^2 \cdot g^{-1}$, thus, providing a large surface chemistry platform [64]

(b) its ability to serve as either a reduction agent (electron donor) or an oxidizing agent (electron acceptor) [65] and

(c) low Johnson noise, even at low load densities and at room temperature as a result of its high conductivity.

Considering these factors together, functionalized graphene-based sensors has potential to detect changes in the nearby load concentration of less than single electron, and thus the involvement of small quantities of functional surface groups and adsorbents [66].

Graphene adhibitions are highlighted in the analysis of the Cd^{2+}, Hg^{2+} and Pb^{2+} toxic heavy metal ions. In 2015, Yuan *et al.* proposed light- nanofibrous films based on electrospinning, like conjugated microporous polymers (CMPs) / polylactic acid (PLA). High porosity, excellent versatility and high surface-ratio were demonstrated by the revolutionary chemical sensor, that improved the ability to sense nitroaromatic, oxidizing heavy metal ions etc. [67].

b. Detection of pesticides: Contamination of pesticides, in particular residues of pesticides in foodstuffs appears because of its excessive use in agriculture. The contamination of pesticides has become a major concern due to their long persistence, toxic biological effects, deep-distance transport, and trophic cascades along the food chain[68]. Functionalized graphene-based nanocomposite, in the form of sensor, has shown high specificity in the ampherometric detection of vast varieties of sulfuratedorganophosphorous (SOP) pesticides, malathione, parathione, and fenitrothion [69]. In another study, functionalized graphene based SPE (solid phase extraction) cartridges coupled with GC-MS (gas chromatography-mass spectrometry) exhibited excellent potential in clean-up of organophosphorus pesticides (OPPs) residues from apple juices. Thus, in addition to its ability to serve as sensor for detecting the pesticide, functionalized graphene also finds application in the removing the pesticide contaminants [70].

c. Detection of warfare agents/chemicals: Graphene and its functionalized forms are the basis of nanoelectromechanical (NEMS) mass resonators, which have been used experimentally in recent days to detect different gas molecules in air and in chemical warfare agents (CWA's) identification [71]. Graphene sheets can be classified as Pristine Graphene (PG), Graphene Oxide (GO), and Reduced Graphene Oxide (RGO), based on their structural features. Each of these groups of graphene sheets has unique capabilities for gas sensing. To date, graphene-based NEMS resonators have been used to investigate hypothetical and experimental detection of simple chemical substances, such as H_2, H_2O, O_2, CO_2, CO, NO_2, Fe, Co, Ni, Ru, Rh, Pd, Os, Ir, and Pt [72]. Even though the graphene as a gas sensor building block is highly sensitive to gas molecules, the comparison studies showed that graphene sensing capabilities could be improved by using a nanoporous substratum [73].

d. Detection of inorganic/organic ions and compounds: Graphene is powerful in the absorption of organic contaminants from aqueous solutions through $\pi-\pi$ and other electrostatic interactions. In many cases small organic molecules or surfactants were used in organic media or water to disperse graphene. Such surfactants / molecules mount permanently on the graphene sheets, thus preventing aggregation. This type of dispersion may be defined as a non-covalent form of chemical functionalization due to the immobilization of these molecules on the graphene surface by stacking, H-bonds, electrostatic forces etc. A study shown that carbon nanotube-graphene hybrid aerogels produced from their hydrogel precursors by supercritical CO_2 drying exhibit extremely high desalination efficiency, high binding capacities to certain heavy metal ions, and high adsorption capacities for dye materials. This composite material is promising in water purification because of its high flexibility, which can be attributed to its low weight, high conductivity, large BET surface area and hierarchically porous structure [74]. Engineered graphene has been used for detection of vast varieties of inorganic/organic compounds (hydrogen peroxide, nitroaromatics, hydrazine, nitriteand sulfite) as discussed below.

i. Sensing Hydrogen Peroxide: Due to pretty simplistic molecule in existence, hydrogen peroxide (H_2O_2) has been frequently utilized in the pharmacy, biomedical, environmental, mine, fabric, print, nutrition and chemical industries due to its powerful oxidization and reduction capabilities [75]. Its widespread application of underlies the necessity for concentration analysis of H_2O_2 for industrial and academic motives and creation of fast, simple, sensitive and low-cost H_2O_2 sensors. Till now, quite enough work has been applied to evaluating H_2O_2 concentration quickly and sensitively depending on various methodological concepts and therefore, a number of approaches were adapted and enforced to H_2O_2 assessment such as fluorimetry, colorimetry, chemiluminescence, titrimetry, spectrophotometry, electron spin resonance spectroscopy, chromatography and electrochemical methods [76]. The electrochemical detectors used for analysis of H_2O_2 are mainly categorized into enzymatic and nonenzymatic sensors. Non-enzymatic sensors comprised of transition metal nanoparticles (NPs), such as Pt, Ag, Au, Ni, etc.have established a fantastic degree of curiosity compared to enzymatic sensors, because of their exceptional deep stability and sensitivity, irrespective of the underlying function of enzymes. Table 1 presents different modification of graphene-based electrodes for sensing H_2O_2.

Table 1:Comparison of different graphene dependent electrodes for sensing H_2O_2

Electrode	Sensitivity ($\mu A\ cm^{-2}\ mmol^{-1}$)	LOD ($\mu mol/L$)	Linear Range ($\mu mol/L$)	References
Au@PBNPs graphene paper	5.0	0.1	1-30	[77]
AuNPs/graphene paper	0.24	2	5-8600	[78]
PtNPs/graphene paper	0.56	0.2	0-2500	[79]
PtNPs/MnO$_2$ nanowires/graphene paper	0.13	1	2.0-13.3	[80]
HRP/graphene paper	0.56	1.7	3.5-329	[81]
AgNCs/graphene film	0.18	3.0	20-10,000	[82]
AuNPs-Porous graphene film	0.08	0.1	0.5-4900	[83]
PtNPs/graphene paper	67.51	0.1	0.2-2000	[84]

ii. Sensing Nitroaromatics: Taking into account the environmental consequences, nitroaromatics sensing with certainty and sampling (from soil and water) cause problems for detection mechanisms. 2,4,6-trinitrotoluene (TNT) is one of the main traditional aromatic explosives. Moreover, highly volatile and soluble 2,4-dinitrotoluene (DNT) operates as a precursor in the industrial manufacture of TNT and hence frequently identified as an impurity suggesting the existence of TNT in contamination areas [85]. It is likely that undetonated explosives and ammunition flow into groundwater, which eventually passes and contaminates the ocean. This kind of harmful substances emerge as undesirable health effects in human populations, like blood, skin and nail discoloration, aplastic anemia and liver dysfunction caused by bioamplification and bioaccumulation [86].

The combination of traditional experimental approaches along with the use of functionalized graphene based sensing platforms for the quantitative and qualitative study of DNT and TNT involve gas chromatography, high-performance liquid chromatography, Raman spectroscopy, infrared absorption spectroscopy, mass spectrometry, immunoassay techniques, and electrochemical techniques [87-89]. Further improvisations in sensing platform has been made by equating electrochemical and fluorescent sensing techniques for TNT evaluation based on carbon quantum dots with the aim to obtain greater linear detection abilities [90]. DNT and TNT electrochemical analysis is made feasible by a 4-electron progressive reduction of each $-NO_2$ group to a $-NHOH$ group accompanied by the 2-electron reduction of $-NHOH$ group to $-NH_2$ group. Research results showing strong sensitivity and higher specificity as well as rapid response times with graphene and doped graphene demonstrate the substance's capacity as a detector for detecting trinitrotoluene (TNT) [91]. Several electrochemical approaches, such as nitrogen and sulphur co-doped graphene nanoribbons (LOD: 0.1 ppb), carbon nanotubes, oligomer-coated carbon nanotubes (LOD: 95 ppb), glassy carbon electrodes equipped with multi-wall carbon nanotubes (LOD: 0.6 ppb) and their configurations, have been documented for detecting TNT with a strong sensing capacity [92].

iii. Sensing Hydrazine: Hydrazine (HDZ) is a highly toxic chemical, cancer causing, neurotoxins, and hepatotoxic agent which impacts glutathione in the liver and brain. The fast water solubility as well as broad variety of industrial processes often makes HDZ a dangerous water and soil contaminant [93]. Hydrazine (HDZ) is an essential substance widely employed in a number of industries, like fuel cells, farming and medical. The hydrazine intermediates were also used for the manufacture of fabric dyes, explosive materials, aviation oils, antioxidants and visual designers. Functionalized graphene-based nanocomposites for designing electrochemical sensors possessed excellent sensitivity towards the detection of carcinogenic HDZ. Further such sensors exhibited outstanding recovery of HDZ and the limit of detection and quantification was 0.0045 and 0.015 µM respectively [94].

Several analytical approaches, particularly spectrophotometry, titrimetry, chemiluminescence, chromatography and electrochemical processes, have been evaluated for identification of HDZ. Recently, magnetite nanoparticle functionalized graphene nanohybrid has shown promising detection potential for HDZ and an extended linear detection range as compared to existing electrosensors. The large surface area of graphene along with the mesoporous nature of magnetite enhanced the electrocatalytic activity, stability, reproducibility and re-usability potential. It further improved the limit of detection (59nM), sensitivity ($27µAµM^{-1}$) and linear detection range (1-4400µM) for detection of HDZ thus served as efficient electrochemical sensing platform [95].

Nanomaterials based on carbon particularly graphene oxide (GO), reduced graphene oxide (RGO) and carbon nanomaterials like carbon nanofiber, carbon nanotube (CNT), carbon nano-horn are suitable candidates for electrochemical sensing implementations due to their excellent electrical conductivity, broad surface area and great structural versatility and low noise. The exploitation of unique graphene nanostructures such as ultra-thin graphene nanoflakes, graphene nanoribbons as an electrode substance has been of considerable relevance due to their wide specific surface area comparison to single exfoliated graphene sheets and carbon nanotube [96].

There are numerous studies on detection of HDZs using various metal and metal oxide nanoparticles such as G/GCE, cobalt hexacyanoferrate/RGO/GCE (CoHCF-RGO/GCE), Au/single-walled carbon nanohorns/GCE (Au/SWCNHs/GCE), RGO-Co_3O_4@Au/GCE, Co_3O_4/graphic-carbon nitride (Co_3O_4/g-C_3N_4), Ag–Ni/RGO, Au@Pd/carbon black-dihexadecylphosphate/GCE (Au@Pd/CB-DHP/GCE),Pt nanoclusters (PtNCs), Cu/Cu_2O@carbon/GCE, Au nanocage/chemically modified graphene (Au nanocage/CMG/GCE), cobalt-G/GCE and GCE/CNTs/catechol. Much of this HDZ detection analyses were performed by utilizing extremely costly catalysts such as Au, Ag, Pd, Pt [95] and to overcome this problem, Fe3O4 nanocomposites centered on G/GO/RGO/CNTs may be a suitable candidate towards the expense-effective electrochemical sensors of HDZ and Fe_3O_4 related nanohybrids. The electrochemical detector Fe_3O_4/polypyrrole/GO/GCE produced for HDZ monitoring displayed a LOD as well as a linear range of 1 μM and 5 μM–1.3 mM, respectively [97].

iv. Sensing Nitrite: Nitrite as being an essential oxynitride serves a massive role in the biological environment's nitrogen cycle. Furthermore, it is utilized extensively as a fertilizer and even as a preservative in the sectors of agriculture and the food sector. Although, nitrite consumption in behavioral environments is prohibited because unnecessary nitrite intake can result to harmful implications on physical health, particularly for children, aged women and pregnant mothers. For instance, haemoglobin oxidation in the bloodstream, the development of cancerous nitrosamines in the digestive tract and several conditions including oesophageal cancer and blue baby syndrome have all been induced by over nitrite intake [98]. According to World Health Organization report, the average nitrite content in drinkable water does not surpass 3 mg l^{-1}.

Numerous strategies to determining NO_2^- including spectrophotometry, chemiluminescence, ionic chromatography, high-performance liquid chromatography, gas chromatography-mass spectrometry, capillary electrophoresis, spectrofluorimetry and electrochemical methods have been established [99]. In comparison to all of these methods, related to its strong sensitivity, durability, convenience of use, inexpensive, better specificity, quick process, small detection limits and excellent efficiency, the electrochemical approach has received considerable importance [100]. Li et al. have found the electrochemical nitrite detector focused on graphene quantum dots coated with composite N-doped carbon nanofibers [101] and another research group used Ag@AgCl-graphene oxide@Fe_3O_4 nanocomposite to measure nitrite by chemiluminescent [102]. In addition, various chemical and electrochemical reduction approaches were employed for the development of graphene composites coated with metal nanoparticles like AuNPs-rGO compound. Moreover, AuNPs decorated with rGO in the possession of chitosan were often widely known [103]. In the table below (Table 2), we classified different strategies used for detection of Nitrate and Nitrite based on functionalized graphene, graphene oxide and reduced graphene oxide:

Table 2: *Different electrochemical sensing platforms for detection of nitrate and nitrite.*

Sensor type	Material	Detection range (µM)	LOD (µM)	Sensitivity (µA µM^{-1}cm^{-2})	References
GO based	Cube-like Pt/GO	0.5-227780	0.2	0.358	[104]
GO based	Ag-Fe_3O_4-GO	0.5-720 720-8150	0.17	1.966 0.426	[105]
GO based	GO-MWCNTs-PMA-Au	2-10000	0.67	0.484	[106]
GO based	Au-HNTs-GO	0.1-6330 6330-61900	0.03	0.0231 0.0865	[107]
GO based	Fe_3O_4/GO/COOH/GCE	1-85 90-600	0.37	0.192	[108]
RGO based	CuNDs/RGO	1.25-13000	0.4	0.214	[109]
RGO based	AuNP/RGO/MCNT/GCE	0.05-2200	0.014	1.201	[110]
RGO based	Fe_2O_3-RGO/GCE	0.05-780	0.015	0.204	[111]
RGO based	Fe_2O_3/H-C_3N_4/RGO	0.025-3000	0.0186	0.0487	[112]
RGO based	Co_3O_4/RGO	1-380	0.14	29.5	[113]
RGO based	Co_3O_4/RGO/CNTs	0.1-8000	0.016	0.408	[114]
RGO based	ERGO/β-CD/CdS/SPCE	0.05-447	0.021	0.00336	[115]

v. Sensing Sulphite: Sulphite (SO_3^{2-}) broadly utilized as a preservative in the nutrition and beverage sectors and because of owing to its possible toxicity, research centered on the identification of sulphite. The guidelines of the US Food and Drug Administration specifically reported that the sulphite ions in food and drink do not surpass 10 ppm. Sulphite is often employed in dissolved oxygen control boilers and boiler-feed waters therefore, a quick and accurate process for analyzing sulphite in these items is needed.

Numerous experimental procedures were documented for the analysis of sulphite, including high-performance liquid chromatography, capillary electrophoresis, chemiluminescence, and spectrophotometry. But these methods are not effective in comparison with electrochemical detection method due to its strength of simplicity, quick processing time, cheap, strong reproducibility and sensitivity [116]. As a recent electrode component, potassium doping (K-doped graphene) has been used to manipulate glassy carbon electrode (GCE) for the synchronized identification of sulphite (SO_3^{2-}) beneath neutral environments and it shows 2.5 μM to 10.3 mM linear range, with a detection level of 1.0 μM (S/N=3) for SO_3^{2-} [117]. Benzoylferrocene (BF)/ionic liquid adjusted graphene nano-sheet paste electrode (BF/IL/GPE) was being employed to develop as a modern electrode in electrocatalysis and sulphite evaluation in aqueous buffer fluid. Hydrophilic ionic liquid (n-hexyl-3-methylimidazolium hexafluoro phosphate) was also utilized for the preparation of the transformed electrode as a binder. Square wave voltammetry (SWV) shows a linear dynamic scale from $5.0*10^{-8}$ to $2.5*10^{-4}$ M and a 20.0 nMsulphite detection range [118].

e. Detection of fire retardants: In aqueous solutions, the chemically exfoliated single-layer graphene oxide (GO) sheets can be well distributed, carrying the advantages of minimal cost and wide quantity of manufacturing [119]. Thus, enormous efforts have been committed to GO reduction work, with the focus on the reduction mechanism, to boost their conductivity efficiently for large and versatile electronic implementations [120]. GO sheet reduction results in reduced graphene oxide (r-GO), also known as chemical-modified graphene. Various methods to GO reduction have been implemented like solution / chemical reduction, thermal heating, radical high temperature reduction, electrochemical reduction, photocatalytic reduction, and photothermal heating triggered by flash-light. The massive amount of heat produced during GO reduction can actually be used to promote other useful chemical processes, such as nanoparticles development [121].

Graphene is an extremely effective fire retardant and is non-toxic. Graphene is synergistic and can be used in coatings, resins, and foams to strengthen current fire-retardant products. This ensures that graphene flame retardants can work by themselves or add additional value to existing sources.

f. Detection of biologically harmful compounds:

Biochemical sensors are remarkably attractive due to their high accuracy, rapidity, accessibility, low cost and power consumption for mobile health monitoring. A standard biochemical sensor's basic sensing functions are as follows: a receptor such as an enzyme, antibody, DNA, or even whole cell is used in the specimens to precisely identify the desired analyte, and produces physicochemical indications. The transducer, such as an electrochemical, optical and mechanical transducer, then transforms the signal into an electrical, optical signal which can be computed [122].

Biochemical sensors commercially available on the market, such as blood glucose test strips, have been frequently used for blood testing and need blood sampling via an intrusive, painful process, especially for children, the old people and diabetics, which often bears the probable risk of infection or is inappropriate for large sampling rates in active monitoring.Biofluids like sweat, saliva, tears, and interstitial fluid (ISF) can be conveniently sampled as an alternative to blood in a non-invasive, user-friendly manner without penetrating the protective layer of the skin and touching blood, which provides a range of biochemical targets relevant to health and shows the potential for health monitoring. For example, chloride, lactate and glucose concentrations in sweat can be used to test for cystic fibrosis in children, identify the initiation of pressure-induced ischemia and the change between an aerobic and anaerobic state during physical activity [123]. The volatile biomarker gases still contain essential details, such as identification of metabolites and electrolytes. For example, analyzing ammonia from human breath gassesmay prescribe infections in the abdomen of helicobacter pylori; assessing exhaled NO can detect asthma. Approximately 2600 volatile organic compounds (VOCs) through air, skin, urine and blood were identified to be essential diagnostic supplementary references [124]. There have recently been significant biochemical graphene-based sensors monitoring markers of dimethyl methyl phosphonate (DMMP), ethanol, NO_2, SO_2 and tumor. A group of researchers introduced a revolutionary DMMP gas sensor with high versatility and accessibility, manufactured onto graphene by drop-coating polypyrrole. This gas sensor displayed excellent strain capability and exceptional variability independent of acetone, methanol, water and tetradecane [125].

Given the versatile graphene-based sensors for human health status, graphene and its derivatives have major applications in the detection of the natural environment which may pose a threat to human health. These can be used for the manufacture of gas sensors to detect massive harmful gasses such as NO_2, NH_3 and hazardous gases, as well as for ultraviolet (UV) rays. This form of reaction, particularly the doping phase, can trigger a change in GO's electrical resistance which could be detected using electrical methods. In a research, scientists implemented an advanced, layer-by-layer (LBL) flexible NO_2 gas sensor that succumbs GO to a gold electrode [126]. However considerable improvements have been made in recent years toward graphene-based sensors, a range of engineering and science hurdles need to be overcome unless practical implications may start. To begin with, risks to human health such as biocompatibility, biological toxicity, including the environmental effect of graphene and its variants, should have to be further evaluated, particularly in deep-term in vivo studies. Entire sensors are also expected to be properly verified, with graphene as its core. Certain applications include the impact of abnormalities, dysfunction, chemical functionality to incapacitate molecular receptors on the graphene oxide (GO) surface, reduced graphene oxide (RGO) and graphene- quantum dots (GQDs). So many awesome reports recently centered on the connections between graphene, GO and RGO- biosensors and their biomaterials targets [127].

Biosensors dependent on the GO, RGO and GQD functionalized graphene are commonly used in medicine, biomedicine and bioimaging procedures. This is prior to their incredibly wide surface area and ability to interact with different molecular forms. The excellent characteristics of solubility, biocompatibility, and functionalization also play a crucial role in the frameworks of sensing. Actually, the GO- FET has easily demonstrated a detection of glucose without an enzymatic solution [128]. GO is used as the preferential material for glucose in this module, whereas the sensor sensitivity is increased up to 1 μM by attaching CuNPS and AgNPs. It is important to note that functionalized graphene oxide (GO) dye has been printed on a pentacene FET todetect synthetic DNA and diffuse tumor cells. The negative charge causes gaps at the grain border of the pentacene layer as the DNA is caught by its phosphate component and creates clash or dispersion in the pentacene layer area. Consequently, the flexibility of the FET improves tremendously to reach the highest sensitivity of 0.1 pM and might this be enhanced for the manufacturing of printed biosensors [129].

In addition to providing an ideal medium for the preparation of usable sensor materials functionalized graphene based nanomaterials also enhances sensors' electrode efficiency [57]. Therefore, these sensors possess excellent electrochemical properties further, larger surface area of graphene facilitates accommodation large number of active sites, reproducibility, interference-free electron conduction, highly sensitive and specific detection of chemical analytes.

10.   Market trend for the development of electrochemical sensors

Global market trend for the electrochemical sensors valued 21 billion USD in 2019 and is expected to reach a value of 32.96 billion USD by 2025, at annual growth rate of 7.51 percent during the forecast year for 2020-2025. The increasing levels of chemical toxins in the ecosystem has instigated the demand for the electrochemical sensors. Further, the user-friendly interface, low-costs and portable nature of the sensors fuels the market demand. The chemical toxins are major cause for deterioration of human health and the environment thus, increasing the disease vulnerability in all age groups. Electrochemical sensors based on molecularly imprinted functionalized graphene and gold nanoparticle has been potent at determining pesticides [130]. In 2018, a Korean institute developed chemical nanosensor for detecting trace elements from the exhaled breath that associated with certain diseases. Recently, functionalized graphene-based nanocomposite has shown potency in determining organophosphate pesticide.

The chemical sensor market is driven by continuous environmental, research laboratories and healthcare concerns. Further, technical advances accelerate the need for chemical sensors in the food, pharmaceutical as well as the chemical industries. Additionally, electrochemical sensors also track hazardous chemicals and gases thus, facilitates the identification of man-made and/or natural incidents in terms of homeland security. In the light of abovementioned facts, it is likely that sensor application has been increasing worldwide however, the poor life-span of chemical sensors remains one of the major causes for their limited acceptance by the customers. Likewise, high maintenance costs, requirement of air-conditioned facility, frequent recalibrations have obstructed the growth of chemical sensor market. Obviously, new strategies are been searched for improving the efficiencies of the electrochemical sensors. Currently, the development of high-throughput, highly sensitive electrochemical sensors capable of detecting the analytes in smaller sample volumes (<50 microlitres) have been reported and now are ready to boom the sensor market in coming years.

11. Future perspective

Routine exposure to millions of chemicals in the form of pollutants, food and water contaminants, has been associated with the evolution of chronic human diseases. Graphene based electrochemical sensors with outstanding properties, presents a viable option for detection of chemical toxins, that will help in shaping the healthcare priorities for saving the mankind. Recent advances are discussed and described in electrochemical sensors, bio-sensors and pesticide sensors made of different nanocomposite materials with various morphologies for the reactive monitoring of food additives, human health and environmental protection. Of interest, due to their outstanding durability electrochemical techniques have been given considerable attention, although potential applicants for real time analytes detection in the newly published studies are also eligible. Electrochemical sensing platforms are ideal for advanced nanostructured materials including nanorods, nanocubes, nanoneedles, nanobundles, nano tubes and nanoflowers, as they show high efficacy electrochemicals sensitivity, selectivity and long-term stability. The use of nanotechnology for development of high-throughput, portable, inexpensive, sensitive and flexible electrochemical sensor along with suitable biomonitoring tool offers potential for better understanding the risk factor associated with health adversities. The futuristic goal for improving quality and utility of electrochemical sensors relies on collaboration of physicists, chemists, epidemiologists and policymakers for improving the market demands and quality of lives.

Conclusion

The emphasis on the existence of toxic chemical products in the atmosphere is directing scientific research towards the establishment new and eco-sustainable management strategies. Such technological improvements allow scientists to boost the limit of detection in the atmosphere, food and the human body as well as develop new techniques to eliminate them from our environments. However, further research is required to achieve the goal of a continuous monitoring of the environment and of providing, in real time, information on its current state.

Acknowledgement

This work was supported by Department of Science & Technology (DST) (SR/NM/NS-57/2016) New Delhi (under nano mission). The funder has no role in design, decision to publish or preparation of report.

References

[1] Organization, W.H., The public health impact of chemicals: knowns and unknowns. 2016, World Health Organization.

[2] Prüss-Üstün, A., et al., Preventing disease through healthy environments: a global assessment of the burden of disease from environmental risks. 2016: World Health Organization.

[3] Febbraio, F., Biochemical strategies for the detection and detoxification of toxic chemicals in the environment. World journal of biological chemistry, 2017. **8**(1): p. 13. https://doi.org/10.4331/wjbc.v8.i1.13

[4] Chen, Y., Organophosphate-induced brain damage: mechanisms, neuropsychiatric and neurological consequences, and potential therapeutic strategies. Neurotoxicology, 2012. **33**(3): p. 391-400. https://doi.org/10.1016/j.neuro.2012.03.011

[5] Barnett, L.M. and B.S. Cummings, Nephrotoxicity and renal pathophysiology: a contemporary perspective. Toxicological Sciences, 2018. **164**(2): p. 379-390. https://doi.org/10.1093/toxsci/kfy159

[6] Rim, K.-T., Reproductive Toxic Chemicals at Work and Efforts to Protect Workers' Health: A Literature Review. Safety and health at work, 2017. **8**(2): p. 143-150. https://doi.org/10.1016/j.shaw.2017.04.003

[7] Ribeiro, A.R. and G. de Aragão Umbuzeiro, Effects of a textile azo dye on mortality, regeneration, and reproductive performance of the planarian, Girardia tigrina. Environmental Sciences Europe, 2014. **26**(1): p. 22. https://doi.org/10.1186/s12302-014-0022-5

[8] Leonardi-Bee, J., J. Britton, and A. Venn, Secondhand smoke and adverse fetal outcomes in nonsmoking pregnant women: a meta-analysis. Pediatrics, 2011. **127**(4): p. 734-741. https://doi.org/10.1542/peds.2010-3041

[9] Pressman, P., et al., Food additive safety: A review of toxicologic and regulatory issues. Toxicology Research and application, 2017. **1**: p. 2397847317723572. https://doi.org/10.1177/2397847317723572

[10] Frankos, V.H. and J.V. Rodricks, Food additives and nutrition supplements. Regulatory toxicology. London: Taylor & Francis, 2001: p. 133-166. https://doi.org/10.1201/9781420025187.ch5

[11] Dasgupta, M., Neurotoxicity, Immunotoxicity and Drug Toxicity–A Review.

[12] Schulz, M., et al., Therapeutic and toxic blood concentrations of nearly 1,000 drugs and other xenobiotics. Critical care, 2012. **16**(4): p. R136. https://doi.org/10.1186/cc11441

[13] Harvey, A.L., Toxins and drug discovery. Toxicon, 2014. **92**: p. 193-200. https://doi.org/10.1016/j.toxicon.2014.10.020

[14] Waller, D.G. and T. Sampson, Medical pharmacology and therapeutics E-Book. 2017: Elsevier Health Sciences.

[15] Nriagu, J.O., Encyclopedia of environmental health. 2019: Elsevier.

[16] Wilkinson, C.F., et al., Assessing the risks of exposures to multiple chemicals with a common mechanism of toxicity: how to cumulate? Regulatory Toxicology and Pharmacology, 2000. **31**(1): p. 30-43. https://doi.org/10.1006/rtph.1999.1361

[17] Franco, R., et al., Molecular mechanisms of pesticide-induced neurotoxicity: Relevance to Parkinson's disease. Chemico-biological interactions, 2010. **188**(2): p. 289-300. https://doi.org/10.1016/j.cbi.2010.06.003

[18] Katagi, T., Bioconcentration, bioaccumulation, and metabolism of pesticides in aquatic organisms, in Reviews of environmental contamination and toxicology. 2010, Springer. p. 1-132. https://doi.org/10.1007/978-1-4419-1440-8_1

[19]   Zhang, Y., Why do we study animal toxins? Zoological research, 2015. **36**(4): p. 183.

[20]   Schafer, K.S. and S.E. Kegley, Persistent toxic chemicals in the US food supply. Journal of Epidemiology & Community Health, 2002. **56**(11): p. 813-817. https://doi.org/10.1136/jech.56.11.813

[21]   Hernández, R., et al., Reduced graphene oxide films as solid transducers in potentiometric all-solid-state ion-selective electrodes. The Journal of Physical Chemistry C, 2012. **116**(42): p. 22570-22578. https://doi.org/10.1021/jp306234u

[22]   Norouzi, P., et al., A glucose biosensor based on nanographene and ZnO nanoparticles using FFT continuous cyclic voltammetry. Int. J. Electrochem. Sci, 2011. **6**(11): p. 5189-5195.

[23]   Yuan, M., et al., Bimetallic PdCu nanoparticle decorated three-dimensional graphene hydrogel for non-enzymatic amperometric glucose sensor. Sensors and Actuators B: Chemical, 2014. **190**: p. 707-714. https://doi.org/10.1016/j.snb.2013.09.054

[24]   Council, N.R., Expanding the vision of sensor materials. 1995: National Academies Press.

[25]   Nomenclature, E.T. and A. Terminology, Standard MC6. 1-1975 (ISA S37. 1). Instrument Society of America, Research Triangle Park, NC, 1975.

[26]   Lion, K.S., Transducers: Problems and prospects. IEEE Transactions on Industrial Electronics and Control Instrumentation, 1969(1): p. 2-5. https://doi.org/10.1109/TIECI.1969.229858

[27]   Göpel, W., J. Hesse, and J.N. Zemel, Sensors: a comprehensive survey. 1989.

[28]   Middelhoek, S. and D. Noorlag, Three-dimensional representation of input and output transducers. Sensors and Actuators, 1981. **2**: p. 29-41. https://doi.org/10.1016/0250-6874(81)80026-1

[29]   Khan, S.B., et al., Humidity and temperature sensing properties of copper oxide–Si-adhesive nanocomposite. Talanta, 2014. **120**: p. 443-449. https://doi.org/10.1016/j.talanta.2013.11.089

[30]   Wu, C.H., et al., Ta3N5 Nanowire Bundles as Visible-Light-Responsive Photoanodes. Chemistry–An Asian Journal, 2013. **8**(10): p. 2354-2357. https://doi.org/10.1002/asia.201300717

[31]   Tüysüz, H., et al., Mesoporous Co 3 O 4 as an electrocatalyst for water oxidation. Nano Research, 2013. **6**(1): p. 47-54. https://doi.org/10.1007/s12274-012-0280-8

[32]   Dadashi-Silab, S., et al., Photoinduced atom transfer radical polymerization using semiconductor nanoparticles. Macromolecular rapid communications, 2014. **35**(4): p. 454-459. https://doi.org/10.1002/marc.201300704

[33]   Ling, X.Y., et al., Alumina-coated Ag nanocrystal monolayers as surfaceenhanced Raman spectroscopy platforms for the direct spectroscopic detection of water splitting reaction intermediates. Nano Research, 2014. **7**(1): p. 132-143. https://doi.org/10.1007/s12274-013-0380-0

[34]   Liu, S. and Z. Tang, Nanoparticle assemblies for biological and chemical sensing. Journal of Materials chemistry, 2010. **20**(1): p. 24-35. https://doi.org/10.1039/B911328M

[35]   Marwani, H.M., et al., Cellulose-lanthanum hydroxide nanocomposite as a selective marker for detection of toxic copper. Nanoscale research letters, 2014. **9**(1): p. 466. https://doi.org/10.1186/1556-276X-9-466

[36]   Khan, M.M., et al., Novel Ag@ TiO2 nanocomposite synthesized by electrochemically active biofilm for nonenzymatic hydrogen peroxide sensor. Materials Science and Engineering: C, 2013. **33**(8): p. 4692-4699. https://doi.org/10.1016/j.msec.2013.07.028

[37]   Asif, S.A.B., S.B. Khan, and A.M. Asiri, Efficient solar photocatalyst based on cobalt oxide/iron oxide composite nanofibers for the detoxification of organic pollutants. Nanoscale research letters, 2014. **9**(1): p. 510. https://doi.org/10.1186/1556-276X-9-510

[38]   Khan, S.B., et al., Detection and Monitoring of Toxic Chemical at Ultra Trace Level by Utilizing Doped Nanomaterial. PloS one, 2014. **9**(10). https://doi.org/10.1371/journal.pone.0109423

[39]   Liu, G. and Y. Lin, Biosensor based on self-assembling acetylcholinesterase on carbon nanotubes for flow injection/amperometric detection of organophosphate pesticides and nerve agents. Analytical Chemistry, 2006. **78**(3): p. 835-843. https://doi.org/10.1021/ac051559q

[40]   Wang, J., C. Timchalk, and Y. Lin, Carbon nanotube-based electrochemical sensor for assay of salivary cholinesterase enzyme activity: an exposure biomarker of organophosphate pesticides and nerve agents. Environmental science & technology, 2008. **42**(7): p. 2688-2693. https://doi.org/10.1021/es702335y

[41]   Kim, S.N., J.F. Rusling, and F. Papadimitrakopoulos, Carbon nanotubes for electronic and electrochemical detection of biomolecules. Advanced materials, 2007. **19**(20): p. 3214-3228. https://doi.org/10.1002/adma.200700665

[42]   Chen, S., et al., Amperometric third-generation hydrogen peroxide biosensor based on the immobilization of hemoglobin on multiwall carbon nanotubes and gold colloidal nanoparticles. Biosensors and Bioelectronics, 2007. **22**(7): p. 1268-1274. https://doi.org/10.1016/j.bios.2006.05.022

[43]   Patolsky, F., G. Zheng, and C.M. Lieber, Fabrication of silicon nanowire devices for ultrasensitive, label-free, real-time detection of biological and chemical species. Nature protocols, 2006. **1**(4): p. 1711. https://doi.org/10.1038/nprot.2006.227

[44]   Liu, G. and Y. Lin, Electrochemical sensor for organophosphate pesticides and nerve agents using zirconia nanoparticles as selective sorbents. Analytical chemistry, 2005. **77**(18): p. 5894-5901. https://doi.org/10.1021/ac050791t

[45]   Guo, S. and E. Wang, Synthesis and electrochemical applications of gold nanoparticles. Analytica chimica acta, 2007. **598**(2): p. 181-192. https://doi.org/10.1016/j.aca.2007.07.054

[46]   Pumera, M., et al., Graphene for electrochemical sensing and biosensing. TrAC Trends in Analytical Chemistry, 2010. **29**(9): p. 954-965. https://doi.org/10.1016/j.trac.2010.05.011

[47]   Wang, Q.H. and M.C. Hersam, Room-temperature molecular-resolution characterization of self-assembled organic monolayers on epitaxial graphene. Nature Chemistry, 2009. **1**(3): p. 206. https://doi.org/10.1038/nchem.212

[48]   Malig, J., et al., Wet chemistry of graphene. Electrochemical Society Interface, 2011. **20**(1): p. 53. https://doi.org/10.1149/2.F06111if

[49]   Georgakilas, V., et al., Functionalization of graphene: covalent and non-covalent approaches, derivatives and applications. Chemical reviews, 2012. **112**(11): p. 6156-6214. https://doi.org/10.1021/cr3000412

[50]   Georgakilas, V., et al., Noncovalent functionalization of graphene and graphene oxide for energy materials, biosensing, catalytic, and biomedical applications. Chemical reviews, 2016. **116**(9): p. 5464-5519. https://doi.org/10.1021/acs.chemrev.5b00620

[51]   Young, R.J., et al., The mechanics of graphene nanocomposites: a review. Composites Science and Technology, 2012. **72**(12): p. 1459-1476. https://doi.org/10.1016/j.compscitech.2012.05.005

[52]   Bartlett, N. and B. McQuillan, Intercalation Chemistry ed MS Whittingham and AJ Jacobson. 1982, New York: Academic Press.

[53]   Shi, Y., et al., Work function engineering of graphene electrode via chemical doping. ACS nano, 2010. **4**(5): p. 2689-2694. https://doi.org/10.1021/nn1005478

[54]   Wang, X., et al., N-doping of graphene through electrothermal reactions with ammonia. Science, 2009. **324**(5928): p. 768-771. https://doi.org/10.1126/science.1170335

[55]   Lin, Y.-C., C.-Y. Lin, and P.-W. Chiu, Controllable graphene N-doping with ammonia plasma. Applied Physics Letters, 2010. **96**(13): p. 133110. https://doi.org/10.1063/1.3368697

[56]   McNaught, A.D. and A. Wilkinson, Compendium of chemical terminology. Vol. 1669. 1997: Blackwell Science Oxford.

[57]   Wang, B., et al., Preparation of nickel nanoparticle/graphene composites for non-enzymatic electrochemical glucose biosensor applications. Materials Research Bulletin, 2014. **49**: p. 521-524. https://doi.org/10.1016/j.materresbull.2013.08.066

[58]   Yang, Z., et al., Graphene oxide based ultrasensitive flow-through chemiluminescent immunoassay for sub-picogram level detection of chicken interferon-γ. Biosensors and Bioelectronics, 2014. **51**: p. 356-361. https://doi.org/10.1016/j.bios.2013.07.067

[59]   Cheng, Y., et al., Highly sensitive luminol electrochemiluminescence immunosensor based on ZnO nanoparticles and glucose oxidase decorated graphene for cancer biomarker detection. Analytica chimica acta, 2012. **745**: p. 137-142. https://doi.org/10.1016/j.aca.2012.08.010

[60]   Li, M., et al., Nanostructured sensors for detection of heavy metals: a review. 2013, ACS Publications. https://doi.org/10.1021/sc400019a

[61]   Aragay, G., J. Pons, and A. Merkoçi, Recent trends in macro-, micro-, and nanomaterial-based tools and strategies for heavy-metal detection. Chemical reviews, 2011. **111**(5): p. 3433-3458. https://doi.org/10.1021/cr100383r

[62]   Carter, K.P., A.M. Young, and A.E. Palmer, Fluorescent sensors for measuring metal ions in living systems. Chemical reviews, 2014. **114**(8): p. 4564-4601. https://doi.org/10.1021/cr400546e

[63]   Tian, W., X. Liu, and W. Yu, Research progress of gas sensor based on graphene and its derivatives: A review. Applied Sciences, 2018. **8**(7): p. 1118. https://doi.org/10.3390/app8071118

[64]   Gadipelli, S. and Z.X. Guo, Graphene-based materials: Synthesis and gas sorption, storage and separation. Progress in Materials Science, 2015. **69**: p. 1-60. https://doi.org/10.1016/j.pmatsci.2014.10.004

[65]   Tuantranont, A., Applications of nanomaterials in sensors and diagnostics. Springer series on chemical sensors and biosensors (Springer, Berlin Heidelberg, 2014), 2013. https://doi.org/10.1007/978-3-642-36025-1

[66]   Irudayaraj, J.M., Biomedical nanosensors. 2012: Pan Stanford. https://doi.org/10.1201/b13721

[67]   Yuan, K., et al., Nanofibrous and graphene-templated conjugated microporous polymer materials for flexible chemosensors and supercapacitors. Chemistry of Materials, 2015. **27**(21): p. 7403-7411. https://doi.org/10.1021/acs.chemmater.5b03290

[68]   Smith, A. and S. Gangolli, Organochlorine chemicals in seafood: occurrence and health concerns. Food and Chemical Toxicology, 2002. **40**(6): p. 767-779. https://doi.org/10.1016/S0278-6915(02)00046-7

[69]   Li, Z., et al., A nanocomposite of copper (ii) functionalized graphene and application for sensing sulfurated organophosphorus pesticides. New Journal of Chemistry, 2013. **37**(12): p. 3956-3963. https://doi.org/10.1039/c3nj00528c

[70]   Han, Q., et al., Application of graphene for the SPE clean-up of organophosphorus pesticides residues from apple juices. Journal of separation science, 2014. **37**(1-2): p. 99-105. https://doi.org/10.1002/jssc.201301005

[71]   Anđelić, N., Z. Car, and M. Čanađija, NEMS Resonators for Detection of Chemical Warfare Agents Based on Graphene Sheet. Mathematical Problems in Engineering, 2019. **2019**. https://doi.org/10.1155/2019/6451861

[72]   Varghese, S.S., et al., Recent advances in graphene based gas sensors. Sensors and Actuators B: Chemical, 2015. **218**: p. 160-183. https://doi.org/10.1016/j.snb.2015.04.062

[73]   Yang, C.-S., et al., Enhancing gas sensing properties of graphene by using a nanoporous substrate. 2D Materials, 2016. **3**(1): p. 011007. https://doi.org/10.1088/2053-1583/3/1/011007

[74]   Sui, Z., et al., Green synthesis of carbon nanotube–graphene hybrid aerogels and their use as versatile agents for water purification. Journal of Materials Chemistry, 2012. **22**(18): p. 8767-8771. https://doi.org/10.1039/c2jm00055e

[75]   Chen, W., et al., Recent advances in electrochemical sensing for hydrogen peroxide: a review. Analyst, 2012. **137**(1): p. 49-58. https://doi.org/10.1039/C1AN15738H

[76]   Zhang, R. and W. Chen, Recent advances in graphene-based nanomaterials for fabricating electrochemical hydrogen peroxide sensors. Biosensors and Bioelectronics, 2017. **89**: p. 249-268. https://doi.org/10.1016/j.bios.2016.01.080

[77]   Zhang, M., et al., Free-standing and flexible graphene papers as disposable non-enzymatic electrochemical sensors. Bioelectrochemistry, 2016. **109**: p. 87-94. https://doi.org/10.1016/j.bioelechem.2016.02.002

[78]   Xiao, F., et al., Coating graphene paper with 2D-assembly of electrocatalytic nanoparticles: a modular approach toward high-performance flexible electrodes. ACS nano, 2012. **6**(1): p. 100-110. https://doi.org/10.1021/nn202930m

[79]   Liang, B., et al., Fabrication and application of flexible graphene silk composite film electrodes decorated with spiky Pt nanospheres. Nanoscale, 2014. **6**(8): p. 4264-4274. https://doi.org/10.1039/C3NR06057H

[80]   Xiao, F., et al., Growth of metal–metal oxide nanostructures on freestanding graphene paper for flexible biosensors. Advanced Functional Materials, 2012. **22**(12): p. 2487-2494. https://doi.org/10.1002/adfm.201200191

[81]   Zhang, Q., et al., Direct Electrochemistry and Electrocatalysis of Horseradish Peroxidase Immobilized on Water Soluble Sulfonated Graphene Film via Self-assembly. Electroanalysis, 2011. **23**(4): p. 900-906. https://doi.org/10.1002/elan.201000614

[82]   Zhong, L., et al., Electrochemically controlled growth of silver nanocrystals on graphene thin film and applications for efficient nonenzymatic H2O2 biosensor. Electrochimica Acta, 2013. **89**: p. 222-228. https://doi.org/10.1016/j.electacta.2012.10.161

[83]   Xi, Q., et al., Gold nanoparticle-embedded porous graphene thin films fabricated via layer-by-layer self-assembly and subsequent thermal annealing for electrochemical sensing. Langmuir, 2012. **28**(25): p. 9885-9892. https://doi.org/10.1021/la301440k

[84]   Song, R.-M., et al., Flexible hydrogen peroxide sensors based on platinum modified free-standing reduced graphene oxide paper. Applied Sciences, 2018. **8**(6): p. 848. https://doi.org/10.3390/app8060848

[85]   Luning Prak, D.J. and D.W. O'Sullivan, Solubility of 2, 4-dinitrotoluene and 2, 4, 6-trinitrotoluene in seawater. Journal of Chemical & Engineering Data, 2006. **51**(2): p. 448-450. https://doi.org/10.1021/je050373l

[86]   Letzel, S., et al., Exposure to nitroaromatic explosives and health effects during disposal of military waste. Occupational and environmental medicine, 2003. **60**(7): p. 483-488. https://doi.org/10.1136/oem.60.7.483

[87]   Walsh, M.E., Determination of nitroaromatic, nitramine, and nitrate ester explosives in soil by gas chromatography and an electron capture detector. Talanta, 2001. **54**(3): p. 427-438. https://doi.org/10.1016/S0039-9140(00)00541-5

[88]   Liu, M. and W. Chen, Graphene nanosheets-supported Ag nanoparticles for ultrasensitive detection of TNT by surface-enhanced Raman spectroscopy. Biosensors and Bioelectronics, 2013. **46**: p. 68-73. https://doi.org/10.1016/j.bios.2013.01.073

[89]   Mullen, C., et al., Detection of explosives and explosives-related compounds by single photon laser ionization time-of-flight mass spectrometry. Analytical chemistry, 2006. **78**(11): p. 3807-3814. https://doi.org/10.1021/ac060190h

[90]   Zhang, L., et al., Simple and sensitive fluorescent and electrochemical trinitrotoluene sensors based on aqueous carbon dots. Analytical chemistry, 2015. **87**(4): p. 2033-2036. https://doi.org/10.1021/ac5043686

[91]   Amin, K.R. and A. Bid, High-performance sensors based on resistance fluctuations of single-layer-graphene transistors. ACS applied materials & interfaces, 2015. **7**(35): p. 19825-19830. https://doi.org/10.1021/acsami.5b05922

[92]   Dettlaff, A., et al., Electrochemical determination of nitroaromatic explosives at boron-doped diamond/graphene nanowall electrodes: 2, 4, 6-trinitrotoluene and 2, 4, 6-trinitroanisole in liquid effluents. Journal of hazardous materials, 2020. **387**: p. 121672. https://doi.org/10.1016/j.jhazmat.2019.121672

[93]   Vernot, E., et al., Long-term inhalation toxicity of hydrazine. Toxicological Sciences, 1985. **5**(6part1): p. 1050-1064. https://doi.org/10.1016/0272-0590(85)90141-1

[94]   Vellaichamy, B., P. Periakaruppan, and S.K. Ponnaiah, A new in-situ synthesized ternary CuNPs-PANI-GO nano composite for selective detection of carcinogenic hydrazine. Sensors and Actuators B: Chemical, 2017. **245**: p. 156-165. https://doi.org/10.1016/j.snb.2017.01.117

[95]   Vinodha, G., P. Shima, and L. Cindrella, Mesoporous magnetite nanoparticle-decorated graphene oxide nanosheets for efficient electrochemical detection of hydrazine. Journal of materials science, 2019. **54**(5): p. 4073-4088. https://doi.org/10.1007/s10853-018-3145-z

[96]   Martín, A., et al., Graphene nanoribbon-based electrochemical sensors on screen-printed platforms. Electrochimica Acta, 2015. **172**: p. 2-6. https://doi.org/10.1016/j.electacta.2014.11.090

[97]   Yang, Z., et al., One-pot synthesis of Fe 3 O 4/polypyrrole/graphene oxide nanocomposites for electrochemical sensing of hydrazine. Microchimica Acta, 2017. **184**(7): p. 2219-2226. https://doi.org/10.1007/s00604-017-2197-0

[98]   Mirvish, S.S., Role of N-nitroso compounds (NOC) and N-nitrosation in etiology of gastric, esophageal, nasopharyngeal and bladder cancer and contribution to cancer of known exposures to NOC. Cancer letters, 1995. **93**(1): p. 17-48. https://doi.org/10.1016/0304-3835(95)03786-V

[99]   Jaiswal, N., et al., Highly sensitive amperometric sensing of nitrite utilizing bulk-modified MnO2 decorated Graphene oxide nanocomposite screen-printed electrodes. Electrochimica Acta, 2017. **227**: p. 255-266. https://doi.org/10.1016/j.electacta.2017.01.007

[100] Fu, L., et al., Carbon nanotube and graphene oxide directed electrochemical synthesis of silver dendrites. Rsc Advances, 2014. **4**(75): p. 39645-39650. https://doi.org/10.1039/C4RA06156J

[101] Li, L., et al., Quantitative detection of nitrite with N-doped graphene quantum dots decorated N-doped carbon nanofibers composite-based electrochemical sensor. Sensors and Actuators B: Chemical, 2017. **252**: p. 17-23. https://doi.org/10.1016/j.snb.2017.05.155

[102] Abdolmohammad-Zadeh, H. and E. Rahimpour, Utilizing of Ag@ AgCl@ graphene oxide@ Fe3O4 nanocomposite as a magnetic plasmonic nanophotocatalyst in light-initiated H2O2 generation and chemiluminescence detection of nitrite. Talanta, 2015. **144**: p. 769-777. https://doi.org/10.1016/j.talanta.2015.07.030

[103] Fang, Y., et al., Simple one-pot preparation of chitosan-reduced graphene oxide-Au nanoparticles hybrids for glucose sensing. Sensors and Actuators B: Chemical, 2015. **221**: p. 265-272. https://doi.org/10.1016/j.snb.2015.06.098

[104] Bai, W., Q. Sheng, and J. Zheng, Morphology controlled synthesis of platinum nanoparticles performed on the surface of graphene oxide using a gas–liquid interfacial reaction and its application for high-performance electrochemical sensing. Analyst, 2016. **141**(14): p. 4349-4358. https://doi.org/10.1039/C6AN00632A

[105] Li, B.-Q., et al., An electrochemical sensor for sensitive determination of nitrites based on Ag-Fe 3 O 4-graphene oxide magnetic nanocomposites. Chemical Papers, 2015. **69**(7): p. 911-920. https://doi.org/10.1515/chempap-2015-0099

[106] Rao, D., Q. Sheng, and J. Zheng, Self-assembly preparation of gold nanoparticle decorated 1-pyrenemethylamine functionalized graphene oxide–carbon nanotube composites for highly sensitive detection of nitrite. Analytical Methods, 2016. **8**(24): p. 4926-4933. https://doi.org/10.1039/C6AY01316C

[107] Zhang, S., Q. Sheng, and J. Zheng, Synthesis of Au nanoparticles dispersed on halloysite nanotubes–reduced graphene oxide nanosheets and their application for electrochemical sensing of nitrites. New Journal of Chemistry, 2016. **40**(11): p. 9672-9678. https://doi.org/10.1039/C6NJ02103D

[108] Rostami, M., et al., Nanocomposite of magnetic nanoparticles/graphene oxide decorated with acetic acid moieties on glassy carbon electrode: A facile method to detect nitrite concentration. Journal of Electroanalytical Chemistry, 2019. **847**: p. 113239. https://doi.org/10.1016/j.jelechem.2019.113239

[109] Zhang, D., et al., Direct electrodeposion of reduced graphene oxide and dendritic copper nanoclusters on glassy carbon electrode for electrochemical detection of nitrite. Electrochimica Acta, 2013. **107**: p. 656-663. https://doi.org/10.1016/j.electacta.2013.06.015

[110] Yu, H., R. Li, and K.-l. Song, Amperometric determination of nitrite by using a nanocomposite prepared from gold nanoparticles, reduced graphene oxide and multi-walled carbon nanotubes. Microchimica Acta, 2019. **186**(9): p. 624. https://doi.org/10.1007/s00604-019-3735-8

[111] Radhakrishnan, S., et al., A highly sensitive electrochemical sensor for nitrite detection based on Fe2O3 nanoparticles decorated reduced graphene oxide nanosheets. Applied Catalysis B: Environmental, 2014. **148**: p. 22-28. https://doi.org/10.1016/j.apcatb.2013.10.044

[112] Wang, S., et al., Protonated carbon nitride induced hierarchically ordered Fe2O3/HC3N4/rGO architecture with enhanced electrochemical sensing of nitrite. Sensors and Actuators B: Chemical, 2018. **260**: p. 490-498. https://doi.org/10.1016/j.snb.2018.01.073

[113] Haldorai, Y., et al., An enzyme-free electrochemical sensor based on reduced graphene oxide/Co3O4 nanospindle composite for sensitive detection of nitrite. Sensors and Actuators B: Chemical, 2016. **227**: p. 92-99. https://doi.org/10.1016/j.snb.2015.12.032

[114] Zhao, Z., et al., Synthesis and electrochemical properties of Co3O4-rGO/CNTs composites towards highly sensitive nitrite detection. Applied Surface Science, 2019. **485**: p. 274-282. https://doi.org/10.1016/j.apsusc.2019.04.202

[115] Balasubramanian, P., et al., A single-step electrochemical preparation of cadmium sulfide anchored erGO/β-CD modified screen-printed carbon electrode for sensitive and selective detection of nitrite. Journal of The Electrochemical Society, 2019. **166**(8): p. B690-B696. https://doi.org/10.1149/2.0981908jes

[116] Zainudin, N., M.M. Yusoff, and K.F. Chong. A promising electrochemical sensing platform based on a graphene nanomaterials for sensitive sulfite determination. in 2015 2nd International Conference on Biomedical Engineering (ICoBE). 2015. IEEE. https://doi.org/10.1109/ICoBE.2015.7235886

[117] Li, X.-R., et al., Potassium-doped graphene for simultaneous determination of nitrite and sulfite in polluted water. Electrochemistry communications, 2012. **20**: p. 109-112. https://doi.org/10.1016/j.elecom.2012.04.014

[118] Beitollahi, H., S. Tajik, and P. Biparva, Electrochemical determination of sulfite and phenol using a carbon paste electrode modified with ionic liquids and graphene nanosheets: application to determination of sulfite and phenol in real samples. Measurement, 2014. **56**: p. 170-177. https://doi.org/10.1016/j.measurement.2014.06.011

[119] Gao, W., et al., New insights into the structure and reduction of graphite oxide. Nature chemistry, 2009. **1**(5): p. 403. https://doi.org/10.1038/nchem.281

[120] Eda, G., G. Fanchini, and M. Chhowalla, Large-area ultrathin films of reduced graphene oxide as a transparent and flexible electronic material. Nature nanotechnology, 2008. **3**(5): p. 270-274. https://doi.org/10.1038/nnano.2008.83

[121] Shi, Y. and L.-J. Li, Chemically modified graphene: flame retardant or fuel for combustion? Journal of Materials Chemistry, 2011. **21**(10): p. 3277-3279. https://doi.org/10.1039/C0JM02953J

[122] Kim, J., et al., Wearable biosensors for healthcare monitoring. Nature biotechnology, 2019. **37**(4): p. 389-406. https://doi.org/10.1038/s41587-019-0045-y

[123] Ray, T.R., et al., Bio-integrated wearable systems: A comprehensive review. Chemical reviews, 2019. **119**(8): p. 5461-5533. https://doi.org/10.1021/acs.chemrev.8b00573

[124] Tricoli, A., N. Nasiri, and S. De, Wearable and miniaturized sensor technologies for personalized and preventive medicine. Advanced Functional Materials, 2017. **27**(15): p. 1605271. https://doi.org/10.1002/adfm.201605271

[125] HyungáCheong, W., J. HyebáSong, and J. JoonáKim, Wearable, wireless gas sensors using highly stretchable and transparent structures of nanowires and graphene. Nanoscale, 2016. **8**(20): p. 10591-10597. https://doi.org/10.1039/C6NR01468B

[126] Su, P.-G. and H.-C. Shieh, Flexible NO2 sensors fabricated by layer-by-layer covalent anchoring and in situ reduction of graphene oxide. Sensors and Actuators B: Chemical, 2014. **190**: p. 865-872. https://doi.org/10.1016/j.snb.2013.09.078

[127] Li, D., et al., When biomolecules meet graphene: from molecular level interactions to material design and applications. Nanoscale, 2016. **8**(47): p. 19491-19509. https://doi.org/10.1039/C6NR07249F

[128] Said, K., et al., Fabrication and characterization of graphite oxide–nanoparticle composite based field effect transistors for non-enzymatic glucose sensor applications. Journal of Alloys and Compounds, 2017. **694**: p. 1061-1066. https://doi.org/10.1016/j.jallcom.2016.10.168

[129] Lee, D.-H., et al., Highly selective organic transistor biosensor with inkjet printed graphene oxide support system. Journal of Materials Chemistry B, 2017. **5**(19): p. 3580-3585. https://doi.org/10.1039/C6TB03357A

[130] Song, Y., et al., A simple electrochemical biosensor based on AuNPs/MPS/Au electrode sensing layer for monitoring carbamate pesticides in real samples. J Hazard Mater, 2016. **304**: p. 103-9. https://doi.org/10.1016/j.jhazmat.2015.10.058

Chapter 3

Heteroatom-Doped Graphene-Based Electrochemical Sensors for Toxic Chemicals

Chinnathambi Suresh[1], Sankararao Mutyala[2*], Jayaraman Mathiyarasu[1]

[1]Electrodics and Electrocatalysis Division, CSIR- Central Electrochemical Research Institute, Karaikudi, Tamil Nadu, India - 630 003.

[2]Department of Chemistry, Koneru Lakshmaiah Education Foundation (KLEF), Vaddeswaram, Guntur, Andhra Pradesh. India - 522 502.

*mutyala.sankararao@gmail.com

Abstract

Toxic chemicals are the group of compounds which including nitrogenous compounds, phenols, sulfurs, quinolines, and toluene, etc. These compounds were utilized for various applications in daily life, but it is highly hazardous to the human and living organism. The frequent uptake through inhalation leads to acute poisoning of the human body which causes a serious illness like cancer, asthma, etc. Hence, identification and determination of the toxic chemicals are highly essential. There are many methods like chromatography, spectroscopy, electrochemical sensors, fluorescence sensors, and surface-enhanced Raman scattering (SERS) sensors are techniques in vogue to identify the trace concentration of toxic chemicals. Among all, electrochemical sensors are an important analytical technique for the detection of toxic chemicals due to low cost, high portability, and precision. Heteroatom doped graphene materials play an important role in the fabrication of electrochemical sensors. Further, the fabrication sensors using different hetero atom doped graphene materials improve the sensing selectivity, stability, and sensitivity due to its exception physic-chemical properties. Hence, hetero atom doped graphene is served as a finite electrochemical sensors platforms for accurate determination of various toxic chemicals in the environment and food.

Keywords

Electrochemical Sensors, Heteroatom Doped Graphene, Selectivity, Sensitivity, Toxic Chemicals

Contents

1.   Introduction

Since the use of charcoal electrodes by Michael Faraday, carbon has been widely used and extensively studied as an electrode material for electrochemical applications. Carbon is a remarkable element and it has been defined as one of the key elements of living organisms [1]. Carbon is having the ability to form strong covalent bonds and produces a variety of stable allotropes with a changeable dimensionality [2]. Because of its bonding property carbon can also yield numerous structures. Carbon is having two best-known allotropes namely diamond and graphite. Diamond has a large network of sp^3 hybridization involved in σ- bonding and is a transparent, electronic insulator [3]. Graphite is an excellent electrical conductor, comprises pure sp^2 carbon forming an σ-bonded network, alongside π-bonding from p_z orbitals [4]. This facility to make all carbon containing nanomaterials, in particular those containing C=C double bonds, has been one of the key events that have led to the current nanostructured materials based on novel technology revolution. In particular, developments in low-dimensional carbon materials, such as 0D-fullerenes, 1D-CNTs, 2D-graphene etc., have further boosted the research for high performance and active materials for electrochemical biosensor applications. In comparison with traditional carbon supports, nanostructured materials have several unique nanometer sized effects that can add attractive electrochemical features to the resulting hybrid materials [5-6].

The remarkable properties of carbon include:

- Carbon is a lighter element and structures made from carbon tend to be light weight.
- Carbon can act as conductor, semi-conductor and insulator; this depends on the carbon-carbon bonding and its structure.
- The thermal conductivity of carbon is a variable.
- The surface of carbon materials can be chemically modified and doped and which leads to developing novel composite materials.

1.2   Graphene

Graphene is defined by IUPAC as "a single carbon layer of graphite structure, describing its nature by analogy to a polycyclic aromatic hydrocarbon of quasi-infinite size" Graphene is the name given to a flat monolayer of carbon atoms tightly packed as a 2D honeycomb lattice [7-8]. Graphene is the basic structural element of some carbon allotropes such as graphite, charcoal, CNTs, and fullerenes (Fig.1).

An ideal sheet of graphene consists of a 2D honeycomb structure of carbon atoms in a single layer and considered a graphene structure cannot have more than ten layers [9]. One of the simple and first methods to produce graphene is achieved through micromechanical cleavage [7-8]. By repeatedly exposing a piece of graphite to the mechanical stress of a piece of tape, the weaker bindings between planes are destroyed. After creating a sufficient number of flakes, the tape can be dissolved and graphene material is collected. However, if the graphene sheets are re-stacked on top of another, the crystalline structure of graphite should not reappear. Because of the size, the graphene sheets can have a wide range of lateral sizes depending on the method that creates the nanomaterials [10]. Graphene is having remarkable properties such as high stiffness and breaking strength. The values measured corresponds to Young's modulus of E = 1.0 TPa and intrinsic strength of 130 GPa, which would suggest that graphene is the strongest material ever measured [11].The most intriguing feature of this material is its exceptional electronic quality and transport properties of individual graphene sheets. Due to its 2D nature, it is theoretically a zero-bandgap semiconductor with excellent room temperature electrical conductivity and it has a very high thermal conductivity [12-14]. Moreover, the pronounced carrier mobility is well maintained at the highest electron or hole concentrations of up to 10^{13} cm^{-2} in ambipolar field-effect devices which are suggesting a ballistic transport on graphene [15]. The exceptional electronic properties of graphene also refer to unusual room temperature half-integer quantum Hall effect [16]

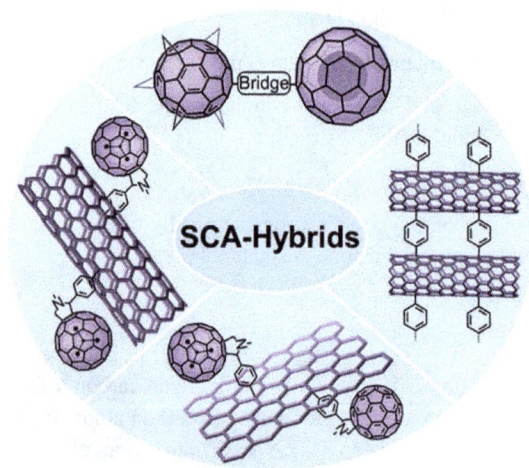

Figure.1. synthetic-carbon-allotrope (SCA) architectures are constructed by hybridizing different types of carbon allotropes such as 0D-fullerene, 1D-carbon nanotubes (CNTs), and 2D-graphene. Reproduced from ref. [7]

1.2.1 Oxidation of graphite to graphene oxide

To produce graphene from graphite the reaction has to proceed through two steps. The first step is the oxidation of graphite into GO and the second step is the reduction of GO to graphene. Two main reasons are behind for using this procedure. The first one is graphite flake has low solubility in water whereas oxidized graphite is soluble to a large extent. The second reason is the methods that are turning graphite into graphene in one step require a lot of force on the sheets. This often occurs when an extremely high temperature or pressure is used. By dividing up the process in two steps less stress required with minimal defects in the final product than one step approach [17-18]. The structure of GO is complicated and dependent on many parameters of the process under which it is formed. Graphene in its oxidized form show two unwanted properties; (i) it is not conductive and (ii) it is very sensitive to the higher temperature. By removing the oxygen functionalities (reducing it), these properties can be restored. In the second step, various routes are existing to perform the reduction of GO. But the most common way is a chemical route by adding a reducing agent. This reaction can be enhanced by adjusting pH and by adding a promotor to the reaction mixture [19]. Within a short time graphene and its related materials are known as novel scientific revolutionary materials.

1.3 Influence of dopants on graphene

Heteroatom doping is mainly classified into withdrawing (i.e., halogens) or donating (i.e., nitrogen) free electrons to the carbon atoms, has been an effective way to give graphene with tunable and enhanced properties. Doping of graphene with various elements such as N, B, S, O, and P is developed to improve/changes their physical and chemical properties of the resulted material, which signifies the electrocatalytic activity. From the past decade, heteroatom-doped carbon materials, like porous carbon, carbon nanotubes and graphene, have been widely explored in various fields [20]. The finite catalytic activity, structural stability and high electrical conductivity make them as a special class of material in biosensors applications. Amongst them, heteroatom-doped graphene (H-G) exhibited excellent catalytic activity due to its 2D ultrathin open structure, which endows it with large specific surface area (theoretically 2630 m^2 g^{-1}) to expose abundant accessible active sites [21].

1.3.1 Synthesis of heteroatom doped graphene

Several methods have been developed for the synthesis of graphene, in which a choice of doping strategies could be derived. As the preparation methods of graphene can be generally classified into bottom-up and top-down routes [22] similar classification is used there to introduce the preparation methods for the H-G. In the past decade, several heteroatom's have been explored to modify graphene such as boron, nitrogen, oxygen, phosphor, sulfur, halogen and metal atoms [23]. Here, mainly focus on the introduction of oxygen, boron, nitrogen, phosphor, and sulfur-doped graphene (denoted as O-G, B-G, N-G, P-G, and S-G), these are used as electrocatalysts in an electrochemical biosensor.

Top-down methods

A The top-down route involves the breakage of the stacked layers of graphite or graphite oxide to produce hetero atom doped graphene, including thermal exfoliation, wet chemical, arc discharge, plasma treatment and ball milling methods [24-26].

Bottom-up approaches

Another most important chemical route for the synthesis of H-G is the bottom-up approach which involves the preparation of H-G from the assembly of carbon-containing precursors in presence of dopants. The typical methods are chemical vapor deposition (CVD) and solvothermal reaction [27-28].

2. Elemental doping of Boron

Boron is an element with unique and incomparable properties within the periodic table. It is one of the most interesting elements for the doping of carbon materials, owing to their comparable atomic sizes and valance electron numbers [29]. Compared to N doping, B doping is a comparable way to alter the graphene properties [30]. The B doping induces the p-type conducting behavior in graphene lattice with a change in work function, bandgap, and enhanced chemical activity [31]. However, unlike N doping, preparation of B-G by using CVD techniques is still rare and only recently has reached the same ripeness in terms of synthesis procedures, the study of physical and chemical properties, and technological implementation in practical applications [32]. The most commonly exploited synthesis of B-G is based on the solid-state reaction between graphite or GO combine with boron precursor, mainly H_3BO_3, B_2O_3 or B_4C. Subsequently, boron doped graphite can be mechanically exfoliated to obtain a single layer of B-G. GO can be doped by gas-solid reactions, Pumera et al. [33] have used thermal exfoliation of GO in the presence of boron trifluoride diethyl etherate (BF_3Et_2O) as a B precursor. The doping concentration of B can be tuned by changing the temperature and composition of the gaseous atmosphere (i.e. H_2/N_2 ratio). Umrao et.al. proposed a rapid method for a cost-effective synthesis of B-G has been developed. A GO suspension has been mixed with a B_2O_3 followed by stirring for 8 h at 60 °C to promote the reaction. The resulted materials are purified and dried in a microwave oven (700 W) for 45 s, leading to the formation of B-G material [34]. For the first time Yan et al. synthesized B-G on the successful construction of a vertically grown B-G on boron doped diamond electrode (B-G/BDD). Dark-field transmission electron microscopy (TEM) image and the corresponding energy-dispersive X-ray spectroscopy (EDX) elemental maps of a B-G showed in Figure 2. Further C, B, and O elemental distributions signify that boron is uniformly distributed throughout the B-G sheet. X-ray photoelectron spectroscopy (XPS) (Figure 2b) confirms the elemental boron doping concentration between 1.04 and 2.45 atom % at B-source-gas flow rates in the 20–100 sccm range this is also in good agreement with the Raman spectrum (Figure 2d). The presence of a highly intense sharp D peak in the spectrum of the B-G sample is consistent with the existence of large numbers of exposed sides and faces that are ascribable to the vertical growth of BG sheets and the existence of multiple defects, such as B-doping, in the hexagonal carbon lattice.

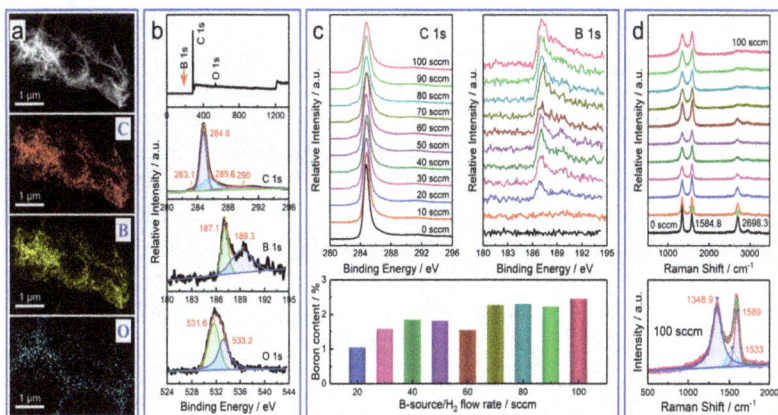

Figure 2. (a) TEM image and corresponding EDX elemental C, B, and O maps of B-G prepared at a B-source-gas flow rate of 50 sccm. (b) XPS survey spectrum and C 1s, B 1s, and O 1s spectra of B-G grown at a B-source-gas flow rate of 50 sccm. (c) C 1s and B 1s XPS spectra of B-G layers grown at various B-source-gas flow rates and the dependence of the B content in the B-G layer on the B-source-gas flow rate during the growth of the B-G layer. (d) Raman spectra of B-G layers prepared at various B-source-gas flow rates and an enlarged spectrum of the BG prepared at 100 sccm. The BG samples were grown for 10 min. [reproduced from ref. 35]

3. Elemental doping of Nitrogen

Amongst hetero atoms, N doping in graphene lattice is one of the most extensively studied one, due to its comparable atomic size. The N atom can be easily introduced into graphene because of its open structure. The N doping has three different bonding configurations in the graphene network such as pyridinic N, pyrrolic N, and graphitic N (or quaternary N) [36]. The N doping atom could be introduced into graphene lattice directly during the graphene preparation or post-treatment of GO or graphene. Among several reported methods CVD, solvothermal and arc-discharge are frequently used for *in situ* growth of N-G. Compared with *in situ* synthesis, post-treatment methods which include thermal annealing, plasma or solution treatment are simpler and likely closer to commercialization [37]

3.1   In situ doping

CVD is one of the promising synthetic routes for N-G preparation [38]. For the first time single-layer graphene is detected by Wei et al. using Cu/Si as the CH_4 as the C source and Ammonia is used as N source to produce N-G under 800 °C. However by using the sole source that contains both C and N e.g., acetonitrile [39] and pyridine [40]. The doping of N can be tune in the range of 1.2–16 at.% by adjusting the flow gas rate and C to N ratio source [41]. Further Deng *et al.* proposed a simple one-pot direct solvothermal synthesis is developed to prepare the N-G through the reaction between tetrachloromethane and lithium nitride under the N_2 atmosphere at below 350 °C [42]. This method allows scalable synthesis with a doping level of 4.5-16.4 wt. % of N. Panchakarla et al. is also employed arc-discharge technique in the presence of pyridine vapor or NH_3 as an N source for N-G synthesis [43]. The resulted technique produced N-G with the N content level around 0.5–1.5 at.%. However, this process requires complicated purification steps with low yield due to the formation of excessive by-products.

3.2   Post-treatment

In general thermal technique is used in the presence of the NH_3 atmosphere is an easy and commonly used method to obtain N-G by post-modification. Since N incorporation reactions occur mostly at the defect sites and the edges of graphene with a low N level approximately 2.8 at.%. [44]. To increase the N doping level, researchers bowed their attention with rich oxygen functional groups and more GO in order to provide a more active deposition. Dai et al group reported the doping of N under the NH_3 atmosphere in GO by using the thermal annealing method [45]. The N-doping process imitates at 300 °C and highest doping level approximately 5 at. % N was attained at 500 °C. The further atomic percentage of N can reach up to 10.1 at.% N by using melamine as the N source to synthesize N-G [46]. In addition to the above techniques, some physical methods such as plasma treatment or ion implantation are used to provoke chemical defects for N doping up to 8.5 at.% N by adjusting the plasma density or exposure time [47]. For example, Guo *et al.* used Ammonium ion irradiation to induce the defects in grapheme lattice followed by annealed under NH_3 atmosphere for N-G synthesis [48]. In hydrothermal route NH_3 will release and react with GO contain oxygen functional groups and results N-G. Researchers are tuning the doping level of N by changing the experimental parameters e.g., the mass ratio between GO and the reducing agent, or the reaction temperature.

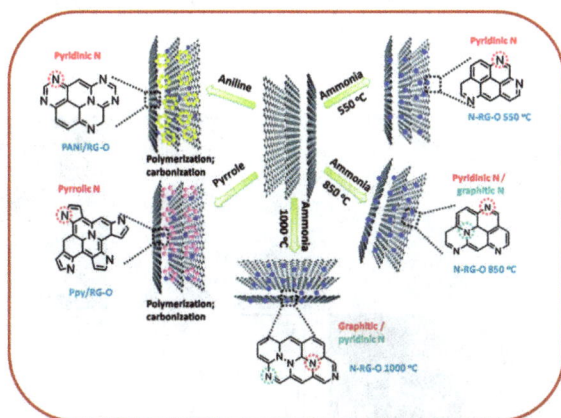

Figure 3. N-doping configuration depends on the precursor and temperature. Reproduced from ref. [49].

4.   Elemental doping of sulfur

In comparison to B and N, S doping in graphene lattice is quite rare and is also one of the emerging fields within carbon material research.   Denis et al. is theoretically predicted the common viability of S-doping in graphene and its doping is quite difficult than N doping, taking into account the size and different binding behavior of S atoms. Nevertheless it was calculated that S doping should allow for a targeted tuning of the graphene bandgap, depends on the quantity S atoms incorporated into the sheets [50]. Practically it is possible to synthesis S-G by CVD approach using a solution of elemental S in hexane as the precursor [51]. Further Yang et al. developed another way to obtain S-G is to blend GO homogeneously with benzyl disulfide and subsequent annealing [52]. Gao and co-workers proposed the formation of S-G on Cu foil using liquid S-hexane mixture vapor at 950 °C. Figure 4 showed that resulted material is composed of graphene sheets with partially wrinkled and folded morphological motifs. XPS is indicating covalent bonding between C and S in the synthesized S-G method. However, the HR-TEM image taken from S-G reveals a well-resolved lattice fringe with a lattice spacing of 0.34 nm (Fig. 4b), corresponding to the (002) plane of graphene. In Fig. 4c, the selected area electron diffraction (SAED) pattern exhibits two diffraction rings indexed to the (002) and (100) planes of the graphene phase, respectively [52]. Energy-dispersive X-ray (EDX) elemental mapping images of S-G (Fig. 4d) indicated that S atom with a concentration of 1.5 wt % is consistently distributed throughout these graphene sheets and edges were confirmed by elemental mapping data.

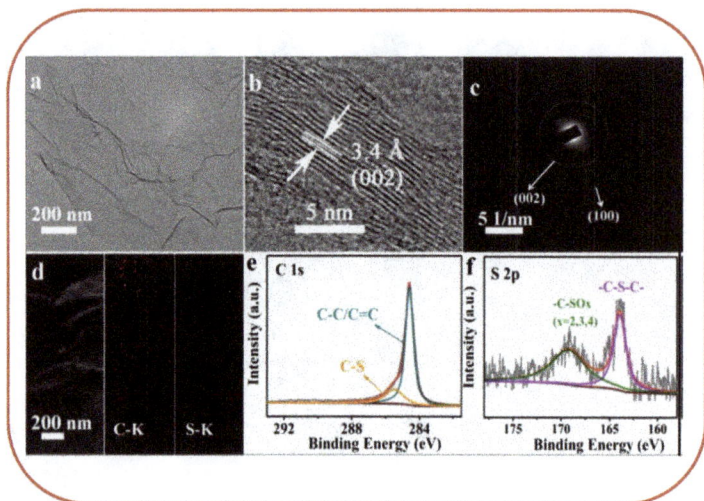

Figure 4. (a) TEM image for S-G. (b) HRTEM image and (c) SAED pattern taken from S-G. (d) TEM and corresponding EDX mapping images of S-G. XPS spectra for S-G in the (e) C 1s and (f) S 2p regions. Reproduced from ref. [53].

5.   Elemental doping of Phosphorus

Phosphorus is an element in the N group, has the same number of valence electrons as N and often exhibit similar chemical properties. This verdict confirms that doping of P into the hexagonal network of graphene and changes the electronic structure to some extent. Li et al synthesized P-G nanosheets by thermal annealing of a homogeneous mixture of GO and an ionic liquid 1-butyl-3-methlyimidazoliumhexafluorophosphate (BmimPF6), which was employed as a mild phosphorus source for in situ doping the thermally reduced GO. Because BmimPF6 is one of the most commonly used ionic liquid with non-volatility and high thermal stability. The proposed ionic liquid assisted one-step synthetic route is very simple and scalable for simultaneous P doping and reduction of GO. Moreover, a high concentration of GO easily dispersed in the solution of BMIMPPF$_6$–ethanol–water. After removal of the water and ethanol solvents results in a homogenous composite of GO and BmimPF6. In the course of the thermal annealing process, PF6 strongly reacts with oxygen-containing groups on the basal plane and edge of GO to form a carbon framework followed by simultaneous P doping and reduction of GO. The resultant P-G nanosheets exhibited high surface area 496.67 m2 g2 with a P-doping level of 1.16 at.%,. [54]. Wen et al. prepared P-G by using simple annealing method with the P doping level of 1.30 at% was synthesized by the mixture of graphene and phosphoric acid Zhang et al. also prepared P-G by thermal treatment of GO and triphenylphosphine mixture and the resulted is evidenced by improved the electrochemical activity.

Lin et.al proposed the doping of P infraphene in vacancies of graphene sheets by with change in the P content of 2.86-6.40 at. % by thermal annealing of fluorographite in P vapor. A large area fluorographite film has a relatively flat surface transforms into the transparent graphene sheets with obvious wrinkles and folds which may be due to the structural defects generated during the P-doping and defluorination process. as shown in figure.5 [55]."

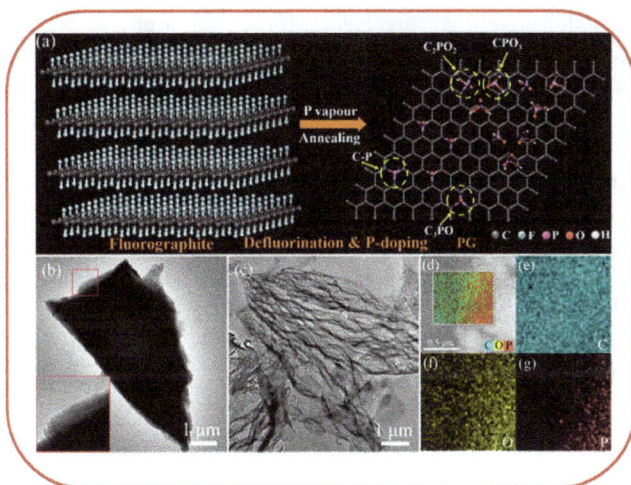

Figure.5. (a) Schematic of defluorination and P-doping of a fluorographite lattice. TEM images of fluorographite (b) and P-doped graphene (c). Elemental mapping of C + O + P (d), C (e), O (f), and P (g) for P-doped graphene. Reproduced from .ref. [55].

6.   Elemental doping of halogen

It is a well known, halogens acquire higher reactivity than group B and N group elements. As halogen-doping in carbon lattice changes its hybridization from sp2 carbon bonding to sp3 state, it consequences in extreme distortions in the geometric and electronic structures of the graphene network. Amongst halogens F is one of the high electronegative and most reactive elements and its bonding to carbon is strong and inert. However after doping of F in the graphene network stretches the C–C bond length from 1.57 to 1.58 Å. In fluorine-doped graphene (F-G), F atoms are covalently bonded to sp3 C and the graphene sheet is buckled as the F attachment alternates on both sides (i.e., a basal plane is sandwiched by two F-layers) of the graphene sheet. F-G has received great attention across researchers from the globe is mainly due to its outstanding mechanical, thermodynamic,and superb chemical properties and it also increases the hydrophobic property of graphene [56]. Robinson et al. obtained a single sided F-G (25% F coverage,C4F) by exposing only one-side of CVD graphene to xenon difluoride and it is optically transparent with a bandgap of 2.93 eV with the increase in resistance when compared graphene [57]. Halogen-doped graphene mainly synthesized by using the liquid-phase exfoliation of graphite halide. Though the doping level of halogens cannot be controlled by using this method and it is very difficult to make a large quantity of halogenated graphite. Zheng et al. identified an interesting simple microwave-spark method for incorporation of Cl- and Br- into graphite in the presence of chlorine and bromine solution, later this could be easily exfoliated into single-layered Cl⁻/Br⁻doped graphene sheets via simple sonication technique [58]. Under the luminous microwave-sparks, active graphiteflakes generated by a short temperature shock can react with halogen precursor's solutions and a consequent sharp decrease in temperature quenches the reaction and prevents the thermal decomposition of the resultant materials. The resulted from halogen doped graphene materials containa high amount of covalently bonded Cl and Br with a 21 at% and 4 at%, respectively with a strong hydrophobic nature and disperses well in organic solvents.

7. Co-doping

The co-doping of multiple species of hetero atoms may generate new properties or create synergistic effects. Gendey et al. developed a reliable one-pot and green synthetic route for N, P co-doped graphene by a reaction between GO and guanosine monophosphate as a green reagent. This is mainly due to P being a more electron-rich atom than compared to carbon and it is also easily doped into the graphene basal plane. In addition to this larger atomic radius and greater electron donating properties of P than N is signified to enhance the electrocatalytic properties of graphene in electrochemical sensor applications [59]. First time Zhou et al. present a 3D N and co-doped graphene grown on Ni form (3DSNG/NF) by using the CVD method. Further the doping concentration and growth parameters of N and S can be easily controlled.

8.   Basic electrochemistry of H-graphene

The basic electrochemical behavior of doped graphene materials is highly interesting to be studied to determine several important parameters of H-G modified electrodes such as potential window, electron transfer rate, redox potentials, etc. This will be helpful for their applications in the fields of biosensors and electrocatalysis [61]. Graphene and heteroatom doped graphene exhibits a wide potential window which is comparable with graphite, glassy carbon [62]. Further the charge transfer resistance (Rct) of H-G is determined from electrochemical impedance spectra. It is much lower than that of graphite and GCE [63]. The electron transfer behavior studies of H-G determined using cyclic voltammetry using standard redox probes, such as $[Fe(CN)_6]^{3-/4-}$ and $[Ru(NH_3)_6]^{3+/2+}$, which exhibited well-defined redox peaks [64]. With finite peak currents in the CV and low peak-to-peak potential separations (ΔEp) in CV with a close to the ideal value of 59 mV [61]. The ΔEp separation is connected to the electron transfer coefficient [65], and a low ΔEp value signifies a fast electron transfer for one electron electrochemical reaction on H-G [66-67]. These studies were made to know the electrochemical response/activity of H-G towards different kinds of toxic chemicals in Environment and Water samples.

9.   Electrochemical biosensors

A biosensor can be defined as a device that incorporates a biological sensing element that is connected with a transducer to convert the observed response into a measurable signal, whose magnitude is proportional to the concentration of a specific chemical or set of chemicals [68]. Depending on the type of signal transduction, biosensors can be differentiated as optical, bioluminescent, electrochemical, piezoelectric, and calorimetric. Among these various kinds of biosensors, electrochemical biosensors are the class of the most widespread, numerous, and successfully commercialized devices of biomolecular electronics [69]. This is due to their capability to be miniaturized, can be operated in turbid media (unlike optical ones), the short response time (unlike bioluminescent ones), and lower detection limit and cheaper, compared to the other types of biosensors. For these reasons, electrochemical biosensors are widely used for monitoring the products of industrial bioprocesses (amino acids, yeast, lactic acid, ethanol, etc.), the pollutants in the environment (pesticides, Nitrobenzene, Phenols, fertilizers, substances estrogenic, etc...), the relevant substances in clinical diagnostics (glucose, alcohol, DNA, hormones, etc...) and the forensic field (cocaine, anthrax, nerve agents, etc...) [70]. Electrochemical biosensors may be classified according to the type of signal transduction (conductometric, potentiometric and amperometric) or depending on the nature of the biological recognition process: biocatalytic devices (for example, based on enzymes as immobilized biocomponents) and affinity sensors (based on antibodies, membrane receptors, or nucleic acids) [71].

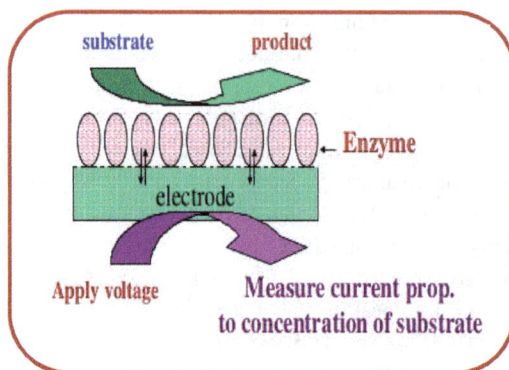

Figure 6. Principle of electrochemical biosensors.

Working principle of an electrochemical biosensor

- Analyte diffuses from solution to the surface of the biosensor.
- Analyte reacts specifically & efficiently with the biological component of the biosensor.
- This reaction changes the physicochemical properties of the transducer surfaceleads to a change in optical/electronic properties of the transducer surface.
- The change in the optical/electronic properties are measured /converted into an electrical signal which is detected.

Heteroatom doped graphene based electrochemical sensors for toxic chemicals detection

With the rapid focus towards the environmental pollution, there have been urgent need to develop fast, sensitive, low-cost and multiplexed sensing devices for the detection of environmental pollutants. Two-dimensional (2D) nanomaterials hold great promise as a electrochemical sensing platfor due to their unique chemical and physical properties, which have been extensively employed to monitor the environmental pollutants combined with different detection techniques. In this view, a general electrochemical detection methods have been adopted for the determination of some of the important toxic chemicals were discussed in details.

10.   Nitrobenzene (NB):

NB is one of the important and hazardous organic materials and used different fields such as petroleum refineries, pesticides, herbicides, insecticides, dyes, shoe polishes, soap, etc. The existence of NB in human bodies causes methemoglobinemia by acute (short-term) and chronic (long-term) diseases. Several symptoms like fatigue, weakness, dyspnea, headache, and dizziness occurred at low nitrobenzene concentrations and depressed respiration, bluish-gray skin, disturbed vision, and coma may occur at higher concentrations. Nitrobenzene (NB), as a toxic and carcinogenic compound is widely used in dyes and drug synthesis [72]. According to the strickt environmental concern, the NB content in drinking water should be in the range of 1 mg/L [73]. The possible conversion of NB through chemical and electrochemical approaches will leads to protect the environment and human health. In early reports, the quantitative determination of NB was done by analytical techniques like ultraviolet spectrophotometry (UV), high-performance liquid chromatography (HPLC), fluorescence quenching, gas chromatography (GC) and electrochemical methods. Among all methods, the electrochemical method is a much familiar and reliable technique [74]. The simple, cost-effective, highly sensitive, and easy to operate for long term stability features are the interesting significances of the electrochemical method. Nanomaterials based electrochemical detection of NB is the versatile approach due to the rapid and accurate determinations. Weiqing et al report that, graphene doped with monometal nitrogen (N), sulfur (S), and bi-metal (NS) via low temperature methods were utilized for NB detections. They conclude that, by changing the synthesis sequence of S and N over the graphene, the electronic properties significantly varied. The developed sensors exhibit a very good limit of deduction value (0.216 µM) in optimized experimental conditions [74].

11.   Organo pesticides

Wide varieties of fertilizers are routinely used in agricultural land to accelerate the productivity. The rapid growth of global population also demands surplus food materials within short duration. Amongst various fertilizers, organophosphorus (OP) compounds are the commonly used pesticides because of their high efficiency in controlling insects. However, the overuses of OP pesticides resulted in environmental contaminations and eventually insecure the healthiness of humans as well as the agro-products and consumers. The very common pesticides methyl patathion is highly toxic and have the half-life periods of 110-144 days [75]. It is mainly used for the controlling of millets on cotton, aphids, boll weevils, artichokes, etc. Also, the over usage of the methyl parathion is causing many problems in humans as it affects the nervous system [76]. Hence, the detection and removal of methyl parathione is the most important task to maintain the environment, but the development of simple, rapid, low-cost and user-friendly detection techniques are attracted more and more attention and is a challenging task. Various chromatography (GC, HPLC and GC-MS) and spectroscopy methods have been used for the detection of methyl parathion [77]. In addition to the above methods, the electrochemical approach is also performed to determine the pesticide concentration. The unique nature of electrochemical methods like easy, low cost and fast sensing maks many nanomaterials based electrochemical sensors being utilized for pesticide determination [78].

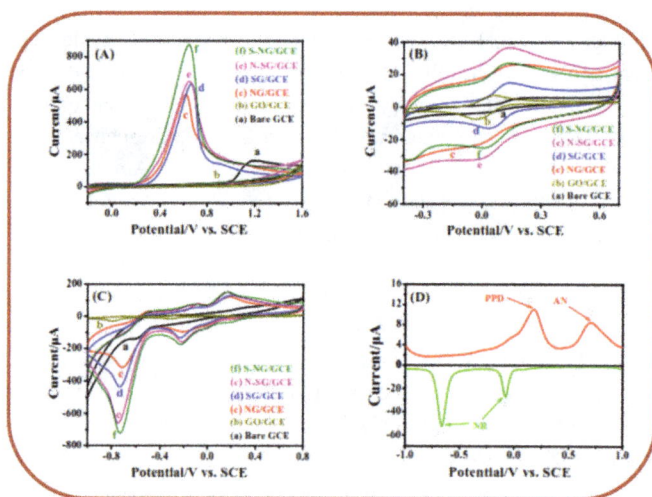

Figure 7. CVs of bare GCE, GO/GCE, NG/GCE, SG/GCE, N-SG/GCE and S-NG/GCE in presence of (A) 0.2 mM AN, (B) 0.1 mMPPD and (C) 0.5 mM NB (pH = 5.0). Scan rate: 50 mV s-1.(D) DPVs obtained for the oxidation of mixed solution containing 20 µM AN and 10 µM PPD, the reduction of 20 µM NB at S-NG/GCE in pH 5.0 citrate-phosphate buffer. Reproduced from [74].

The organophosphorus (OP) pesticide inhibits the activity of enzyme acetylcholine esterase (AChE) present in our body and essential for the normal functioning of neurotransmission system [76]. Within the framework of electrochemical sensors, the nanomaterials based detection is attracted much attention. The detection of phoxim

Materials Research Foundations **82** (2020) 61-90   https://doi.org/10.21741/9781644900956-3

pesticide was studied using Poly(3 methylthiophene)/nitrogen doped graphene modified glassy carbon electrode. The developed sensors exhibit inherent electrochemical sensor activity towards the pesticide. In a typical experimental condition, the electrode exhibit a wide liner rage of 0.02–0.2 µM and 0.2-2.0 µM and a good detection limit of 6.4 nM. The prepared electrode tested with 1000 fold excess metal ions and 100 fold excess biomolecules and phenolic compounds. However, the sensor does not show any significant interference effect [79]. The molecularly imprinted polymer (MIP) and nitrogen-doped graphene nanosheet have been prepared and deployed for pesticide parathion detection. The unique cavity nature of the developed MIP sensors, it preferably senses the parathion rather than the verity of other pesticides. The sensor showed excellent linearity in the rage of 0.1 to 10 µg mL^{-1} with the LOD value of 0.01 µg mL^{-1} [80]. Electrochemical determination of carbendazim (CBZ) exists in the food samples have been reported by Yu et al. using the modified electrode consisting of electrochemically reduced graphene oxide on glassy carbon. The prepared (ERNGO/GCE) electrode showed excellent activity with good selectivity towards the selected pesticides. The ERNGO/GCE exhibited wide linearity of 5.0~850 µg/L with a detection limit of 1.0 µg/L (S/N= 3) under optimized conditions [81]. A layer by layer assembled of hexachlorobenzene (HCB) sensor has been demonstrated using the nitrogen-doped graphene(NG) and chitosan modified GCE [82-83]. The study pointed out that, the graphene structure makes very good adsorption affinity towards HCB and enhances the sensor activity. The sensor showed wide linear range (3 µg/Lto10 µg/L) and excellent low detection limit (1.72 µg/L).

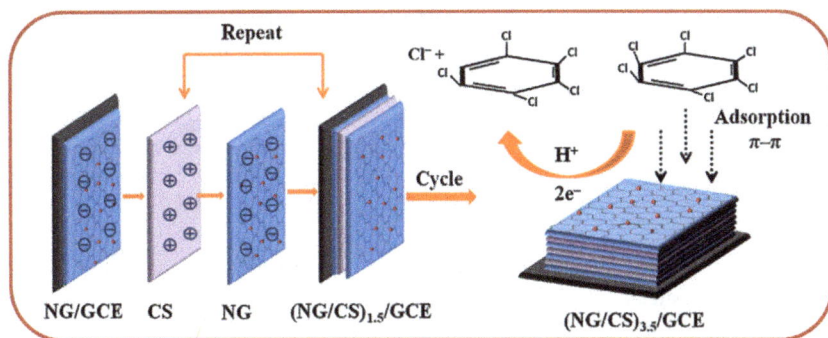

Figure 8. Schematic drawing of the assembly of (N-G/CS)3.5 /GCE and schematic representation of the electrochemical responses of HCB at (N-G/CS)3.5/GCE. Reproduced from ref. [82].

A nonelectroactive pesticide dimethoate has been determined by using oxime-functionalized gold nanoparticles(AuNPs) and nitrogen-doped graphene (NG) composites (NG/AuNPcs). The surface functionality of (-SH group) will play a significant role to interact with the gold nanoparticles and make a better sensing matrix. The sensor showed a linear range from 1×10^{-12} to 4×10^{-8} M with a low detection limit of $8.7 \times 10-13$ M (S/N = 3) [83].

Nitrogen doped graphene modified with molecularly imprinted polymer (MIP) has been developed for detection of pesticide methyl parathion [84]. The specific cavity of the MIP will lead to enhance the sensing performance of the developed sensor. Under the optimized experimental condition, the sensor showed a wide liner range value of 0.1 to 10 $\mu g\ mL^{-1}$ with very low detection limit of 0.01 $\mu g\ mL^{-1}$.

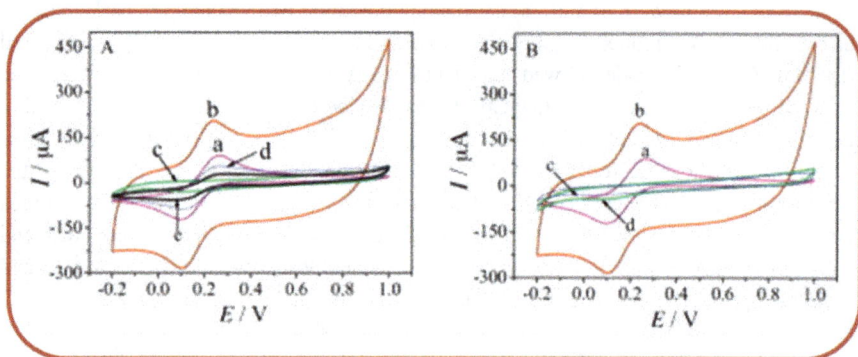

Figure.9. Cyclic voltammetry results of the MIP (A) and nMIP (B) Au electrodes in 5 mmol L^{-1} $K_3[Fe(CN)_6]$ solution containing 0.1 mmol L^{-1} KCl: (A) (a) bare Au electrode, (b) N-GS-Au electrode, (c) MIP-Au electrode before template removal, (d) MIP-Au electrode after template removal, (e) MIP-Au electrode after rebinding template; (B) (a) bare Au electrode, (b) N-GS-Au electrode, (c) nMIP-Au electrode, and (d) nMIP-Au electrode after the elution step. Scan rate: 0.1 $V s^{-1}$ [Reproduced from ref. 84].

A versatile nitrogen-doped graphene modified glassy carbon electrode was fabricated for sensing the pesticide methyl parathion. To facilitate the sensing ability due to nitrogen doping, systematic experiments have been carried out for doping various concentrations of nitrogen. By altering the graphene oxide to glycine ratio, the catalysts with a 1:8 ratio exhibit better performance. In the experimental condition, the developed sensors show a very good linear range (5.0×10^{-8}–4×10^{-5}, $molL^{-1}$) and LOD of 1.7×10-8 mol L^{-1}.

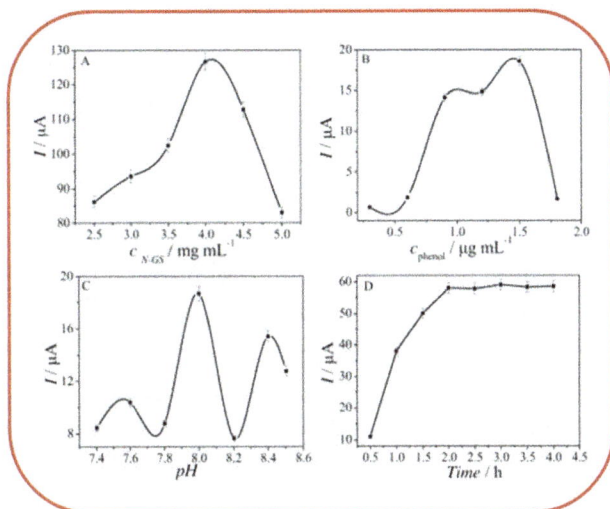

Figure. 10. Effect of (A) N-GS concentration, (B) the ratio of template to monomer, (C) pH and (D) elution time on respond signals (The current is the difference of MIP-modified Au electrode after and before removal of the imprinted molecules). Reproduced from ref. [84]

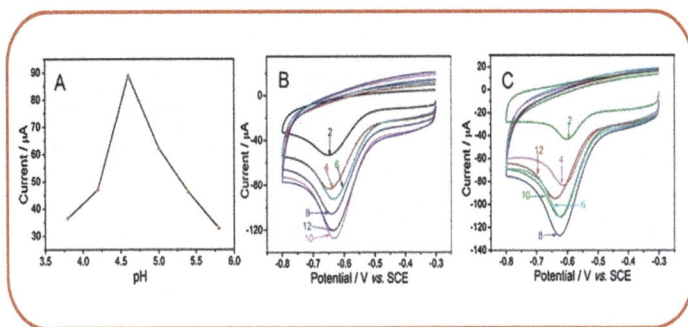

Figure 11 . Effect of pH of 0.1 M ABS (10 μL 0.25 mg mL^{-1} N-GE/DMF, 8 cycles PCZ) (A), the content of N-GE loading (pH 4.6, 8 cycles PCZ) (B) and the cycle number of PCZ polymerization (pH 4.6, 10 μL 0.25 mg mL^{-1} N-GE/DMF) (C) on the current response of 1 × 10^{-4} M 4-NP at PCZ/N-GE/GCE. Scan rate: 100 mV s^{-1}. Reproduced from ref. [86].

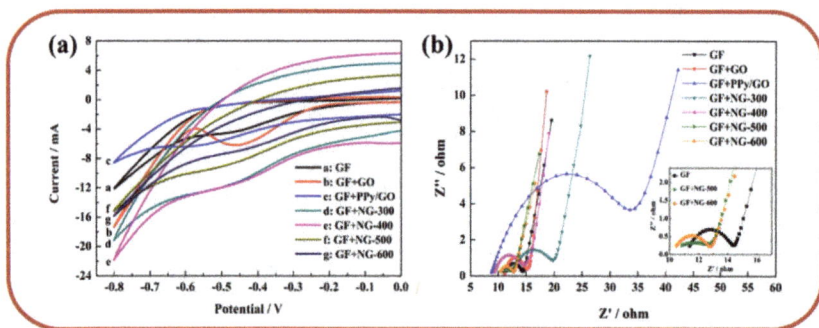

Figure 12. (a) *CV curves and (b) Nyquist plots generated from EIS measurements of GO, PPy/GO and NG modified electrodes obtained at different pyrolysis temperatures (300, 400, 500, and 600 °C, respectively. Reproduced from ref. [87].*

12. Para-Nitrophenol

Nitrophenols are the essential chemicals in many important industrial process like manufacturing of pesticides, explosive compouns and pharma products etc. The over utilization and unsafe disposal mechanism, the trace quantity of these compound will have a significant environmental effect [85]. The para-nitrophenol or 4-nitrophenol (4-NP) are highly toxic for human, that lead to huge adverse effect to human parts like kidney, liver and some time in the nervous system. The existence of these chemical will reduce the function of immune system of human, which causes nausea, headaches and cyanosis etc. The United States Environmental Protection Agency (EPA) identified the maximum allowable level of 4-NP in drinking water is 60 ppb [85]. Hence, the determination of 4-NP in the water as well in the environment is an important task. Numerous methods like spectroscopic, electrophoresis and fluorescence based techniques are used to estimate the concentration of 4-NP. However, due to the high cost per sample analysis, requirement for trained workforce and more sophisticated operation condition, these methods fails in routine analysis. Meanwhile, the electrochemical methods is highly suitable for 4-NP determination due to the fast, easy operation and being made to be portable etc. Yuehua et al. report a facile polymerization synthesis of Polycarbazole(PCZ)/nitrogen-dopedgraphene (N-GE) composite for electrochemical determination of 4-NP [86]. The electrodes exhibit well-defined conductivity due to the existence of PCZ polymer and graphene heterostructures. The improved sensor characteristics were correlated with the synergic effect of PCZ and NGE. The developed sensor selectively detects the 4-NP molecule without and interference in the real water sample. The sensors has wide linear range (8×10^{-7} to 2×10^{-5} M) with low detection value of [(0.062 μM(S/N=3)].

Table 1 List of the heteroatom-doped graphene methods and its presursors.

Method for synthesis of	Precursors	Doping element	Reference
Microwave oven	B_2O_3	B	32
Thermal exfoliation	Trifluoride diethyl etherate	B	33
Hot-filament Chemical Vapor Deposition	B_2H_6	B	34
Hot-filament chemical vapor deposition	B(OCH3)3	B	35
Chemical Vapor Deposition	Dicyandiamide	N	38
Solvothermal	Ammonia	N	42
Thermal treatment	Ammonia	N	44
Electrothermal Method	Ammonia	N	45
Plasma Treatment	N2-Plasma	N	47
Chemical Vapor Deposition	Liquid S liquid precursor	S	51
Chemical Vapor Deposition	Benzyl disulfide	S	52
Thermal Annealing	Benzyl disulfide	S	53
Annealing	phosphoric acid	P	55
Microwave processing	Ammonium polyphosphate	N, P	88
	ammonia,borontrifluoride (NH3BF3)	B,N	89
arc discharge	B_2H_6	B, N	43

13. Future perspectives

Though there has been significant growth in the development of suitable sensor plotforms for electrochemical biosensor applications toward the development of Point of care (POC) with a full automation and miniaturization. So far acroos the globe huge number of efforts have engrossed especially on the methodology rather than application for future commercialization. It can be predictable that, with further development and resources, such POC systems are going to speed up in electrochemical biosensor applications and make analytical results available at patient bedside or physician office within only a few minutes

References

[1] P. Levi, The periodic table, New York, Shocken Books, 1984, pp 227

[2] C. Srinivasan, Graphene-: Mother of all graphitic materials, Curr. Sci. 92 (2007) 1338-1339

[3] K. Bobrov, A. J. Mayne, G. Dujardin, Atomic-scale imaging of insulating diamond through resonant electron injection, Nature, 413 (2001) 616-619. https://doi.org/10.1038/35098053

[4] K. S. Novoselov, A. K. Geim, S. V. Morozov, D. Jiang, Y. Zhang, S. V. Dubonos, I.V. Grigorieva, A. A. Firsov, Electric field effect in atomically thin carbon films Science, 306 (2004) 666-669. https://doi.org/10.1126/science.1102896

[5] D. S. Su, S. Perathoner, G. Centi, Nanocarbons for the Development of Advanced Catalysts Chem. Rev. 113 (2013) 5782-5816. https://doi.org/10.1021/cr300367d

[6] D. Jariwala, V. K. Sangwan, L. J. Lauhon, T. J. Marks, M. C. Hersam, Carbon nanomaterials for electronics, optoelectronics, photovoltaics and sensing Chem. Soc. Rev. 42 (2013) 2824-2860. https://doi.org/10.1039/C2CS35335K

[7] T.Wei, F. Hauke, H. Andreas, Covalent Inter-Synthetic-Carbon-Allotrope Hybrids, Acc. Chem. Res. 52 (2019) 2037–2045. https://doi.org/10.1021/acs.accounts.9b00181

[8] Y. B. Zhang, Y. W. Tan, H. L. Stormer, P. Kim, Experimental observation of the quantum Hall effect and Berry's phase in graphene Nature. 438 (2005) 201-204. https://doi.org/10.1038/nature04235

[9] E. Fitzer, K.H. Kochling, H.P. Boehm, H.MarshRecommended terminology for the description of carbon as a solid, Pure Appl. Chem. 67 (1995) 473-506. https://doi.org/10.1351/pac199567030473

[10] S. Stankovich, D. A.Dikin, G.H. B. Dommett, K. M. Kohlhaas, E.J. Zimney, E. A. Stach, R. D. Piner, S.B. T. Nguyen, R.S. Ruoff, Graphene-based composite materials Nature, 442 (2006) 282-286. https://doi.org/10.1038/nature04969

[11] C. Lee, X. Wei, J. W. Kysar J. Hone, Measurement of the Elastic Properties and Intrinsic Strength of Monolayer Graphene, Science. 321 (2008) 385-388. https://doi.org/10.1126/science.1157996

[12] J. Q. Liu, Z. Y. Yin, X. H. Cao, A. Lin, L. Xie, Q. Fan, F. Boey, H. Zhang, W. Huang, Bulk Heterojunction Polymer Memory Devices with Reduced Graphene Oxide as Electrodes ACS Nano, 4 (2010) 3987-3992. https://doi.org/10.1021/nn100877s

[13] J. Q. Liu, Z. Y. Lin, T. J. Liu, Z. Y. Yin, X. Zhou, S. Chen, L. Xie, F. Boey, H. Zhang, W. Huang, Multilayer Stacked Low-Temperature-Reduced Graphene Oxide Films: Preparation, Characterization and Application in Polymer Memory Devices, Small. 6 (2010) 1536-1542. https://doi.org/10.1002/smll.201000328

[14] A. A. Balandin, S. Ghosh, W. Z. Bao, I. Calizo, D. Teweldebrhan, F. Miao, C. N. Lau, Superior Thermal Conductivity of Single-Layer Graphene, Nano Lett. 8 (2008) 902-907. https://doi.org/10.1021/nl0731872

[15] K. I. Bolotin, K. J. Sikes, Z. Jiang, M. Klima, G. Fudenberg, J. Hone, P. Kim, H. L. Stormer, Ultrahigh electron mobility in suspended graphene Solid State Commun.146 (2008) 351-355. https://doi.org/10.1016/j.ssc.2008.02.024

[16] K. S. Novoselov, Z. Jiang, Y. Zhang, S. V. Morozov , H. L. Stormer, U. Zeitler , A.K. Geim, Room- Temperature quantum hall effect in graphene.315 (2007) 1379. https://doi.org/10.1126/science.1137201

[17] A. Lerf, H.He, M. Forster, J. Klinowski,Structure of Graphite Oxide Revisited J Phys Chem. B,102 (1998) 4477-4482. https://doi.org/10.1021/jp9731821

[18] H. He, J. Klinowski, M. Forster, A. Lerf, A new structural model for graphite oxide Chem. Phys. Lett, 287 (1998) 53-56. https://doi.org/10.1016/S0009-2614(98)00144-4

[19] P.G. Ren, X. Ji, T.Chen, Z. M Li,Temperature dependence of graphene oxide reduced by hydrazine hydrate, Nanotechn. 22 (2010) 055705. https://doi.org/10.1088/0957-4484/22/5/055705

[20] Y. Deng, Y. Xie, K. Zou and X. Ji, Review on recent advances in nitrogen-doped carbons: preparations and applications in supercapacitors J. Mater. Chem. A, 2016, **4**, 1144-1173. https://doi.org/10.1039/C5TA08620E

[21] W. Wang, X. Wang, j. Xing Q. Gong H. Wang j. Wang, Z. Chen, Y. Ai, X.Wang Multi-heteroatom doped graphene-like carbon nanospheres with 3D inverse opal structure: a promising bisphenol-A remediation material. Environ. Sci. Nano, 6 (2019) 809-819. https://doi.org/10.1039/C8EN01196F

[22] A. Ambrosi, C. K. Chua, A. Bonanni, M. PumeraElectrochemistry of Graphene and Related MaterialsChem. Rev. 114 (2014) 7150-7188. https://doi.org/10.1021/cr500023c

[23] X. Wang, G. Sun, P. Routh, D.H. Kim, W. Huang, P. Chen, Heteroatom doped graphene materials: syntheses, properties and applications, Chem. Soc. Rev, 2014, 43, 7067-7098. https://doi.org/10.1039/C4CS00141A

[24] X. L. Li, X. R. Wang, L. Zhang, S. W. Lee, H. J. Dai, Chemically Derived, Ultrasmooth Graphene Nanoribbon Semiconductors, Science, 319 (2008) 1229-1232. https://doi.org/10.1126/science.1150878

[25] X. Li, G. Zhang, X. Bai, X. Sun, X. Wang, E. Wang, H. Dai, Highly conducting graphene sheets and Langmuir–Blodgett films, Nat. Nanotechnol. 3 (2008) 538-542. https://doi.org/10.1038/nnano.2008.210

[26] J. H. Lee, D. W. Shin, V. G. Makotchenko, A. S. Nazarov, V. E. Fedorov, Y. H. Kim, J. Y. Choi, J. M. Kim, J. B. Yoo, One-Step Exfoliation Synthesis of Easily Soluble Graphite and Transparent Conducting Graphene Sheets Adv. Mater. 21 (2009) 4383-4387. https://doi.org/10.1002/adma.200900726

[27] C. Berger, Z.M. Song, T.B. Li, X.B. Li, A.Y. Ogbazghi, R. Feng, A.N. Marchenkov, E. H. Conrad, W.A. de Heer, Ultrathin Epitaxial Graphite: 2D Electron Gas Properties and a Route toward Graphene-based Nanoelectronics, J. Phys. Chem. B 108 (2004) 19912-19916. https://doi.org/10.1021/jp040650f

[28] K. Novoselov, V.Falko, L. Colombo, P. Gellert, M. Schwab, K. Kim, A roadmap for Graphene Nature, 490 (2012) 192-200. https://doi.org/10.1038/nature11458

[29] T. Wu, H. Shen, L. Sun, B. Cheng, B. Liu, J. Shen, Nitrogen and Boron Doped Monolayer Graphene by chemical vapor deposition using polystyrene urea and boric acid. New J. Chem, 36 (2012) 1385-1391. https://doi.org/10.1039/c2nj40068e

[30] S. Agnoli, M. Favaro, Doping graphene with boron: A review of synthesis methods, physicochemical characterization, and emerging applications. J. Mater. Chem. A, 4 (2016) 5002-5025. https://doi.org/10.1039/C5TA10599D

[31] H. Wang, T. Zhou, D. Wu, L. Liao, S. Zhao, H. Peng, Z. Liu, Synthesis of boron doped graphene monolayers using the sole solid feedstock by chemical vapor deposition. Small 9 (2013) 1316-1320. https://doi.org/10.1002/smll.201203021

[32] S. Agnoli, M. Favaro, Doping graphene with boron: A review of synthesis methods, physicochemical characterization and emerging applications. J. Mater. Chem. A 4 (2016) 5002-5025. https://doi.org/10.1039/C5TA10599D

[33] L. Wang, Z. Sofer, P. Simek, I. Tomandl, M. Pumera, Boron-doped graphene: scalable and tunable p-type carrier concentration doping J. Phys. Chem. C, 117 (2013) 23251-23257. https://doi.org/10.1021/jp405169j

[34] S.Z. Zhai, H. Shen, J. Chen, X. Li Y. Li. Metal-free synthesis of boron-doped graphene glass by hot-filament CVD for wave energy harvesting ACS Appl. Mater. Interfaces 12 (2020) 2805–2815. https://doi.org/10.1021/acsami.9b17546

[35] D. Cui, H. Li, M. Li, C. Li, L. Qian, B. Zhou, B. Yang, Boron doped graphene directly grown on boron doped diamond for high voltage aqueous supercapacitors ACS Appl. Energy Mater, 2 (2019) 1526-1536. https://doi.org/10.1021/acsaem.8b02120

[36] Q. Wei, X. Tong, G. Zhang, J. Qiao, Q. Gong, S. Sun, Nitrogen-doped carbon nanotube and graphene materials for oxygen reduction reactions, Catalysts, 5 (2015) 1574-1602. https://doi.org/10.3390/catal5031574

[37] K.N. Wood, R. Hayre, S. Pylypenko, Recent progress on nitrogen/carbon structures designed for use in energy and sustainability applications. Energy Environ. Sci, 7 (2014) 1212-1249. https://doi.org/10.1039/C3EE44078H

[38] D. Wei, Y. Liu, Y. Wang, H. Zhang, L. Huang, G. Yu, Synthesis of N-doped graphene by chemical vapor deposition and its electrical properties. Nano Lett, 9 (2009) 1752-1758. https://doi.org/10.1021/nl803279t

[39] A.L.M. Reddy, A. Srivastava, S.R. Gowda, H. Gullapalli, M. Dubey, P.M. Ajayan, Synthesis of nitrogen doped graphene films for lithium battery application. ACS Nano, 4 (2010) 6337-6342. https://doi.org/10.1021/nn101926g

[40] Z. Jin, Y. J. Yao, C. Kittrell, J. M. Tour, Large-scale growth and characterizations of Nitrogen doped monolayer graphene sheets, ACS Nano, 5 (2011) 4112-4117. https://doi.org/10.1021/nn200766e

[41] Z. Luo, S. Lim, Z. Tian, J. Shang, L. Lai, B. MacDonald, C. Fu, Z. Shen, Z. T. Yu, J. Lin, Pyridinic N doped graphene: Synthesis, electronic structure, and electrocatalytic property. J. Mater. Chem. 21 (2011) 8038–8044. https://doi.org/10.1039/c1jm10845j

[42] D. Deng, X. Pan, L. Yu, Y. Cui, Y. Jiang, J. Qi, W.X. Li, Q. Fu, X. Ma, Q. Xue, Toward N- doped graphene via solvothermal synthesis. Chem. Mater. 23 (2011)1188-1193. https://doi.org/10.1021/cm102666r

[43] L.S. Panchakarla, K.S. Subrahmanyam, S.K. Saha, A. Govindaraj, H.R. Krishnamurthy, U.V. Waghmare, C.N.R. Rao, Synthesis, structure, and properties of boron and nitrogen doped graphene. Adv. Mater, 21 (2009) 4726-4730. https://doi.org/10.1002/adma.200901285

[44] D. Geng, Y. Chen, Y. Chen, Y. Li, R. Li, X. Sun, S. Ye, S. Knights, High oxygen reduction activity and durability of nitrogen-doped graphene. Energy Environ. Sci, 4 (2011) 760-764. https://doi.org/10.1039/c0ee00326c

[45] X. Wang, X.X. Li, L. Zhang, L. Yoon, P.K. Weber, H. Wang, J. Guo, H. Dai, N-doping of graphene through electrothermal reactions with ammonia. Science 324 (2009)768-771. https://doi.org/10.1126/science.1170335

[46] X. Li, H. Wang, J.T. Robinson, H.S Sanchez, G. Diankov, H. Dai, Simultaneous nitrogen doping and reduction of graphene oxide. J. Am. Chem. Soc, 131 (2009) 15939-15944. https://doi.org/10.1021/ja907098f

[47] Y. Shao, S. Zhang, M.H. Engelhard, G. Li, G. Shao, Y. Wang, J. Liu, I.A. Aksay, Y. Lin, Nitrogen doped graphene and its electrochemical applications. J. Mater. Chem. 20 (2010) 7491-7496. https://doi.org/10.1039/c0jm00782j

[48] B. Guo, Q. Liu, E. Chen, H. Zhu, L. Fang, J. R. Gong, Controllable N-doping of graphene. Nano Lett. 10 (2010)4975-4980. https://doi.org/10.1021/nl103079j

[49] L. F. Lai, J. R. Potts, D. Zhan, L. Wang, C. K. Poh, C. H. Tang, H. Gong, Z. X. Shen, L. Y. Jianyi, R. S. Ruoff, Exploration of the active center structure of nitrogen-doped graphene- based catalysts for oxygen reduction reaction Energy Environ. Sci. 5 (2012) 7936-7942. https://doi.org/10.1039/c2ee21802j

[50] P. A. Denis, Density functional investigation of thioepoxidated and thiolated graphene J. Phys. Chem. C. 113 (2009) 5612-5619. https://doi.org/10.1021/jp808599w

[51] H. Gao, Z. Liu, L. Song, W. Guo, W. Gao, L. Ci, A. Rao, W. Quan, R. Vajtai, P. M. Ajayan, Synthesis of S-doped graphene by liquid precursor, Nanotechnology. 23 (2012) 275605. https://doi.org/10.1088/0957-4484/23/27/275605

[52] J. Yang, D. Voiry, S. J. Ahn, D. Kang, A. Y. Kim, M. Chhowalla, H. S. Shin, Two-dimensional hybrid nanosheets of tungsten disulfide and reduced graphene oxide as ctalysts for enhanced hydrogen evolution Angew. Chem.Int. Ed. 125 (2013) 13996-13999. https://doi.org/10.1002/ange.201307475

[53] L. Xia, J. Yang, H. Wang, R. Zhang, H. Chen, W. Fang, M. Abdullah F. Xie, G. Cuib, X. Sund, Sulfur doped graphene for efficient electrocatalytic N2-to-NH3 fixation,Chem. Commun. 55 (2019) 3371-3374. https://doi.org/10.1039/C9CC00602H

[54] R. Li, Z. Wei, X. Gou, W. Xu Phosphorus-doped graphene nanosheets as efficient metal- free oxygen reduction electrocatalysts. RSC Adv. 3 (2013) 9978-9984. https://doi.org/10.1039/c3ra41079j

[55] L. Lin, L. Fu, K. Zhang,J. Chen,W. Zhang, S. Tang,Y. Du, N. Tang P-Superdoped Graphene: Synthesis and Magnetic Properties, ACS Appl. Mater. Interfaces 42 (2019), 39062–39067.https://doi.org/10.1021/acsami.9b11505

[56] M. A. Ribas, A. K. Singh, P. B. Sorokin, B. I. Yakobson, Patterning nanoroads and quantum dots on fluorinated graphene Nano Res, 4 (2011) 143-152. https://doi.org/10.1007/s12274-010-0084-7

[57] J. T. Robinson, J. S. Burgess, C. E. Junkermeier, S. C. Badescu, T. L. Reinecke, F. K. Perkins, M. K. Zalalutdniov, J.W. Baldwin, J. C. Culbertson, P. E. Sheehan E. S. Snow, Properties of fluorinated graphene films Nano Lett, 10 (2010) 3001-3005. https://doi.org/10.1021/nl101437p

[58] J. Zheng, H. T. Liu, B. Wu, C. A. Di, Y. L. Guo, T. Wu, G. Yu, Y. Q. Liu, D. B. Zhu, Production of Graphite Chloride and Bromide Using Microwave Sparks Sci. Rep, 2 (2012) 662. https://doi.org/10.1038/srep00662

[59] D.M.E. Gendy, N.A.A. Ghany, N. K. AllamGreen single-pot synthesis of functionalized Na/N/P co-doped grapheme nanosheets for high-performance supercapacitors, J. Electroanal. Chem. 837 (2019) 30-38. https://doi.org/10.1016/j.jelechem.2019.02.009

[60] J. Zhou, Z. Wang, D. Yang, W. Zhang Y.Chen, Free-standing S, N co-doped graphene/Ni foam as highly efficient and stable electrocatalyst for oxygen evolution reaction Electrochimica Acta 317 (2019) 408-415. https://doi.org/10.1016/j.electacta.2019.06.015

[61] R. L. McCreery, Advanced carbon electrode materials for molecular electrochemistry, Chem. Rev. 108 (2008) 2646-2687. https://doi.org/10.1021/cr068076m

[62] J. B. Jia, D. Kato, R. Kurita, Y. Sato, K. Maruyama, K. Suzuki, S. Hirono, T. Ando, O. Niwa, Structure and electrochemical properties of carbon films prepared by a electron cyclotron resonance sputtering method. Anal. Chem, 79 (2007) 98-105. https://doi.org/10.1021/ac0610558

[63] O. Niwa, J. Jia, Y. Sato, D. Kato, R. Kurita, K. Maruyama, K. Suzuki, S. Hirono, Electrochemical Performance of Angstrom Level Flat Sputtered Carbon Film Consisting of sp^2 and sp^3 Mixed Bonds, J. Am. Chem. Soc. 128 (2006) 7144-7145. https://doi.org/10.1021/ja0606091

[64] S. L. Yang, D. Y. Guo, L. Su, P. Yu, D. Li, J. S. Ye, L. Q. Mao, A facile method for preparation of graphene film electrodes with tailor-made dimensions with Vaseline as

the insulating binder, Electrochem. Commun.11 (2009) 1912-1915.
https://doi.org/10.1016/j.elecom.2009.08.020

[65] R. S. Nicholson, Theory and application of cyclic voltammetry for measurement of electrode reaction kinetics, Anal. Chem. 37 (1965) 1351-1355.
https://doi.org/10.1021/ac60230a016

[66] N. G. Shang, P. Papakonstantinou, M. McMullan, M. Chu, A. Stamboulis, A. Potenza, S. S. Dhesi, H. Marchetto, Catalyst-free efficient growth, orientation and biosensing properties of multilayer graphene nanoflake films with sharp edge planes, Adv. Funct. Mater.18 (2008) 3506-3514. https://doi.org/10.1002/adfm.200800951

[67] A. E. Fischer, Y. Show, G. M. Swain, Electrochemical performance of diamond thin-film electrodes from different commercial sources, Anal. Chem. 76 (2004) 2553-2560. https://doi.org/10.1021/ac035214o

[68] B.R. Eggins, Biosensors: an introduction, Wiley-VCH (1996).
https://doi.org/10.1007/978-3-663-05664-5

[69] S. V. Dzyadevych, V. N. Arkhypova, A. P. Soldatkin, A. V. Elskaya, C. Martelet, J. Renault. Amperometric enzyme biosensors: Past, present and future, IRBM. 29 (2008)171- 180. https://doi.org/10.1016/j.rbmret.2007.11.007

[70] D. R. Thevenot, K. Toth, R. A. Durst, G. S. Wilson, Electrochemical biosensors: recommended definitions and classification, Anal Lett.34(2001) 635-659.
https://doi.org/10.1081/AL-100103209

[71] J. Wang "Analytical Electrochemistry", 3rd edition, Wiley-VCH, (2006)201.
https://doi.org/10.1002/0471790303

[72] N. J. Renault New Trends in Biosensors for Organophosphorus Pesticides, Sensors 2 (2001) 60-74. https://doi.org/10.3390/s10100060

[73] Y. Zhang, X. Bo, A. Nsabimana, C. Luhana, G. Wang, H. Wang, M. Li, L. Guo, Fabrication Of 2D ordered mesoporous carbon nitride and its use as electrochemical sensing platform for H_2O_2, Nitrobenzene, and NADH Detection. Biosens. Bioelectron. 53 (2014) 250-256. https://doi.org/10.1016/j.bios.2013.10.001

[74] W. Zhu, J. Gao, H. Song, X. Lin, S. Zhang, Nature of the synergistic effect of N and S Co- Doped graphene for the enhanced simultaneous determination of toxic pollutants. ACS Appl. Mater. Interfac. 11 (2019) 44545-44555.
https://doi.org/10.1021/acsami.9b13211

[75] M. Trojanowicz, Determination of pesticides using electrochemical enzymatic biosensors. Electroanalysis. 14 (2002) 19-20. https://doi.org/10.1002/1521-4109(200211)14:19/20<1311::AID-ELAN1311>3.0.CO;2-7

[76] D. M. Quinn. Acetylcholinesterase: enzyme structure, reaction dynamics, and virtual transition states, Chem. Rev. 87 (1987) 955-979. https://doi.org/10.1021/cr00081a005

[77] R. Su, X. Xu, X. Wang, D. Li, X. Li, H. Zhang, A. Yu, Determination of organophosphorus pesticides in peanut oil by dispersive solid phase extraction gas chromatography-mass spectrometry, J. Chromatogr. B Analyt. Technol. Biomed. Life Sci. 879 (2011) 3423-3428. https://doi.org/10.1016/j.jchromb.2011.09.016

[78] C. Wu, H. Liu, W. Liu, Q. Wu, C. Wang, Z. Wang, Determination of organophosphorus pesticides in environmental water samples by dispersive liquid-liquid microextraction with solidification of floating organic droplet followed by highperformance liquid chromatography, Anal. Bioanal. Chem. 397 (2010) 2543-2549. https://doi.org/10.1007/s00216-010-3790-9

[79] L.Wu, W. Lei, Z. Han, Y. Zhang, M. Xia, Q. Hao, A novel non-enzyme amperometric poly(3-methylthiophene)/nitrogen platform based on electrode doped graphene modified for determination of trace amounts of pesticide phoxim. Sensor Actuat B Chem. 206 (2015) 495–501. https://doi.org/10.1016/j.snb.2014.09.098

[80] X. Xue, Q. Wei, D.Wu, H. Li, Y.Zhang, R. Feng, B. Du. Determination of methyl parathion by a molecularly imprinted sensor based on nitrogen doped graphene sheets. Electrochimica Acta. 116 (2014) 366–371. https://doi.org/10.1016/j.electacta.2013.11.075

[81] Y. Ya, C. Jiang, L. Mo, T. Li, L. Xie, J. He, L Tang, D. Ning, F. Yan, Electrochemical determination of carbendazim in food samples using an electrochemically reduced nitrogen doped graphene oxide modified glassy carbon electrode. Food. anal. meth. 10 (2017) 1479–1487. https://doi.org/10.1007/s12161-016-0708-y

[82] G. Yu, W Zhang, Q. Zhao, W. Wu, X. Wei, Q. Lu. Enhancing the sensitivity of hexachlorobenzene sensor electrochemical based on nitrogen–doped graphene. Sensor Actuat B Chem. 235 (2016) 439-446. https://doi.org/10.1016/j.snb.2016.05.072

[83] Y. Zhang, H. B. Fa, B. He, C. Hou, D. Huo, T. Xia, W. Yin, Electrochemical biomimetic sensor based on oximegroup-functionalized gold nanoparticles and nitrogen- dopedgraphene composites for highly selective and sensitive dimethoatedetermination. J Solid State Electr. 21 (2017) 2117–2128. https://doi.org/10.1007/s10008-017-3560-0

[84] X. Xue,Q.Wei,D. Wu,H. Li,Y. Zhang,R. Feng,B.Du, Determination of methyl parathion by a molecularly imprinted sensor based on nitrogen doped graphene shee, Electrochemical Acta, 116 (2014), 366-371, https://doi.org/10.1016/j.electacta.2013.11.075

[85] T. Vincent, E. Guibal, Chitosan-supported palladium catalyst influence of experimental parameters on nitrophenol degradation, Langmuir 19 (2003) 8475–8483. https://doi.org/10.1021/la034364r

[86] Y. Zhang, L. Wu, W. Lei, X. Xia, M. Xia, Q. Hao. Electrochemical determination of 4- nitrophenolpolycarbazole/N-doped at graphene modified glassy carbon electrode. Electrochimica Acta 146 (2014) 568–576. https://doi.org/10.1016/j.electacta.2014.08.153

[87] X. Jiang, H. Shi, J. Shen, W. Han, X. Sun, J. Ji, L. Wang, Synergistic effect of pyrrolic N and graphitic N for the enhanced nitrophenol reduction of nitrogen doped graphene modified cathode in the bioelectrochemical system, J. Electroanal. Chem.823 (2018) 32-39. https://doi.org/10.1016/j.jelechem.2018.05.036

[88] B.S. Suresh, A. Elavarasan, M. Sathish, High performance supercapacitor using Ndoped graphene prepared via supercritical fluid processing with an oxime nitrogen source, Electrochim. Acta 200 (2016) 37-45. https://doi.org/10.1016/j.electacta.2016.03.150

[89] Z. S. Wu, A. Winter, L. Chen, Y. Sun, A. Turchanin, X. L. Feng and K. Mullen, Three-dimensional nitrogen and boron co-doped graphene for high-performance all-solid-state supercapacitors Adv. Mater., 2012, 24, 5130–5135. https://doi.org/10.1002/adma.201201948

Materials Research Forum LLC
https://doi.org/10.21741/9781644900956-4

Chapter 4

Graphene-Metal Modified Electrochemical Sensors for Toxic Chemicals

S. Vinodha[*,1], L. Vidhya[1], T. Ramya[2], R. Jeba Beula[3], P. Jegathambal[4]

[1] Department of Chemical Engineering, Sethu Institute of Technology, Kariapatti, Virudhunagar District

[2] Department of Environmental Science, Bharathiar University, Coimbatore

[3] Department of Physics, Karunya University, Coimbatore

[4] Water Institute, Karunya University, Coimbatore

vinodha.harris@gmail.com, vidhuram236@gmail.com rtramya1@gmail.com

Abstract

The extent of amplified wastes produced and discharged into the environment contains wide variety of libelous chemicals, contaminants and toxic substances which are carcinogenic in nature. Toxic chemicals require special consideration in view of the risks posed to ecosystem. Comprehensive investigations on development of diverse variety of sensing materials are versatile and highly adaptive and have proven to be promising. A new class of electrochemical sensors has emerged where chemical functionalisation, hybridisation have led to improved performance, stability or versatility. Graphene which is a novel, fascinating type of nanocarbon possess positive contribution to the electrode properties and exhibits beneficial behaviour for a wide variety of electrochemical applications. Graphene based metal frameworks have received great interest and opened new gates in scientific communities due to their enhanced physiochemical properties, excellent stability, higher flexibility, and very good electrical conductivity. Graphene metal hybrid electrochemical sensors lay an enhanced platform for electro analytical applications as it effectively accelerates the transfer of electrons which provides a fast and highly sensitive current response. Owing to the enhanced electronic transport property and high electrocatalytic activity of graphene, the electrochemical reactions of analyte are greatly promoted on graphene film resulting in enhanced voltammetric response. This chapter discloses the advances in the field of graphene metal modified electrochemical sensors with particular focus on toxic chemicals. The main emphasis of

Graphene-Based Electrochemical Sensors for Toxic Chemicals   Materials Research Forum LLC
Materials Research Foundations **82** (2020) 91-124   https://doi.org/10.21741/9781644900956-4

the chapter is on the electrochemical sensing application, summarizing the advantages, disadvantages, and challenges offered in the field.

Keywords
Graphene, Modified Electrode, Toxic Chemicals, Electrochemical Sensors

Contents

1.   Preamble

Technological advancements have resulted in the generation of additional wastes proving destructive effects to ecology. Engineering developments are ensuing in reduction of resources and stern environmental damage. Henceforth, if the nearby century strikes socio-economic, scientific and technological developments on one hand, it is overwhelmed to face serious harms on the other hand due to the release of ample quantity of wastes (pollutants, insecticides, pesticides, and fertilizers) into the environment.

Moreover, the extent of amplified wastes produced and discharged into the environment contains wide variety of libelous chemicals, contaminants and toxic substances which are carcinogenic in nature. Contamination of biota with prospective chemicals that are noxious has crucial inference for individual wellbeing [1]. Therefore upon improper management, these toxic substances will lead to ecological catastrophe. Managing toxic chemicals released into the environment has become an extensive task, demanding incessant planning, stable legislative policies, community contribution and above all through effectual industrial techniques. Although existing remediation practices for sequestration of toxic contaminants such as precipitation, adsorption, bio-sorption, and membrane separation [2] etc. are on the way in recent decades, comprehensive investigation on development of diverse variety of sensing materials are versatile and highly adaptive and have proven to be promising [3]. User friendly electronic devices, particularly sensors, have become superior with the aim to resolve tremendous real-world challenges, such as in situ sensing for health and environmental problems. Development of convenient and transportable sensors is an extensive dynamic vicinity of research in the electro-analytical ground with major focus on the development of inexpensive electrodes to employ in such sensors. Traditional electrodes such as glassy carbon electrodes (GCEs) are expensive and incur time for preparation. Mechanical pencil lead electrodes (MPEs) which were an alternative to GCE owned rich drawbacks predominantly structural vulnerability [3]. Subsequent to the progress of numerous viable electrochemical sensors, presently there is a remarkable evolution and mounting claim for the development of flexible electrochemical sensors with the usage of new elements.

Graphene has revealed enormous prospective as sensing element due to its distinctive advantages such as huge surface area, remarkable electrical conductivity, rapid electron carrying capacity, simplicity of functionalisation, and mass manufacture [4,5,6]. Graphene is a collection of single layer carbon atoms attached by sp^2 hybridization forming a honeycomb lattice [7]. This material shows superior electrical, mechanical, thermal, and catalytic properties that are appropriate for applying in sensors [8]. Explicit properties of graphene like single layer thickness and high electron carrying capacity facilitates the detection of chemical species [9,10]. Technologists and scientists have grasped the assets of graphene and are investigating possible modes of incorporating graphene as electrochemical sensors. Cost effective and novel graphene electrochemical sensors have been devised by various methods such as mechanical stripping, chemical disoxidation of graphene oxide, chemical vapor deposition, electrodeposition for the detection of bimolecular compounds (glucose, dopamine, ascorbic acid and uric acid), heavy metal ions (Pb^{2+}, Hg^{2+}, Cd^{2+} and As^{3+}) and environmental contaminants (hydrazine, nitrobenzene, nitrophenols and pesticides) [11] (Figure 1). In addition, graphene into combination with materials possesses large interfacial surface area much better than carbon nanotubes (CNTs) and has proved a wide range of application. The modern rise of synthesis and processing of two-dimensional (2D) graphene and its derivatives present a new opportunity and shows great potential for developing a novel sensor most likely with essential physiochemical properties, as every atom in a graphene sheet is a surface atom, molecular interaction and thus electron transport through graphene can be highly sensitive to adsorbed molecules. This chapter deals with the contemporary progress of graphene metal modified electrochemical sensor for detection of toxic chemicals.

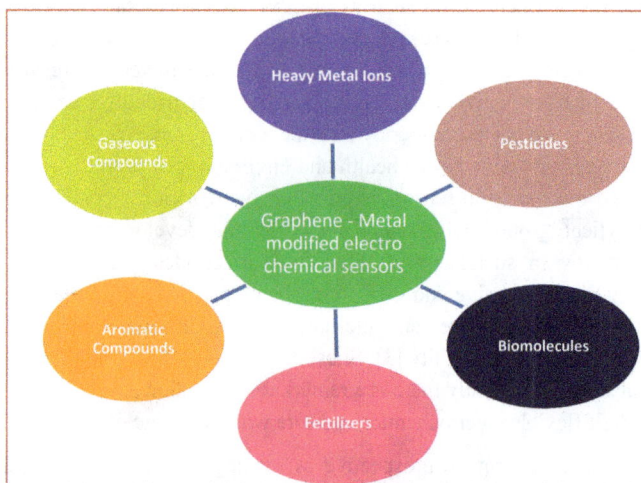

Figure 1 Application of graphene metal modified electro chemical sensors.

2.　　Hybridisation of graphene – metal

Nano materials find a wider range of application as it possesses special properties such as less inter distance of the particle, capability to create good adsorption and surface/volume ratio (that can enhance the voltammetric responses) being high compared to micro particles. In addition, materials in nanoscale have unique magnetic, photonic, electronic, electro-catalytic, and bio-sensing features, which have been exploited effectively in recent times [12]. The transport of charge between graphene layers is one major issue to deal with to ensure high performance of graphene-based sensors. To overcome this problem, different materials have been decorated onto graphene surfaces such as metal nanoparticles. A variety of nanomaterials with well-controlled physicochemical properties are used to modify electrodes in order to improve their analytical performances. Henceforth, development of composite based electrochemical sensors paves way for enhanced applications. The interaction of graphene with transition metals in particular uncovers practical applications. Hybridisation of graphene with metals and other nano-composites finds greater significance in steady, rapid and sharp sensing techniques. This section outlines the possibilities of creating hybrid metallic nanoparticles and graphene based electrochemical sensor in addition with its analytical performance. Also, graphene (or its derivatives) hybridization with nanomaterials enhances the operational properties of components and generate new properties via supportive interface. Electrochemical exfoliation of ionic liquid-functionalized graphite [13] and electrical arc discharge between two graphitic electrodes [14,15] are approaches for preparation of graphene nanosheets. For synthesizing high-quality, large-area graphene nanosheets of different layers and sizes the process of Chemical Vapour Deposition is believed to be the prime means. Physico-chemical characteristics and surface qualities of graphene can be altered using nitrogen, boron, phosphorus, and sulfur (S) heteroatom doping [16]. In recent years, number of investigations in sensing applications (especially electrochemical sensors) have been appropriated to the hybrid of graphene with metallic nanoparticles such as gold nanoparticles, nickel nanoparticles, titanium di oxide nanoparticles, ferrous nanoparticles, copper nanoparticles, and palladium nanoparticles which is discussed herewith.

2.1   Graphene - gold

Gold nanoparticles are one of the maximum considered nanomaterials because of their remarkable properties. Gold nanoparticles also known as colloidal gold, constitute a suspension of submicrometer-sized gold metal particle in a fluid and can be obtained with diameters between 3 and 200 nm. Electrochemical and optical study of the Au_{25} nanoparticles has revealed that Au_{25} exhibited highest occupied molecular orbital, lowest unoccupied molecular orbital, depicting properties of molecule wherein its structure, fundamental, physical and chemical properties also proves promising. These nanoparticles have received considerable attention recently because of their unique size-dependent electrochemical, optical, and high electrochemical catalytic properties, strong adsorption capability, biocompatibility, sound suitability and high conductivity [17]. Chemical functionality added with easy synthetic techniques makes gold as an interesting coating material to be employed as electrochemical sensors successfully [18,19,20,21]. Electrodeposition is the most common method for creating nanoparticles [22]. A graphene gold nanoparticles hybrid composite (Figure 2) is made via layer-by-layer strategy, using an intermediary as linkage [23,24], utilizing reducing agents [25] and without utilizing any reducing agents [26].

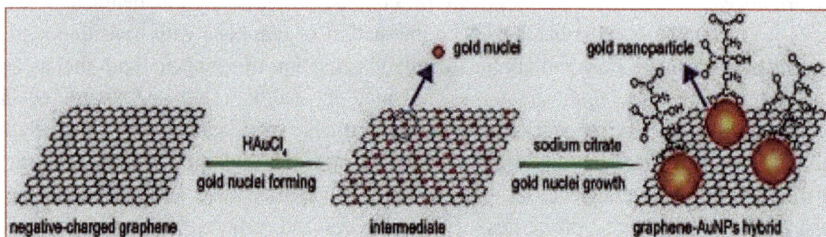

Figure 2 Schematic representation of hybrid graphene gold nanoparticles.

Material used as insulator decreases the conductivity which is the main drawback. Also, difficulty in uniform distribution of nanoparticle still remains as a challenge. There is also a possibility of absence of intercalation of metal nanoparticles between reduced graphene sheets in the above said process [27] however; the stability of ionic strength is under study.

2.2   Graphene - titanium

Titanium has become a part of day to day life, with wide range of world wide application possessing superior stability, reactivity, biocompatibility, non-toxic and environmentally friendly nature [28,29,30]. Due to its high conductivity and cost effectiveness, TiO_2 is being used by material engineers as an attractive electrode material in diverse forms such as nanoparticles (represented in Figure 3), nanoneedles, nanofibers, nanosheets, and nanotubes [31,32,33,34] and accessible for the fabrication of sensors. In combination with the technologies of today's age, TiO_2 nanomaterials are extensively used for the construction of low cost, environmental benign, competent, sensors. They can be decked on graphene or other carbon material to be designed into electrochemical based sensors [35,36,37,38], photocatalytic and electrocatalytic applications [39,40,41] as TiO_2 has the capability of increasing active sites and electron transfer mechanisms [42] thereby used for the detection and analysis of toxic chemicals and its compounds in aqueous media.

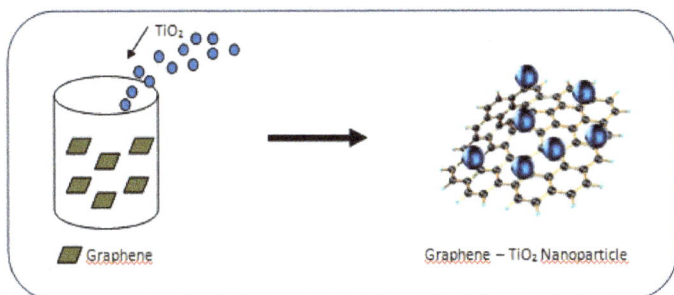

Figure 3: Illustration of ex situ hybridization of graphene TiO_2 nanoparticle

2.3   Graphene - ferrous nanoparticle

Ferrous nanoparticles are inorganic nanoparticles and are classified based on the size of the iron oxide, i.e., standard size of superparamagnetic iron oxide is 60–150 nm, ultra-small size of superparamagnetic iron oxide is 5 to 40 nm, and subset size of monocrystalline iron oxide is 10 to 30 nm. The standard-size nanoparticles are composed of ferric (Fe^{3+}) and ferrous (Fe^{2+}) iron nanoparticles which are a type of magnetic materials that have great potential due to their low toxicity, paramagnetic property, biocompatibility, convenient preparation techniques [43] and more notably, the close contact among the nanoparticles, substrates, and catalysts [44]. In the present decades, iron nano particles have consideration for their potential applications in ferro fluids [45], rechargeable batteries [46], magnetic resonance imaging [47], targeted drug delivery [48,49] and catalysis [50]. To add with, they have been commonly utilized for biosensing because they possess the property of enhancement of the electrode surface area, electrical conductivity, and kinetics of electron transfer. For assessing biomolecules for biosensing these are considered as an effective carrier [51]. The most vital aspect taken into account for fabrication of sensors is the ability to disperse the iron nanoparticles uniformly in apt medium such as chitosan, carbon, and graphene to foil the agglomeration process [52]. It is of great interest to modify electrodes with the combination of graphene and iron nanoparticles for the fabrication of new electrochemical sensors. Recently, iron oxide nanoparticle and graphene composites have been prepared to construct a mediator free H_2O_2 sensor with the Fe_3O_4 particles size of 200-250 nm in diameter. The ferrous oxide nanoparticle composite is produced by in situ reduction of ferric hydroxide by microwave heating [53], whereas larger particle size and poor particle dispersion inhibit its properties. Hence for uniform dispersion of ferrous nanoparticles on graphene sheets, it is reported to disperse nanoparticles of 10 nm in diameter on the graphene sheet [54]. Though various other known chemical methods such as ion exchange, dialysis, precipitation and conjugation chemistry are available, they own the disadvantages of being complicated due to series of procedures and use of expensive raw materials [54]. Therefore, the major challenge that prevails is economically viable fabricating technique and to achieve uniform distribution of nanoparticles.

2.4   Graphene - copper

Copper nanoparticles (1 to 100 nm in size) has become a promising subject since it holds good optical, catalytic, mechanical and electrical properties that lead to new applications and devices such as lubricants, sensors or catalysts. With small size and great porosity, the nanoparticles are able to achieve a higher reaction yield and a shorter reaction time when utilized as reagents in organic and organo-metallic synthesis. Like many other forms of nanoparticles, a copper nanoparticle can be formed by natural processes or through chemical synthesis. Copper nanoparticles enhance the response current and possess better biocompatibility and relatively inexpensive cost [55]. Graphene copper nanoparticles hybrid was set by encapsulating the nanoparticles with graphene. Hybrids have also developed by depositing graphene layers onto copper nanoparticles on a larger scale with the support of reducing flame technique [56]. These hybrid electrodes were also modified through electrostatic adsorption and have been used for the non-enzymatic detection of analytes.

2.5   Graphene - palladium

Palladium nanoparticles have high electron conductivity with small sizes typically 20 - 100 nanometers (nm) with specific surface area (SSA) in the 1 - 3 m^2/g range [57]. Remarkable catalytic efficiency of palladium based nanostructures has attracted and drawn substantial notice in diversified areas. For improving the performance of sensors and grab the benefits of the synergistic properties of surface compositions, palladium nanoparticles have been hybridized with low dimensional carbon materials like graphene. Biosensors were fabricated by immobilizing graphene onto nanocomposites of palladium nanoparticles and chitosan, through covalent crosslinking which displayed electrocatalytic performance towards analysis of H_2O_2 [58] and toxic chemicals.

2.6   Graphene- nickel

Nanoscale nickel particles are typically 10 - 40 nanometers (nm) with specific surface area (SSA) in the 30 - 50 m^2/g range. Nickel oxide (NiO) nanostructured based materials have been extensively applied in super capacitor biosensors and electrochemical sensors, owing to their tremendous electrocatalytic and low-cost properties [59]. Assorted morphological nickel nanostructures have been manufactured by different methods in recent decades such that, nickel oxide and hydroxide have great interest for sensing progress due to their properties including biocompatibility, cost effectiveness, non-toxicity, high chemical stability, and high electrocatalytic effect [60]. Some researchers have investigated the electrocatalytic effect of graphene nickel modified electrodes towards the electro-redox reaction of H_2O_2 and pesticides. Works have been carried out for in-situ electrodeposition of nickel onto graphene using series of steps after dropping of Myoglobin (Mb) solution on the electrode and the use of nafion for the prevention of its leakage from the electrode surface [61].

Other metals such as silver, platinum, cobalt, zinc, manganese etc. have also been used in combination with graphene for devising electrochemical sensors in detection of toxic chemicals.

2.7   Selectivity of hybridized graphene for electrochemical sensors

Although there is a considerable progress in the synthesis, processing, characterization and applications of graphene and metallic nanomaterials for the design of electrochemical sensors, most importantly these sensors have to be specific, reusable and to have high sensitivity towards specific analyte. Thus current challenges comprise the development of low-cost, ultrahigh, and user-friendly sensors, which have high selectivity, fast response and recovery times, and small dimensions. As already discussed, graphene in its metal modified form with nanoparticles has demonstrated ideal properties to overcome these challenges and have emerged as one of the most popular sensing platforms for diverse applications. Electrochemical sensors and biosensors with high sensitivity are extremely important not only in biomedical applications, clinical diagnosis but also in environmental protection. Highly selective graphene based sensor arrays was reported for the detection of volatile organic compounds (VOCs), gases, organophosphate pesticides [61], phenolic compounds [62,63] biomolecules such as cancer biomarkers [64], proteins [65,66], dopamine (DA) [67,68] and amino acid tyrosine [69] and various toxic chemicals listed in [70].

3.   Sensing application of toxic chemicals

Toxic chemicals are harmful to human health and when their contents exceed the limit in soil or water environment generate ecological and environmental harm. These include heavy metal ions, pharmaceuticals, pesticides, plastics, fuels, solvents, explosives, surface coatings, adhesives, disinfectants, and other organic and inorganic compounds. There is a paramount need for detection of such toxic chemicals by a simple, fast and sensitive method. In comparison with the conventional detection methods, graphene metal modified electrochemical sensors encompass the advantages of high sensitivity, selectivity, rapid detection and high stability and found to be useful for toxic chemical analysis.

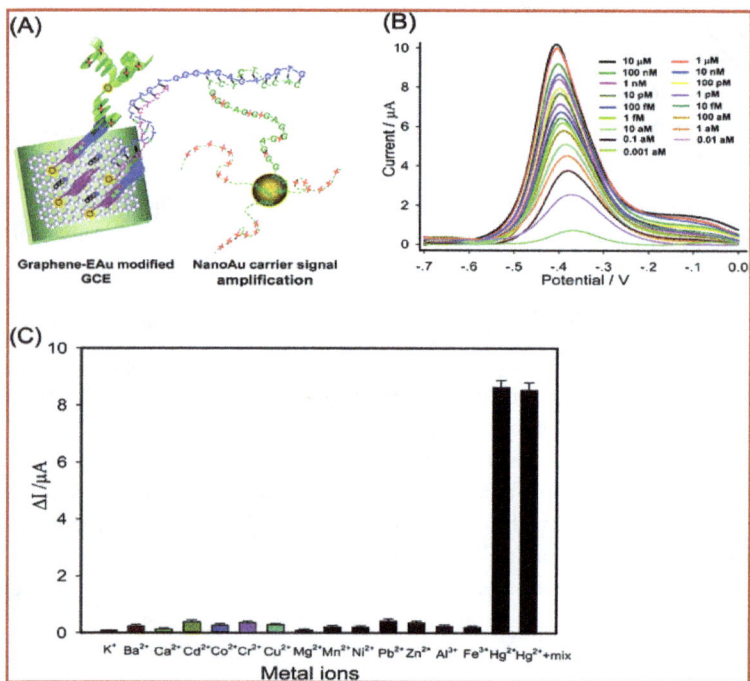

Figure 4 A) Schematic illustration of the sensing of Hg^{2+} using a graphene/nano-Au composite for signal amplification. (B) Square wave voltammograms (SWV) as a function of Hg^{2+} concentration (C) Selectivity of Hg^{2+} ions among various interfering metal ions.

3.1   Ions of heavy metals

Heavy metals due to their toxicity have received a paramount attention among all the pollutants. Heavy metals are usually present in trace amounts in environment but many of them are toxic even at very low concentrations. Their toxicity depends on several factors including the dose, route of exposure, and chemical species, as well as the age, gender, genetics, and nutritional status of exposed individuals. Though certain heavy metals such as Manganese, Copper etc, in small quantities are nutritionally essential for a healthy life, toxic heavy metal ions such as Lead, Mercury, Cadmium, Copper, Chromium and

Graphene-Based Electrochemical Sensors for Toxic Chemicals   Materials Research Forum LLC
Materials Research Foundations **82** (2020) 91-124   https://doi.org/10.21741/9781644900956-4

Arsenic are few among them which are well known to be severely harmful to human health. Amongst the heavy metals, Pb^{2+} and Cd^{2+} pose severe threat to human health such as reproductive toxicity, damage to the human immune system, respiratory disorders, and negative effects on metabolism. The performance of electrochemical sensors for metal ion detection was effectively improved, when graphene based nanomaterials are employed. In recent years, different graphene metal modified electrodes have been put into practice in diversified ways for the detection and elimination of toxic chemicals. Hydrophilic characteristic and functional groups including oxygen atoms that can bind metal ions to the surface improves the performance of graphene in modified form for the removal of heavy metal ions [71,72] as depicted in Figure 4. Few metal ions of great concern are discussed herewith.

Lead

Exposure of lead throw serious consequences on the health of human especially children attacking the brain and central nervous system. Lead ions were detected using the functionalisation of gold nanoparticles onto graphene. Au nanoparticles were decorated initially on graphene due to DNAzyme molecule immobilization. Low cost, high stability and selectivity compared to protein and RNA enzymes remain as an advantage of the DNAzymes. A single gold nanoparticle has the capability to adhere many DNAzyme molecules. A positive graphene transfer curves shift in a large order is observed when decorated with gold nanoparticles due to p-type doping effect. Conversely using n-type doping effects to graphene stimulated by the DNA molecule, there is a curve shift to the negative direction. As a result, the DNAzyme/Au nanoparticles/graphene curve is in the intermediate of graphene and Au nanoparticles/graphene curves. When lead ions become adjacent to the device, the transfer curves shift to the right side again whereby; the shifting is proportional to the concentration of lead. Similar studies have been reported by Erkang Wang [73] where the graphene sheets were dispersed in nafion solution. A composite structure using modified graphene with 1-noctylpyridinum hexa fluorophosphates [74] is widely useful for the electrochemical analysis of Pb^{2+} along with the simultaneous detection of mercury and cadmium without the necessity of separating three elements from complex mixtures prior to measurements. TiO_2 - Gr nanocomposites have also been used in the analysis of lead as represented in Figure 5.

Cadmium

Cadmium which is mainly found in the earth's crust is an inevitable byproduct of zinc, copper and lead. It can accumulate in liver, bones and kidneys in humans and can be toxic to plants and animals too. Wang et al., [73] used graphene nanosheets dispersed in Nafion (Nafion-G) solution along with bismuth film electrode plated in situ, for the fabrication of the electrochemical sensors with enhanced sensing platform for Cd^{2+} using differential pulse anodic stripping voltammetry. The prepared Nafion-G composite film exhibited improved sensitivity for the metal ion detection as a result of the synergistic effect of Nafion-G. The stripping current signal is greatly enhanced and well distinguished on graphene electrodes with a wide linear range and 0.02 µg L^{-1} detection limit for cadmium [74,75]. Another modified form of electrode was using L-cysteine/graphene-carbon monosulfide. The electrochemical behaviors of cadmium and lead on the proposed electrode were studied by differential pulse anodic stripping voltammetry. Operational parameters, such as the deposition potential and time, pH value of the buffer have been optimized with the detection limits 0.45 µg/L [76]. Alternate to these works, a new sensor using gold nano particles on graphene (AuNPs/GR/cysteine composite) proved promising at the detection range of 0.5 - 40 mg L^{-1} and a limit of detection as 0.1 mg L^{-1}[77]. Zhang et al has attempted TiO_2–graphene hybrid nanostructures for Pb(II) and Cd(II) exposure. The sensor showed linear ranges from 1.0 x 10^{-8} M to 3.2 x 10^{-5} M and 6.0 x 10^{-7} M to 3.2 x10^{-5} M and detection limit value of 1.0 x 10^{-10} M and 2.0 x 10^{-9} M for Pb^{2+} and Cd^{2+} respectively [78].

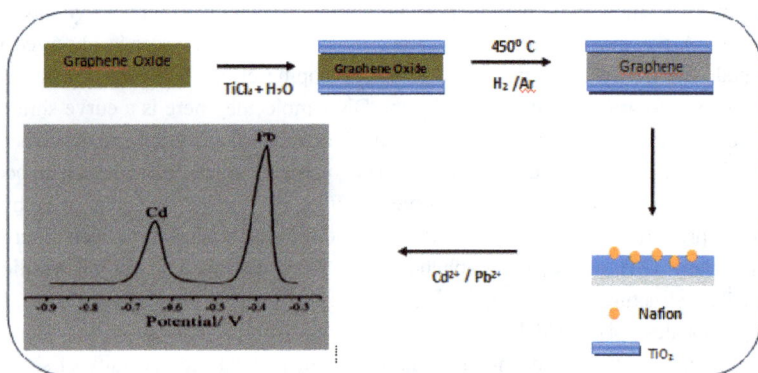

Figure 5: TiO_2-Gr nanocomposites and electrochemical sensing of Pb^{2+} and Cd^{2+}

Mercury

Physiological and neurological toxicity property of mercury (which is one among the predominant heavy metals) demands severe effects on human health and the environment. The global overall annual mercury discharge including both anthropogenic and non anthropogenic sources is approximated 7500 tons per year [79]. Increasing efforts are being made for identifying and understanding mercury exposure and its effects to address mercury elimination. [80] brought to the notice about the electrochemical determination of Hg(II) based on activated graphene modified screen-printed carbon electrodes. The sensor exhibited a linear range of 0.05-14.77 ppm and an limit of detection of 4.6 ppb with a sensitivity of 81.5 mA ppm^{-1} cm^{-2} such that it is found to be considerably below the guideline values for Hg(II) in drinking water. A novel Ionic Liquid/Graphene Modified Electrode has also been used recently for this metal ion detection. Furthermore, the detection of Hg^{2+} with a limit of detection value as 0.001 aM has been achieved with a electrodeposited graphene-Au-modified electrode for signal amplification.

Nickel

Limited research has been carried out in the detection of nickel ions. A type of graphene nanosheet/δ- MnO$_2$ (GNS/MnO$_2$) composite has been synthesized by a microwave-assisted method to be used as an adsorbent for the removal of nickel ions from waste water [81].

Hence there is considerable progress using various choices of graphene metal modified electrochemical sensors for the detection of heavy metals. Table 1 highlights electrochemical sensors for the analysis of different ions under the category of heavy metals.

3.2   Phenolic compounds

Phenols and their derivative compounds are commonly found in the environment wherein dyes, polymers, drugs and other organic substances contain it as a major component. The presence of phenols in the ecosystems is also related with production and degradation of numerous pesticides leading to the generation of industrial and municipal wastes. These compounds when substituted with other atoms such as chlorine are nitrated, methylated or alkylated forming harmful ecotoxins. Phenols and catechols reveal peroxidative capacity, they are hematotoxic and hepatotoxic, provoke mutagenesis and carcinogenesis toward humans and other living organisms [82]. Development of electrochemical sensors has opened new gates for sensing and elimination of these compounds. With this perspective, analysis of few compounds of phenol is discussed below.

Bisphenol

Bisphenol A (BPA), which is extensively used in the production of polycarbonate plastics, phenol and epoxy resins throw undesirable health issues on animals and human beings [83] which necessitated the development of a rapid, simple, sensitive and precise sensor for BPA determination. With the advantage of low cost, simplicity and good sensibility electrochemical sensors have been widely used for BPA detection too. In combination with graphene nanocomposites exhibit excellent synergic effect toward BPA detection. An extensive analysis about the electrochemical sensors and biosensors for BPA determination in real samples has been carried out by Jalalvand et al. [84] which talks about the DNA damage detection induced by BPA. Moreover, graphene modified gold and copper bimetallic nanoclusters/graphene nanoribbons [85] and gold nanoparticles loaded on reduced graphene oxide multiwall carbon nanotubes [86] has been successfully incorporated for bisphenol detection. In addition, gold - palladium nanoparticles disperded in graphene nanosheets [87] also possess the ability detect low BPA concentrations (8 nM) within a broad linear range of 0.01-10 μmol L^{-1}.

Hydroquinone and catechol

Hydroquinone (HQ) and catechol (CC) - isomers of di hydroxybenzene due to their redoxive properties and deleterious effects (mainly DNA damaging ability) fall under the category of toxic chemicals. Graphene has the great potential for distinguishing a diverse range of such aromatic isomers. Graphene metal modified electrode has been used for the simultaneous detection of hydroquinone and catechol. Good potential separation of oxidation peaks of about 110 mV between HQ and CC, wide linear concentration ranges, low detection limits, good stability, and high resolution capacity to isomers [88] have been observed. On the other hand, a highly stable pyridine - graphene was used as an electrochemical sensor for detecting HQ and CC wherein, excellent electrocatalysis of pyridine-NG was attained due to the π–π interactions between the benzene ring of CC and graphene layer and the hydrogen bonds formed between hydroxyl in HQ molecule and pyridinic nitrogen atoms within graphene layers, especially the less density distribution of π electron cloud in pyridinic-NG in acidic condition [89]. A number of several other novel sensors like micro and nanoscale applications monium chloride graphene and laser reduced graphene [90] modified GCE were fabricated for the simultaneous detection of CC and HQ with low detection limit, high sensitivity, excellent potential peak separation, and anti-interference ability.

3.3   Emerging contaminants

These are group of chemical contaminants of emerging concern, requiring additional consideration and a need is demanded to adapt current trends for finding successful solution to eradicate such toxic chemicals. Out of the category of chemicals that are termed to be toxic, pharmaceutical wastes and agro based toxic chemicals are discussed in this segment.

Pharmaceuticals

Pharmaceutical wastes are those derived from unused, damaged and expired medicinal drugs that are regulated as hazardous. Due to its bulk production in recent years all such activities have led to subtle effects on humans and environment. Hence the activity of electrochemical sensors to detect these compounds and trim down pollution has been of great concern. Nanomaterials based sensors is effectively used for the quantification of *rutin* in the pharmaceutical products [91]. A study has reported the use of graphene-gold nanoparticles screen-printed voltammetric sensor for the determination and quantification of *rutin* in pharmaceutics. This sensor allows the detection of rutin on a potential of 0.44 V. The current of the anodic peak varies linearly with the rutin concentration ranging in the domain 0.1×10^{-6} to 15×10^{-6} M, with a detection limit of 1.1×10^{-8} M. Sensors based on the nanohybrid of palladium-reduced graphene oxide modified with gold nanoparticles was developed (Figure 6), which was prepared by electrodepositing Au nanoparticles on Pd/rGO modified on a GCE which exhibited excellent electrocatalytic activity toward the redox of *acetaminophen* and *aminophenol* simultaneously. Study reports that graphene modified electrode obviously promotes the sensitivity of the determination of *paracetamol* with a low detection limit of 32 nM and a satisfied recovery from 96.4% to 103.3% [91].

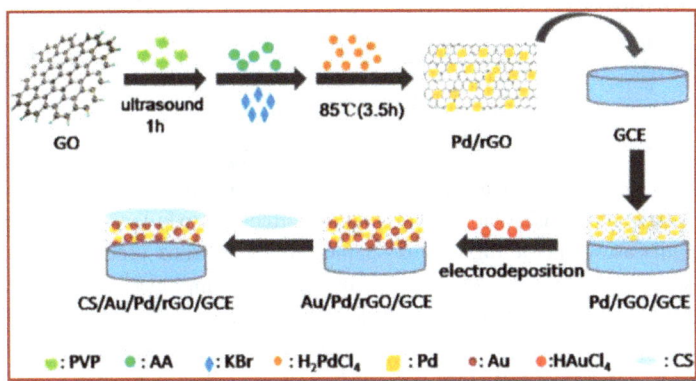

Figure 6 Graphene modified with gold for acetaminophen.

Various other studies also report that, graphene in its modified form has the ability to detect pharma compounds, a few of which are given in Table 1.

Table 1 Examples for modified graphene electrodes for pharmaceutical detection.

S.No	Electrode	Analyte	Linear Range	Detection Limit	Reference
1	Poly(4-aminobenzoic acid)/ electrochemically reduced graphene oxide composite film modified GCE	Acetaminophen	0.1–65 µM	0.01µM	[93]
2	Nafion/TiO_2–graphene Modified	Acetaminophen	1-100 µM	210 nM	[92]
3	Electrochemically reduced and deposited graphene onto GCE	Paracetamol	5.0nM-800 µM	2.13 nM	[91]
4	Nickel oxides loaded (Ni_2O_3–NiO) nanoparticles coated onto graphene	Acetaminophen	0.04-100 µM	0.02 µM	[94]

Agro based toxic chemicals

Agrochemicals or agrichemicals are those used exclusively in the agricultural processes including pesticides & insecticides (to kill bugs), herbicides (to kill weeds), fungicides (to get rid of diseases) and fertilizers (as supplement to enhance production) which are potentially toxic to living beings. Their overuse and improper disposal can cause significant food and environmental contamination as *pesticides* easily dissolve in water and do not decompose easily leading to harmful effects. The most commonly worldwide used organo-phosphorus pesticides are used as insecticides to protect crops, but are equally unsafe to public health due to their high neurotoxicity [95]. Recently, Hondred et al., [96] has developed an electrochemical sensor with Platinum nanoparticles electrodeposited on patterned graphene electrodes for detecting *paraoxons*. Inkjet maskless lithography (IML) has been used to fabricate the electrode primarily, followed by the electrodeposition of platinum nanoparticles to boost up the electrode surface area, electrical conductivity, and electrocatalytic activity. For effective detection of the pesticide, an enzyme phosphotriesterase was conjugated using bifunctional cross-linking molecule glutaraldehyde. The device showed rapid detection of paraoxon pesticide with a response time of 5s, a sensitivity of 370 nA mM, a linear range of 0.1–1 mM, and an LOD of 3 nM. In comparison, the biosensor based on IML– PGE/GA (without PtNPs) showed a sensitivity of 270 nA mM, a linear range of 0.1–1 mM, and an LOD of 12 nM. Thus the sensor showed reusability over 12 repeated operations by retaining 95% sensitivity and 70% stability of anodic current signal to paraoxon after 8 weeks. Also, it exhibited high selectivity toward paraoxon detection against interfering chemical nerve agents including p-nitrophenol, chlorpyrifos methyl, parathion, dichlofenthion, fenitrothion, phoxim, and dimethoate. The main advantage of this sensor is its cost effective bulk production and the feasibility of field pesticide detection. Electrochemical sensors with graphene nanocomposite modified GCEs have been studied for the detection of different types of pesticides such as methyl parathion, carbofuran [97] chlorpyrifos [98,99,100], methyl parathion [101], phoxim with graphene/GCE [102], carbaryl and chlorpyrifos with AgNPs–CGR/NF composite [103], carbaryl and monocrotophos with ionic liquid-functionalized graphene - gelatin [104], paraoxon-ethyl with rGOAuNPs/ polypyrrole, carbaryl with an electrochemically induced porous GO network [105].

3.4   Other toxic chemicals

Cyanide

Cyanide is rapidly acting potentially deadly chemical which exists in various forms. Leung *et al.* has reported that the detection of cyanide can be done by using sensor platform consists of a bio composite layer which are made up of materials like polymers, nano gold sol-gel, and hemoglobin. The anchoring materials such as Au, Pt, and glass carbon electrode are coated with bio composite on the surface. In comparison to other reports in the public domain, this ultra-high sensitive biosensor can detect cyanide at lower concentration levels orders. Further this report discloses the results in twofold: one at low concentration levels above 1×10^{-5} M and another at ultra-low levels of 1×10^{-18}M. The application for routine cyanide monitoring and research exploration represents these two concentration ranges. Owing to its capability, cyanide sensor can detect cyanide at such extensive range of concentrations. It is widely appreciated because of the exclusive applications in the fields of homeland security, biomedical, and environmental monitoring. Hence, this amazing sensitivity and range of detection limit and the development of ultra-high performance sensor fetches its use in several applications in the field of industry, biomedical, and security [106].

Hydrazine

Hydrazine which is an inorganic compound is highly toxic, probably carcinogenic and perilously unstable unless it is hydrated. Dutta *et al*. devised a facile fabrication of noble metal-based bimetallic nanoparticles with reduced graphene oxide (rGO) to detect hydrazine. It proved to be of huge demand due to its utility as a recyclable substrate for electrochemical sensors. A combination of $Au_{core}@Pd_{shell}$ with an average size of ~11.5 nm on an rGO (GAP) through one-step synthetic protocol is synthesized. 2-propanol is used as a precursor which is used as a solvent as well as a reducing agent. This device thus designed is economically desirable for the blend of the chosen GAP material. This active electro catalyst, GAP with optimized elements, has been applied as a substrate to electrochemical sensor to detect toxic chemical hydrazine in trace concentration level with high selectivity and good sensitivity. The GAP material displayed a limit of detection (LOD) as 0.08 μM with a linear range of 2–40 μM from the chrono amperometric (CA) result at −0.15 V *vs*. SCE. The novel concept of detecting N_2H_4 by this device is very well evident and utilized for subsequent practical application, as electrochemical sensor to detect the toxic chemical [107].

Nitro Compounds

Nitro compounds are organic molecules containing at least one nitro group. Nitro aromatics are major pollutants released in the environment almost exclusively from anthropogenic sources posing toxic effects to living organisms. A modified glassy carbon electrode (GCE) based on green synthesized silver nanoparticles (AgNPs) decked with reduced graphene oxide (RGO) for the detection of *nitrobenzene* has been demonstrated by Chelladurai et al. [108]. *Justicia glauca* leaf extract is used as a stabilizing and reducing agent for the synthesis of AgNPs. The nano composite modified electrode was devised by electrochemical reduction of AgNPs dispersed GO solution. The RGO-AgNPs composite endorses that AgNPs are attached on the RGO sheets and the average size of AgNPs is reported to be 40 ± 5 nm. The modified electrode displays an efficient electrode with over potential for electrocatalytic reduction than that of other modified electrodes. They confirm with the reduction peak of nitrobenzene which is linear over the concentrations from 0.5 to 900 μM. The sensitivity of the sensors is reported to be 0.836 $μAμM^{-1}$ cm^{-2} with the detection limit of 0.261 μM. Furthermore, the RGO-AgNPs modified electrode displays a good selectivity in the interfering similar compounds and has a good practical applicability in the waste water samples. Green synthesized AgNPs decked RGO modified GCE has been used for the sensitive detection of nitrobenzene. The devised electrode displays many advantages, such as sensitivity, low LOD, linear response range along with good selectivity for the detection of nitrobenzene [108]. An electrochemical sensing platform made of a reduced graphene oxide-silver (rGO-Ag) nanocomposite is developed for the detection of *4-nitrophenol* [109]. The synthesis of the nanocomposite is monitored at different reaction times (2 h, 6 h, 10 h and 15 h) in modified Tollens test. The nanocomposites prepared with different reaction times have been used for the electrocatalytic reduction of 4-NP, and the rGO-Ag (15 h) nanocomposite-modified glassy carbon (GC) electrode displayed a higher faradic current at an over potential of −0.5 V towards 4-NP reduction. A facile microwave-assisted synthesis of a reduced graphene oxide-silver (rGO-Ag) nanocomposite and its application towards the electrochemical determination of 4-nitrophenol (4-NP) are also reported by [110]. A glassy carbon electrode (GCE) was modified with the rGO-Ag nanocomposite and used for the electrocatalytic reduction and determination of 4-NP. The nanocomposite modified electrode displayed a higher reversible redox peak currents and lower charge transfer resistance.

Table 2 Graphene modified sensors for toxic chemicals detection.

References	Sensors	Analyte	Detection range	Limit of Detection
Heavy metal ions				
111	AuNP/Gr/Cystein	Cd^{2+}, Pb^{2+}	$0.5 - 40$ mgL^{-1}	0.1 mgL-1
112	Gr Nanodots/Au	Cu^{2+}, Pb^{2+}	0.006 -2.5μm	0.8nM
113	Heparin composite CSGR	Pb^{2+}	$0.009 - 4$μm $1.125 - 8.25$μgL^{-1}	1nM 6.03μgL^{-1}
114	Gr/Au	Hg^{2+}	$1aM - 100aM$	0.001aM
Pesticides				
115	PtNP Carboxylic GR Nafion	methyl parathion Carbofuran	$1x10^{-7} - 1x10^{-2}$μm $1x10^{-6} - 1x10^{-7}$μm	$5x10^{-5}$nM $5x10^{-4}$nM
116	SnNP Carboxylic GR Nafion	chloropyris	$1x10^{-6} - 1x10^{-2}$μm	$5x10^{-4}$nM
117	AgNP Carboxylic GR Nafion	carbonyl	$1x10^{-6} - 1x10^{-2}$μm	$5.45x10^{-4}$nM

Phenolic compounds

118	AuNP GR	Hydroquinone	0.06-800 μM	0.018 μM
108	RGOAgNP	Nitrobenzene	-	-

Inorganic Compounds

119	GO-AgNP	Hydrogen Peroxide	100 μM to 11 mM	28.3 μM
120	GO-AgNP	Nitrite ions	2.1 μM - 37 nM	-
121	Nanohybrid of RGO-nafion@Ag NP	Hydrogen Peroxide	-	5.35×10^{-7} M

Conclusion and future perspective

Toxic chemicals are ubiquitous environmental pollutants in both aquatic and terrestrial ecosystems, persistent, bio-accumulative and are more hazardous. Advances in biosensor research provide new and promising platforms for the sensitive detection of various components which can provide potential solutions not only in the biomedical field but also in other applications related to environmental monitoring. A new era of exploration has opened up due to the structural and functional properties of graphene which is considered as the mother of all carbon materials. Rapid acceptance of graphene as a material of interest, owing to its diversified set of unusual properties that happen to match the short comings of other materials paved way for a variety of engineering applications. Development of state of art graphene based precise sensors with better accuracy, elevated selectivity, sensitivity, long term stability and in situ sensing capability is under due consideration as they are featured with good conductivity and large specific surface area. Furthermore, extraordinary properties of graphene composites such as high electrochemical activity, ease of surface functionalization, and excellent electron transfer properties, have enabled to be used with excellent results to detect the analytes with improved sensitivity and selectivity.

Although the progress accompanies new challenges, like biocompatibility, electronic conductivity and stretchable capability are of great concern. To recapitulate, graphene metal modified electrochemical sensors make them extremely competitive in comparison with traditional electrodes, and there are still humpty number of openings for scientific research and applications. Despite the intense research carried out using sensing materials for electronic sensors, the sensing mechanism remains vague, which requires further exploration in order to better interpret the sensing behaviour and optimize the sensing performance. This chapter has showcased electrochemical sensors incorporating metal modification which made it interesting to get a clear picture of different combination of materials for electrochemical sensors pertaining to the detection of toxic chemicals. Furthermore, these modifications may lead to the development of novel sensors for wide range of applications.

References

[1] Hazrat Ali, Ezzat Khan, Ikram Ilahi Environmental Chemistry and Ecotoxicology of Hazardous Heavy Metals: Environmental Persistence, Toxicity, and Bioaccumulation, Journal of Chemistry, Volume 2019, Article ID 6730305, 14 pages https://doi.org/10.1155/2019/6730305

[2] Vhahangwele Masindi and Khathutshelo L. Muedi, Environmental Contamination by Heavy Metals, 2018. https://doi.org/10.5772/intechopen.76082

[3] Md Ashfaque Hossain Khan, Mulpuri V Rao, Quiliang Li, Recent Advances in Electrochemical Sensors for Detecting Toxic Gases: NO_2, SO_2 and H_2S, Sensors, 19(4), 905, 2019, https://doi.org/10.3390/s19040905

[4] Yang G., Zhu C., Du D., Zhu J., Lin Y, Graphene like two-dimensional layered nanomaterials: applications in biosensors and nanomedicine. Nanoscale 7(34), (2015), 14217–14231. https://doi.org/10.1039/C5NR03398E

[5] Varghese S.S., Lonkar S., Singh K., Swaminathan S., Abdala A.: Recent advances in graphene based gas sensors, Sensors Actuators B Chem. 218, (2015), 160–183. https://doi.org/10.1016/j.snb.2015.04.062

[6] Mao S., Lu G., Chen J.: Nanocarbon-based gas sensors: progress and challenges. J. Mat. Chem. A, 2 (16), (2014), 5573–5579. https://doi.org/10.1039/c3ta13823b

[7] Xu M., Liang T., Shi M., Chen H. Graphene-like two-dimensional materials. Chem. Rev. 113 (5), (2013), 3766–3798. https://doi.org/10.1021/cr300263a

[8] Kostarelos K., Novoselov K.S. Graphene devices for life. Nat. Nanotechnol. 9(10), (2014), 744–745. https://doi.org/10.1038/nnano.2014.224

[9] Cheng Z., Li Q., Li Z., Zhou Q., and Fang Y. Suspended graphene sensors with improved signal and reduced noise. Nano Lett. 10, (2010), 1864–1868. https://doi.org/10.1021/nl100633g

[10] Ohno, Y., Maehashi, K., and Matsumoto, K. Label-free biosensors based on aptamer-modified graphene field-effect transistors. J. Am. Chem. Soc. 132, (2010), 18012–18013. https://doi.org/10.1021/ja108127r

[11] Wang X., Zhang X., Electrochemical co-reduction synthesis of graphene/nano-gold composites and its application to electrochemical glucose biosensor. Electrochem. Acta **112**, (2013), 774–782. https://doi.org/10.1016/j.electacta.2013.09.036

[12] Urbanova, V., Magro, M., Gedanken, A., Baratella, D., Vianello, F., Zboril, R.: Nanocrystalline iron oxides, composites, and related materials as a platform for electrochemical, magnetic, and chemical biosensors. Chem. Mater. **26**(23), (2014), 6653–6673. https://doi.org/10.1021/cm500364x

[13] N. Liu, F. Luo, H. Wu, Y. Liu, C. Zhang and J. Chen, One- Step Ionic-Liquid-Assisted Electrochemical Synthesis of Ionic-Liquid-Functionalized Graphene Sheets Directly from Graphite, Adv. Funct. Mater., 18,2008, 1518–1525. https://doi.org/10.1002/adfm.200700797

[14] K. S. Subrahmanyam, L. S. Panchakarla, A. Govindaraj and C. N. R. Rao, Simple Method of Preparing Graphene Flakes by an Arc-Discharge Method, J. Phys. Chem. C, 113, 2009, 4257–4259. https://doi.org/10.1021/jp900791y

[15] Z.-S. Wu, W. Ren, L. Gao, J. Zhao, Z. Chen, B. Liu, D. Tang, B. Yu, C. Jiang and H.-M. Cheng, Synthesis of Graphene Sheets with High Electrical Conductivity and Good Thermal Stability by Hydrogen Arc Discharge Exfoliation, ACS Nano, 3, 2009, 411–417. https://doi.org/10.1021/nn900020u

[16] X. Wang, G. Sun, P. Routh, D.-H. Kim, W. Huang and P. Chen, Heteroatom-Doped Graphene Materials: Syntheses, Properties and Applications, Chem. Soc. Rev., 43, 2014, 7067–7098. https://doi.org/10.1039/C4CS00141A

[17] Huang, K.-J., Li, J., Wu, Y.-Y., Liu, Y.-M.: Amperometric immunobiosensor for α-fetoprotein using Au nanoparticles/chitosan/TiO$_2$–graphene composite based platform. Bioelectrochemistry 90, (2013) 18–23. https://doi.org/10.1016/j.bioelechem.2012.10.005

[18] Lepinay, S., Staff, A., Ianoul, A., Albert, J, Improved detection limits of protein optical fiber biosensors coated with gold nanoparticles. Biosens. Bioelectron. 52, (2014) 337–344. https://doi.org/10.1016/j.bios.2013.08.058

[19] Devi, R.V., Doble, M., Verma, R.S.: Nanomaterials for early detection of cancer biomarker with special emphasis on gold nanoparticles in immunoassays/sensors. Biosens. Bioelectron. 68, (2015), 688–698. https://doi.org/10.1016/j.bios.2015.01.066

[20] Yusoff, N., Pandikumar, A., Ramaraj, R., Lim, H.N., Huang, N.M.: Gold nanoparticle based optical and electrochemical sensing of dopamine. Microchim. Acta 182, (2015), (13–14), 2091–2114. https://doi.org/10.1007/s00604-015-1609-2

[21] Yola, M.L., Eren, T., Atar, N.: A sensitive molecular imprinted electrochemical sensor based on gold nanoparticles decorated graphene oxide: application to selective determination of tyrosine in milk. Sens. Actuators B Chem. 210, (2015), 149–157. https://doi.org/10.1016/j.snb.2014.12.098

[22] Gotti, G., Fajerwerg, K., Evrard, D., Gros, P.: Electrodeposited gold nanoparticles on glassy carbon: correlation between nanoparticles characteristics and oxygen reduction kinetics in neutral media. Electrochim. Acta 128, (2014), 412–419. https://doi.org/10.1016/j.electacta.2013.10.172

[23] Sabury, S., Kazemi, S.H., Sharif, F.: Graphene–gold nanoparticle composite: application as a good scaffold for construction of glucose oxidase biosensor. Mat. Sci. Eng. C 49, (2015) 297–304. https://doi.org/10.1016/j.msec.2015.01.018

[24] Xue, C., Gao, M., Xue, Y., Zhu, L., Dai, L., Urbas, A., Li, Q.: Building 3D layer-by-layer graphene–gold nanoparticle hybrid architecture with tunable interlayer distance. J. Phys. Chem. C 118 (28), (2014), 15332–15338. https://doi.org/10.1021/jp504553w

[25] Dutta, S., Ray, C., Mallick, S., Sarkar, S., Roy, A., Pal, T.: Au@ Pd core–shell nanoparticles-decorated reduced graphene oxide: a highly sensitive and selective platform for electrochemical detection of hydrazine. RSC Adv. **5**(64), (2015), 51690–51700. https://doi.org/10.1039/C5RA04817F

[26] Henry, A.L., Plumejeau, S., Heux, L., Louvain, N., Monconduit, L., Stievano, L., Boury, B.: Conversion of nanocellulose aerogel into TiO_2 and $TiO_2@C$ nano-thorns by direct anhydrous mineralization with $TiCl_4$. Evaluation of electrochemical properties in Li batteries. ACS Appl. Mater. Interfaces 7(27), (2015), 14584–14592. https://doi.org/10.1021/acsami.5b00299

[27] Ma, H., Sun, J., Zhang, Y., Bian, C., Xia, S., Zhen, T.: Label-free immunosensor based on one-step electrodeposition of chitosan-gold nanoparticles biocompatible film on Au microelectrode for determination of aflatoxin B1 in maize. Biosens. Bioelectron. 80, (2016), 222–229. https://doi.org/10.1016/j.bios.2016.01.063

[28] Pasang, T., Namratha, K., Parvin, T., Ranganathaiah, C., Byrappa, K.: Tuning of band gap in TiO_2 and ZnO nanoparticles by selective doping for photocatalytic applications. Mat. Res. Innov. 19(1), (2015), 73–80. https://doi.org/10.1179/1433075X14Y.0000000217

[29] Kapilashrami, M., Zhang, Y., Liu, Y.-S., Hagfeldt, A., Guo, J.: Probing the optical property and electronic structure of TiO_2 nanomaterials for renewable energy applications. Chem. Rev. 114 (19), (2014), 9662–9707. https://doi.org/10.1021/cr5000893

[30] Ghosh, S., Das, A.: Modified titanium oxide (TiO_2) nanocomposites and its array of applications: a review. Toxicol. Environ. Chem. 97(5), (2015), 491–514. https://doi.org/10.1080/02772248.2015.1052204

[31] Liu, J., Song, K., van Aken, P.A., Maier, J., Yu, Y.: Self-supported $Li_4Ti_5O_{12}$–C nanotube arrays as high-rate and long-life anode materials for flexible Li-ion batteries. Nano Lett. 14(5), (2014), 2597–2603. https://doi.org/10.1021/nl5004174

[32] Madian, M., Giebeler, L., Klose, M., Jaumann, T., Uhlemann, M., Gebert, A., Oswald, S., Ismail, N., Eychmuller, A., Eckert, J.: Self-organized TiO_2/CoO nanotubes as potential anode materials for lithium ion batteries. ACS Sustain. Chem. Eng. 3(5), (2015), 909–919. https://doi.org/10.1021/acssuschemeng.5b00026

[33] Kim, S.-J., Cho, Y.K., Seok, J., Lee, N.-S., Son, B., Lee, J.W., Baik, J.M., Lee, C., Lee, Y., Kim, M.H.: Highly branched RuO_2 nanoneedles on electrospun TiO_2 nanofibers as an efficient electrocatalytic platform. ACS Appl. Mater. Interfaces. 7(28), (2015), 15321–15330. https://doi.org/10.1021/acsami.5b03178

[34] Ahmad, K., Mohammad, A., Rajak, R., Mobin, S.M.: Construction of TiO_2 nanosheets modified glassy carbon electrode (GCE/TiO_2) for the detection of hydrazine. Mater. Res. Express 3(7), (2016), 074005. https://doi.org/10.1088/2053-1591/3/7/074005

[35] Zhang, Y., Bai, X., Wang, X., Shiu, K.-K., Zhu, Y., Jiang, H.: Highly sensitive graphene–Pt nanocomposites amperometric biosensor and its application in living cell H_2O_2 detection. Anal. Chem. 86(19), (2014), 9459–9465. https://doi.org/10.1021/ac5009699

[36] Leonardi, S.G., Aloisio, D., Donato, N., Russo, P.A., Ferro, M.C., Pinna, N., Neri, G.: Amperometric sensing of H_2O_2 using Pt–TiO_2/reduced graphene oxide nanocomposites. ChemElectroChem 1(3), (2014), 617–624. https://doi.org/10.1002/celc.201300106

[37] Hong, J., Zhao, Y.-X., Xiao, B.-L., Moosavi-Movahedi, A.A., Ghourchian, H., Sheibani, N.: Direct electrochemistry of hemoglobin immobilized on a functionalized multi-walled carbon nanotubes and gold nanoparticles nanocomplex-modified glassy carbon electrode. Sensors 13(7), (2013), 8595–8611. https://doi.org/10.3390/s130708595

[38] Zhu, J., Liu, X., Wang, X., Huo, X., Yan, R.: Preparation of polyaniline–TiO_2 nanotube composite for the development of electrochemical biosensors. Sens. Actuators B Chem. 221, (2015), 450–457. https://doi.org/10.1016/j.snb.2015.06.131

[39] Wang, J.T.-W., Ball, J.M., Barea, E.M., Abate, A., Alexander-Webber, J.A., Huang, J., Saliba, M., Mora-Sero, I., Bisquert, J., Snaith, H.J.: Low-temperature processed electron collection layers of graphene/TiO_2 nanocomposites in thin film perovskite solar cells. Nano Lett. 14(2), (2013), 724–730. https://doi.org/10.1021/nl403997a

[40] Huang, K.-J., Wang, L., Li, J., Gan, T., Liu, Y.-M.: Glassy carbon electrode modified with glucose oxidase–graphene–nano-copper composite film for glucose sensing. Measurement 46(1), (2013), 378–383. https://doi.org/10.1016/j.measurement.2012.07.012

[41] Yang, N., Liu, Y., Wen, H., Tang, Z., Zhao, H., Li, Y., Wang, D.: Photocatalytic properties of graphdiyne and graphene modified TiO_2: from theory to experiment. ACS Nano 7(2), (2013), 1504–1512. https://doi.org/10.1021/nn305288z

[42] Feng, C., Xu, G., Liu, H., Lv, J., Zheng, Z., Wu, Y.: Facile fabrication of Pt/graphene/TiO_2 NTAs based enzyme sensor for glucose detection. J. Electro chem. Soc. 161(1), (2014), B1–B8. https://doi.org/10.1149/2.025401jes

[43] Iqbal, A., Iqbal, K., Li, B., Gong, D., Qin, W.: Recent advances in iron nanoparticles: preparation, properties, biological and environmental application. J. Nanosci. Nanotechnol. **17**(7), (2017), 4386–4409. https://doi.org/10.1166/jnn.2017.14196

[44] Chatterjee, K., Sarkar, S., Rao, K.J., Paria, S.: Core/shell nanoparticles in biomedical applications. Adv. Colloid Interfaces Sci. **209**, (2014) 8–39. https://doi.org/10.1016/j.cis.2013.12.008

[45] Yang, C., Bian, X., Qin, J., Zhao, X., Zhang, K., Bai, Y.: An investigation of a viscosity-magnetic field hysteretic effect in nano-ferrofluid. J. Mol. Liq. **196**, (2014), 357–362. https://doi.org/10.1016/j.molliq.2014.04.021

[46] Zhao, L., Gao, M., Yue, W., Jiang, Y., Wang, Y., Ren, Y., Hu, F.: Sandwich-structured graphene–Fe_3O_4@Carbon nanocomposites for high-performance lithium-ion batteries. ACS Appl. Mater. Interfaces 7(18), (2015), 9709–9715. https://doi.org/10.1021/acsami.5b01503

[47] Yadav, R.S., Sharma, V., Kuanr, B.K.: Magnetic nanoparticles; synthesis, characterization and application as contrast agent in magnetic resonance imaging

(MRI). Adv. Sci. Lett. **20**(7–9), (2014), 1548–1550.
https://doi.org/10.1166/asl.2014.5562

[48] Gan, Q., Zhu, J., Yuan, Y., Liu, C.: pH-responsive Fe_3O_4 nanoparticles-capped mesoporous silica supports for protein delivery. J. Nanosci. Nanotechnol. **16**(6), (2016), 5470–5479. https://doi.org/10.1166/jnn.2016.11744

[49] Shen, M., Wu, C., Lin, C., Fan, G., Jin, Y., Zhang, Z., Li, C., Jia, W.: Facile solvothermal synthesis of mesostructured chitosan-coated Fe_3O_4 nanoparticles and its further modification with folic acid for improving targeted drug delivery. NANO **9**(7), (2014), 1450081. https://doi.org/10.1142/S1793292014500817

[50] Nasrollahzadeh, M., Sajadi, S.M., Rostami-Vartooni, A., Khalaj, M.: Green synthesis of Pd/Fe_3O_4 nanoparticles using *Euphorbia condylocarpa* M. bieb root extract and their catalytic applications as magnetically recoverable and stable recyclable catalysts for the phosphine-free Sonogashira and Suzuki coupling reactions. J. Mol. Catal. A Chem. **396**, (2015), 31–39. https://doi.org/10.1016/j.molcata.2014.09.029

[51] Gu, T., Wang, J., Xia, H., Wang, S., Yu, X.: Direct electrochemistry and electrocatalysis of horseradish peroxidase immobilized in a DNA/chitosan–Fe_3O_4 magnetic nanoparticle bio-complex film. Materials **7**(2), (2014), 1069–1083. https://doi.org/10.3390/ma7021069

[52] Zhu, S., Guo, J., Dong, J., Cui, Z., Lu, T., Zhu, C., Zhang, D., Ma, J.: Sonochemical fabrication of Fe_3O_4 nanoparticles on reduced graphene oxide for biosensors. Ultrason. Sonochem. **20**(3), (2013), 872–880. https://doi.org/10.1016/j.ultsonch.2012.12.001

[53] Zhou, K., Zhu, Y., Yang, X., Li, C.: Preparation and application of mediator-free H_2O_2 biosensors of graphene–Fe_3O_4 composites. Electroanalysis 23(4), (2011), 862–869. https://doi.org/10.1002/elan.201000629

[54] Hsieh, C.-T., Lin, J.-Y., Mo, C.-Y.: Improved storage capacity and rate capability of Fe_3O_4–graphene anodes for lithium-ion batteries. Electrochim. Acta 58, (2011), 119–124. https://doi.org/10.1016/j.electacta.2011.09.008

[55] Wang, Y., Wei, W., Zeng, J., Liu, X., Zeng, X.: Fabrication of a copper nanoparticle/chitosan/carbon nanotube-modified glassy carbon electrode for electrochemical sensing of hydrogen peroxide and glucose. Microchim. Acta **160** (1–2), (2008), 253–260. https://doi.org/10.1007/s00604-007-0844-6

[56] Chen, Q., Zhang, L., Chen, G.: Facile preparation of graphene-copper nanoparticle composite by in situ chemical reduction for electrochemical sensing of carbohydrates. Anal. Chem. 84(1), (2012), 171–178. https://doi.org/10.1021/ac2022772

[57] Baccar, H., Ktari, T., Abdelghani, A.: Functionalized palladium nanoparticles for hydrogen peroxide biosensor. Int. J. Electrochem. 6, 4, (2011). https://doi.org/10.4061/2011/603257

[58] Rahman, M.M., Ahammad, A., Jin, J.-H., Ahn, S.J., Lee, J.-J.: A comprehensive review of glucose biosensors based on nanostructured metal-oxides. Sensors 10(5), 4855–4886 (2010). https://doi.org/10.3390/s100504855

[59] Nancy, T.E.M., Kumary, V.A.: Synergistic electrocatalytic effect of graphene/nickel hydroxide composite for the simultaneous electrochemical determination of ascorbic acid, dopamine and uric acid. Electrochim. Acta 133, 233–240 (2014). https://doi.org/10.1016/j.electacta.2014.04.027

[60] Sun, W., Gong, S., Deng, Y., Li, T., Cheng, Y., Wang, W., Wang, L.: Electrodeposited nickel oxide and graphene modified carbon ionic liquid electrode for electrochemical myoglobin biosensor. Thin Solid Films 562, 653–658 (2014). https://doi.org/10.1016/j.tsf.2014.05.002

[58] Zeng, Q., Cheng, J.-S., Liu, X.-F., Bai, H.-T., Jiang, J.-H.: Palladium nanoparticle/chitosan-grafted graphene nanocomposites for construction of a glucose biosensor. Biosens. Bioelectron. 26(8), (2011), 3456–3463. https://doi.org/10.1016/j.bios.2011.01.024

[59] C. Kotlowski, M. Larisika, P. M. Guerin, C. Kleber, T. Krober, R. Mastrogiacomo, C. Nowak, P. Pelosi, S. Sch"utz, A. Schwaighofer and W. Knoll, Fine Discrimination of Volatile Compounds by Graphene- Immobilized Odorant-Binding Proteins, Sens. Actuators, B,2018, 256, 564–572. https://doi.org/10.1016/j.snb.2017.10.093

[60] S.-J. Choi, S.-J. Kim and I.-D. Kim, Ultrafast Optical Reduction of Graphene Oxide Sheets on Colorless Polyimide Film for Wearable Chemical Sensors, NPG Asia Mater., 2016, 8, e315. https://doi.org/10.1038/am.2016.150

[61] M. H. M. Facure, L. A. Mercante, L. H. C. Mattoso and D. S. Correa, Detection of Trace Levels of Organophosphate Pesticides Using an Electronic Tongue Based on Graphene Hybrid Nanocomposites, Talanta, 2017, 167, 59–66. https://doi.org/10.1016/j.talanta.2017.02.005

[62] K. S. Kim, J.-r. Jang, W.-S. Choe and P. J. Yoo, Electrochemical Detection of Bisphenol A with High Sensitivity and Selectivity Using Recombinant Protein-

Immobilized Graphene Electrodes, Biosens. Bioelectron., 2015, 71, 214–221.
https://doi.org/10.1016/j.bios.2015.04.042

[63] S. Gupta and R. Wood, Development of FRET Biosensor Based on
Aptamer/Functionalized Graphene for Ultrasensitive Detection of Bisphenol A and
Discrimination from Analogs, Nano-Struct. Nano-Objects, 2017, 10, 131–140.
https://doi.org/10.1016/j.nanoso.2017.03.013

[64] S. Myung, A. Solanki, C. Kim, J. Park, K. S. Kim and K.-B. Lee, Graphene-
Encapsulated Nanoparticle-Based Biosensor for the Selective Detection of Cancer
Biomarkers, Adv. Mater., 2011, 23, 2221–2225.
https://doi.org/10.1002/adma.201100014

[65] D. Khatayevich, T. Page, C. Gresswell, Y. Hayamizu, W. Grady and M. Sarikaya,
Selective Detection of Target Proteins by Peptide-Enabled Graphene Biosensor, Small,
2014, 10, 1505–1513. https://doi.org/10.1002/smll.201302188

[66] S. Tomita, S. Ishihara and R. Kurita, A Multi-Fluorescent DNA/Graphene Oxide
Conjugate Sensor for Signature- Based Protein Discrimination, Sensors, 2017, 17,
E2194. https://doi.org/10.3390/s17102194

[67] P. Si, H. Chen, P. Kannan and D.-H. Kim, Selective and Sensitive Determination of
Dopamine by Composites of Polypyrrole and Graphene Modified Electrodes, Analyst,
136, 2011, 5134–5138. https://doi.org/10.1039/c1an15772h

[68] S. Hou, M. L. Kasner, S. Su, K. Patel and R. Cuellari, Highly Sensitive and Selective
Dopamine Biosensor Fabricated with Silanized Graphene, J. Phys. Chem. C, 114,
2010,14915–14921. https://doi.org/10.1021/jp1020593

[69] S. Dong, Q. Bi, C. Qiao, Y. Sun, X. Zhang, X. Lu and L. Zhao, Electrochemical
Sensor for Discrimination Tyrosine Enantiomers Using Graphene Quantum Dots and
BCyclodextrins Composites, Talanta, 173, 2017, 94–100.
https://doi.org/10.1016/j.talanta.2017.05.045

[70] Siva Kumar Krishnan, Eric Singh, Pragya Singh, Meyya Meyyappan and Hari Singh
Nalwa, A review on graphene-based nanocomposites for electrochemical and
fluorescent biosensors, Reviews, Royal Society of Chemistry, 9, 2019.
https://doi.org/10.1039/C8RA09577A

[71] Sitko, R., Turek, E., Zawisza, B., Malicka, E., Talik, E., Heimann, J., et al.,
Adsorption of divalent metal ions from aqueous solutions using graphene oxide.
Dalton Trans. 42, (2013). https://doi.org/10.1039/c3dt33097d

[72] Zhao, G., Ren, X., Gao, X., Tan, X., Li, J., Chen, C., et al., Removal of Pb (II) ions from aqueous solutions on few-layered graphene oxide nanosheets. *Dalton Trans.* 40, (2011). https://doi.org/10.1039/c1dt11005e

[73] Wang Z, Wang H, Zhang Z, Yang X, Liu G. Sensitive electrochemical determination of trace cadmium on astannum film/poly(p-aminobenzene sulfonic acid)/electrochemically reduced graphene composite modified electrode. Electrochimica Acta, 120, 2014. https://doi.org/10.1016/j.electacta.2013.12.068

[74] H. Bagheri, A. Afkhami, H. Khoshsafar, M. Rezaei, S. J. Sabounchei and M. Sarlakifar, Simultaneous Electrochemical Sensing of Thallium, Lead and Mercury Using a Novel Ionic Liquid/Graphene Modified Electrode, Anal. Chim. Acta, 870, 2015, 56–66. https://doi.org/10.1016/j.aca.2015.03.004

[75] Li J, Guo S, Zhai Y, Wang E. High-sensitivity determination of lead and cadmium based on the nafion graphene composite film. Analytica Chimica Acta 2009. https://doi.org/10.1016/j.aca.2009.07.030

[76] L. Zhu, L. Xu, B. Huang, N. Jia, L. Tan and S. Yao, Simultaneous Determination of Cd(II) and Pb(II) Using Square Wave Anodic Stripping Voltammetry at a Gold Nanoparticle-Graphene-Cysteine Composite Modified Bismuth Film Electrode, Electrochim. Acta, 2014, 115, 471–477. https://doi.org/10.1016/j.electacta.2013.10.209

[77] Shao Y, Wang J, Wu H, Liu J, Aksay I, Lin Y. Graphene based electrochemical sensors and biosensors: A review. Electroanalysis, 22, 2010, 1027–1036. https://doi.org/10.1002/elan.200900571

[78] H. Zhang, S. Shuang, G. Wang, Y. Guo, X. Tong, P. Yang, A. Chen, C. Dong and Y. Qin, TiO2–Graphene Hybrid Nanostructures by Atomic Layer Deposition With Enhanced Electrochemical Performance for Pb(ii) and Cd(ii) Detection, RSC Adv., 5, 2015, 4343–4349. https://doi.org/10.1039/C4RA09779C

[79] Berger, C., Song, Z., Li, X., Wu, X., Brown, N., Naud, C., et al. (2006). Electronic confinement and coherence in patterned epitaxial graphene. *Science* 312, 1191–1196. https://doi.org/10.1126/science.1125925

[80] S. Palanisamy, R. Madhu, S.-M. Chen and S. K. Ramaraj, A Highly Sensitive and Selective Electrochemical Determination of Hg(II) Based on an Electrochemically Activated Graphite Modified Screen-Printed Carbon Electrode, Anal. Methods, 6, 2014, 8368–8373. https://doi.org/10.1039/C4AY01805B

[81] Ren Y, Yan N, Wen Q, Fan Z, Wei T, Zhang M, Ma J. Graphene/δ-MnO2 composite as adsorbent for the removal of nickel ions from waste water. Chemical Engineering Journal, 175, 2010, 1–7. https://doi.org/10.1016/j.cej.2010.08.010

[82] Chen L, Tang Y, Wang Ke, Liu C, Luo S. Direct electrodeposition of reduced graphene oxide on glassy carbon electrode and its electrochemical application. Electrochemistry Communications 13, 2011, doi:10.1016/j.elecom.2010.11.033, 133–137. https://doi.org/10.1016/j.elecom.2010.11.033

[83] S. Almeida,A. Raposo,M. Almeida-González, and C. Carrascosa, Compr. Rev. Food Sci. Food Saf., 17, (2018), 1503. https://doi.org/10.1111/1541-4337.12388

[84] K. Varmira, M. Saed-Mocheshi, and A. R. Jalalvand, Sens. Bio-Sensing Res., 15, (2017), 17. https://doi.org/10.1016/j.sbsr.2017.07.002

[85] H. Yu, X. Feng, X. X. Chen, J. L. Qiao, X. L. gao, N. Xu, and L. J. Gao, Chinese J.Anal. Chem., 45, (2017), 713. https://doi.org/10.1016/S1872-2040(17)61014-4

[86] B. Su, H. Shao, N. Li, X. Chen, Z. Cai, and X. Chen, Talanta, 166, (2017), 126. https://doi.org/10.1016/j.talanta.2017.01.049

[87] J. Michałowicz, W. Duda, Phenols – Sources and Toxicity, Pol. J. Environ. Stud., 16(3), 2007, 347–362

[88] Guo H, Peng S, Xu J, Zhao Y, Kang X. Highly stable pyridinic nitrogen doped graphene modified electrode in simultaneous determination of hydroquinone and catechol, Sensors and Actuators B: Chemical, 193, 2014;623–629. https://doi.org/10.1016/j.snb.2013.12.018

[89] Song D, Xia J, Zhang F, Bi S, Xiang W, Wang Z, Xia L, Xia Y, Li Y, Xia L. Multiwall carbon nanotubes-poly(diallyldimethylammonium chloride)-graphene hybrid composite film for simultaneous determination of catechol and hydroquinone. Sensors and Actuators B: Chemical 206, 2015. https://doi.org/10.1016/j.snb.2014.08.084

[90] Lai T, Cai W, Dai W, Ye J. Easy processing laser reduced graphene: A green and fast sensing platform for hydroquinone and catechol simultaneous determination. Electrochemica Acta 138:48–55. https://doi.org/10.1016/j.electacta.2014.06.070

[91] Kang X, Wang J, Wu H, Liu J, Aksay I, Lin Y. A graphene-based electrochemical sensor for sensitive detection of paracetamol. Talanta, 81. https://doi.org/10.1016/j.talanta.2010.01.009

[92] Zhu W, Huang H, Gao X, Ma H. Electrochemical behavior and voltammetric determination of acetaminophen based on glassy carbon electrodes modified with

poly(4-aminobenzoic acid)/electrochemically reduced graphene oxide composite films. Materials Science and Engineering: C, 2014, 45:21–28. https://doi.org/10.1016/j.msec.2014.08.067

[93] Adhikari B, Govindhan M, Chen A. Sensitive detection of acetaminophen with graphene-based electrochemical sensor. Electrochimica Acta. DOI:10.1016/j.electacta. 2014.10.028

[94] Liu G, Chen H, Lin G, Ye P, Wang X, Jiao Y, Guo X, Wen Y, Yang H. One-step electrodeposition of graphene loaded nickel oxides nanoparticles for acetaminophen detection. Biosensors and Bioelectronics, 56:26–32. 2014, 001-005 145, (2019), 242. https://doi.org/10.1016/j.bios.2014.01.005

[95] D. Gunnell, M. Eddleston, M. R. Phillips and F. Konradsen, The Global Distribution of Fatal Pesticide Self-Poisoning: Systematic Review, BMC Public Health, 2007, 7, 357. https://doi.org/10.1186/1471-2458-7-357

[96] J. A. Hondred, J. C. Breger, N. J. Alves, S. A. Trammell, S. A. Walper, I. L. Medintz and J. C. Claussen, Printed Graphene Electrochemical Biosensors Fabricated by Inkjet Maskless Lithography for Rapid and Sensitive Detection of Organophosphates, ACS Appl. Mater. Interfaces, 10, 2018, 25–11134. https://doi.org/10.1021/acsami.7b19763

[97] L. Yang, G. Wang and Y. Liu, An Acetylcholinesterase Biosensor Based on Platinum Nanoparticles–Carboxylic Graphene–Nafion-Modified Electrode for Detection of Pesticides, Anal. Biochem., 2013, 437, 144–149. https://doi.org/10.1016/j.ab.2013.03.004

[98] L. Yang, G. Wang, Y. Liu and M. Wang, Development of a Biosensor Based on Immobilization of Acetylcholinesterase on NiO Nanoparticles–Carboxylic Graphene– Nafion Modified Electrode for Detection of Pesticides, Talanta, 2013, 113, 135–141. https://doi.org/10.1016/j.talanta.2013.03.025

[99] Q. Zhou, L. Yang, G. Wang and Y. Yang, Acetylcholinesterase Biosensor Based on SnO2 Nanoparticles–Carboxylic Graphene–Nafion Modified Electrode for Detection of Pesticides, Biosens. Bioelectron., 2013, 49, 25–31. https://doi.org/10.1016/j.bios.2013.04.037

[100] X. Tan, Q. Hu, J. Wu, X. Li, P. Li, H. Yu, X. Li and F. Lei, Electrochemical Sensor Based on Molecularly Imprinted Polymer Reduced Graphene Oxide and Gold Nanoparticles Modified Electrode for Detection of Carbofuran, Sens. Actuators, B, 2015, 220, 216–221. https://doi.org/10.1016/j.snb.2015.05.048

[101] T.Jeyapragasam, R. Saraswathi, S.-M. Chen and T.-W. Chen, Acetylcholinesterase Biosensor for the Detection of Methyl Parathion at an Electrochemically Reduced Graphene Oxide-Nafion Modified Glassy Carbon Electrode, Int. J. Electrochem. Sci., 2017, 12, 4768–4781. https://doi.org/10.20964/2017.06.77

[102] M. Chao and M. Chen, Electrochemical Determination of Phoxim in Food Samples Employing a Graphene-Modified Glassy Carbon Electrode, Food Anal. Methods, 2014, 7, 1729–1736. https://doi.org/10.1007/s12161-014-9813-y

[103] Y. Liu, G. Wang, C. Li, Q. Zhou, M. Wang and L. Yang, A Novel Acetylcholinesterase Biosensor Based on Carboxylic Graphene Coated with Silver Nanoparticles for Pesticide Detection, Mater. Sci. Eng., C, 2014, 35, 253–258. https://doi.org/10.1016/j.msec.2013.10.036

[104] Y. Zheng, Z. Liu, Y. Jing, J. Li and H. Zhan, An Acetylcholinesterase Biosensor Based on Ionic Liquid Functionalized Graphene–Gelatin-Modi☐ed Electrode for and Their Applications in Detection of Organophosphorus Pesticides in the Environment, Arch. Toxicol., 2017, 91, 109–130. Detection of Organophosphorus Pesticides in the Environment, Arch. Toxicol., 2017, 91, 109-130.

[105] Y. Yang, A. M. Asiri, D. Du and Y. Lin, Acetylcholinesterase Biosensor Based on a Gold Nanoparticle–Polypyrrole– Reduced Graphene Oxide Nanocomposite Modified Electrode for the Amperometric Detection Of Organophosphorus Pesticides, Analyst, 2014, 139, 3055– 3060. https://doi.org/10.1039/c4an00068d

[106] S. Mozneb, J.C.K. Lai, S.W. Leung, Cyanide detection by highly modified Sol-Gel biocomposite sensor, in Proceeding of the NSTI Nanotechnology Conference & Expo, Diagnostics & Imaging, Washington, D.C., 3 (3) (2015) 155-158.

[107] S. Dutta, C. Ray, S. Mallick, S. Sarkar, A. Roy, T. Pal, AAu@Pd core-shell nanoparticles- decorated reduced graphene oxide a highly sensitive and selective platform for electrochemical detection of hydrazine , Royal Society of Chemistry Advances, 5 (64) (2015) 51690-51700. https://doi.org/10.1039/C5RA04817F

[108] K. Chelladurai, K. Muthupandi, S.M. Chen, M. Ajmal Ali, P. Selvakumar, A. Rajan, P. Prakash, Green synthesized silver nanoparticles decorated on reduced graphene oxide for enhanced electrochemical sensing of nitrobenzene in waste water samples, Royal Society of Chemistry Advances, (2013), 1-3.

[109] Nurul Izrini Ikhsan, Perumal Rameshkumar, Nay Ming Huang, Controlled synthesis of reduced graphene oxide supported silver nanoparticles for selective and sensitive electrochemical detection of 4-nitrophenol. Electrochimica Acta 192, (2016), 392–399. https://doi.org/10.1016/j.electacta.2016.02.005

[110] Perumal Rameshkumar, Norazriena Yusoff, Huang Nay Ming, Mohd Shaiful Sajab, Microwave synthesis of reduced graphene oxide decorated with silver nanoparticles for electrochemical determination of 4-nitrophenol. Ceramics International 42, (2016), 18813–18820. https://doi.org/10.1016/j.ceramint.2016.09.026

[111] L. Zhu, L. Xu, B. Huang, N. Jia, L. Tan and S. Yao, Simultaneous Determination of Cd(II) and Pb(II) Using Square Wave Anodic Stripping Voltammetry at a Gold Nanoparticle-Graphene-Cysteine Composite Modified Bismuth Film Electrode, Electrochim. Acta, 2014, 115, 471– 477. https://doi.org/10.1016/j.electacta.2013.10.209

[112] H. Zhu, Y. Xu, A. Liu, N. Kong, F. Shan, W. Yang, C. J. Barrow and J. Liu, Graphene Nanodots-Encaged Porous Gold Electrode Fabricated via Ion Beam Sputtering Deposition for Electrochemical Analysis of Heavy Metal Ions, Sens. Actuators, B, 2015, 206, 592–600. https://doi.org/10.1016/j.snb.2014.10.009

[113] T. Priya, N. Dhanalakshmi and N. Thinakaran, Electrochemical Behavior of Pb (II) on a Heparin Modified Chitosan/Graphene Nanocomposite Film Coated Glassy Carbon Electrode and its Sensitive Detection, Int. J. Biol. Macromol., 2017, 104, 672– 680. https://doi.org/10.1016/j.ijbiomac.2017.06.082

[114] Y. Zhang, G. M. Zeng, L. Tang, J. Chen, Y. Zhu, X. X. He and Y. He, Electrochemical Sensor Based on Electrodeposited Graphene-Au Modified Electrode and Nano Au Carrier Amplified Signal Strategy for Attomolar Mercury Detection, Anal. Chem., 2015, 87, 989–996. https://doi.org/10.1021/ac503472p

[115] L. Yang, G. Wang and Y. Liu, An Acetylcholinesterase Biosensor Based on Platinum Nanoparticles–Carboxylic Graphene–Nafion-Modified Electrode for Detection of Pesticides, Anal. Biochem., 2013, 437, 144–149. https://doi.org/10.1016/j.ab.2013.03.004

[116] Q. Zhou, L. Yang, G. Wang and Y. Yang, Acetylcholinesterase Biosensor Based on SnO2 Nanoparticles–Carboxylic Graphene–Nafion Modified Electrode for Detection of Pesticides, Biosens. Bioelectron., 2013, 49, 25–31. https://doi.org/10.1016/j.bios.2013.04.037

[117] Y. Liu, G. Wang, C. Li, Q. Zhou, M. Wang and L. Yang, A Novel Acetylcholinesterase Biosensor Based on Carboxylic Graphene Coated with Silver Nanoparticles for Pesticide Detection, Mater. Sci. Eng., C, 2014, 35, 253–258. https://doi.org/10.1016/j.msec.2013.10.036

[118] Hu S, Wang Y, Wang X, Xu L, Xiang J, Sun W. Electrochemical detection of hydroquinone with a gold nanoparticle and graphene modified carbon ionic liquid

electrode. Sensors and Actuators B 2012;168:27–33.
https://doi.org/10.1016/j.snb.2011.12.108

[119] Noor An'amt Mohamed, Shahid Muhammad Mehmood, Rameshkumar Perumal, Huang Nay Ming, A glassy carbon electrode modified with graphene oxide and silver nanoparticles for amperometric determination of hydrogen peroxide. Microchimica Acta 183, (2016), 911–916. https://doi.org/10.1007/s00604-015-1679-1

[120] Nurul Izrini Ikhsan, Perumal Rameshkumar, Alagarsamy Pandikumar, Muhammad Mehmood Shahid, Nay Ming Huang, Swadi Vijay Kumar, Hong Ngee Lim, Facile synthesis of graphene oxide-silver nanocomposite and its modified electrode for enhanced electrochemical detection of nitrite ions. Talanta144, (2015), 908–914. https://doi.org/10.1016/j.talanta.2015.07.050

[121] Norazriena Yusoff, Perumal Rameshkumar, Muhammad Shahid Mehmood, Alagarsamy Pandikumar, Hing Wah Lee, Nay Ming Huang, Ternary nanohybrid of reduced grapheme oxide-Nafion @ silver nanoparticles for boosting the sensor performance in non-enzymatic amperometric detection of hydrogen peroxide. Biosensors and Bioelectronics 87, (2017), 1020–1028. https://doi.org/10.1016/j.bios.2016.09.045

Chapter 5

Graphene-Metal Oxides Modified Electrochemical Sensors for Toxic Chemicals

L. Vidhya [1*], T. Ramya[1], S. Vinodha[2]

[1]Department of Chemical Engineering, Sethu Institute of Technology, Pulloor, Kariapatti, Virudhunagar District, India

[2]Department of Environmental Science, Bharathiar University, Coimbatore

vidhuram236@gmail.com, vinodha.harris@gmail.com, rtramya1@gmail.com

Abstract

This chapter discusses the recent progresses in environmental electrochemistry and its wide capabilities and application towards pollution free environment. Various chemicals including agrochemicals, heavy metals, and other toxic materials polluting the environment can either be treated or transformed to non-toxic elements. Environmental protection and incessant development of people's value of life are found to be the most important areas of the application of electrochemical sensors in future. A sensor, here, is a chemical-play-tool that converts a chemical data like composition, presence of a particular ion, concentration, chemical activity, and partial pressure into a systematically useful signal. Currently, with new challenges and prospects, the electrochemical sensors have new and wide areas of outlook and applications. The electrochemical biosensor is a simple device that measures electronic current either ionic or by change in conductance carried by the bio-electrodes. Generally, carbon materials are widely used as electrode substrates to make different electrodes owing to its soft properties and renewable for exchange of electrons. Befittingly the arrangement of carbon atoms in graphene enhances its promising applications in several fields. On the other hand, nano materials possess good geometric as well as unique mechanical, physical and chemical properties that significantly encourage applications in medicine, electronics, environmental science and biosensors. In this chapter, the application of graphene- ZnO nano composite material is discussed for analysing the toxic chemicals in the environment and biological samples because of its high sensitivity and good reproducibility.

Materials Research Foundations **82** (2020) 125-150   https://doi.org/10.21741/9781644900956-5

Keywords

Electro Chemical Sensors, Graphene, Metal Oxides, Nano Composites, ZnO, Chemical Pollutants

Contents

1. Introduction

Toxicity of pollutants leads to major threats and causes adversarial effects to the environment but their quantum of production increases day-by-day consistently in several industrial and engineering fields [1]. The major reasons for the contamination are due to the increased global industrialization and urbanization. Ultimately, these industrial and engineering fields release pollutants into the water and soil thus ending up the aquatic surroundings and ecosystems. The release of toxic chemicals into the environment is mainly from various anthropogenic activities, agronomic practices and dumping of various types of wastes. According to World Health Organisation (WHO) ranking, the top potential chemicals such as arsenic (As), cadmium (Cd), lead (Pb) and mercury (Hg) and other organic compounds are of major public health concern. They amend the food chain by disturbing the metabolic system of living organisms which leads to life threatening disorders [2]. In detail, the contamination in the environment occurs through natural activities like weathering and volcanic eruptions; and man-made activities such as industrial, agricultural, pharmaceutical, domestic effluents, and atmospheric sources as well. The Industrial pollution in the environment is caused due to mining, foundries and smelters, and other metal-based processes. The toxicity of hazardous substances depends upon the absorbed dose and the direction and period of exposure. For the past few decades, numerous toxic pollutants have been identified through several analytical techniques for environmental monitoring and food safety. Owing to the inaccurate anthropogenic activities in agricultural and industrial processes, some of the toxic substances via agrochemicals (including fertilizers, insecticides, fungicides, bactericides, etc.), nitro aromatic compounds, phenolic derivatives, heavy metal ions, etc., enter the ecosystem and encountered. Several heavy metals such as cobalt, chromium, iron, manganese, copper, nickel, magnesium, selenium, molybdenum and zinc are the major source of nutrients for the metabolic and physiological functions. But the exposure of heavy metals after onset levels will cause lethal effects thereby causing major health issues [3]. The term "emerging" mainly focused the attention in the list of compounds containing toxic chemicals or biological species which disrupts the endocrine activity. Many pollutants are even supposed to be carcinogenic and cause severe health hazards to human beings [4, 5, 6]. Hence, determination and detoxification of the toxic chemicals are paramount importance. Of several options, electrochemical sensors are promising and fastidious.

Electrochemical sensors have a fast progress in the terms of electroactive materials, matrix materials, and size [7, 8]. Owing to their screening capabilities, reliable design, short time of analysis and economic they are very attractive in the determination of toxic chemicals. Some of the sensors have attained the commercial stage and initiated a wide range of applications but electrochemical sensors propose a higher selectivity and sensitivity and offer speedy and replicate reactions to the target substance [9]. The modification in the working electrode's surface remains constantly as an arising area in the electroanalytical chemistry. Varied materials and chemical molecules are created and applied for the emergence of newly modified electrodes as electrochemical sensors with optimum performance in exposing the toxic and harmful chemicals. Literatures reveal that metal oxides are the befitting choice to synergise the electrochemical processes.

Metal Oxides (MO) possess excellent electrical properties. Recent studies focus on various MOs and their applications in engineering fields. MOs have high conductivity, thermal stability, non-hazardous, economic, easier methods of preparation, bulky surface area, amenable morphological characteristics, extraordinary photocatalytic and electro catalytic activity and large binding sites [10]. MOs can act as the layer underneath for other particles to develop into an electrochemical biosensor electrode which can actively be equipped for the applications in biological and pharmaceutical fields [11]. The efficient transfer of electrons and the rapid reaction of the particles can very well be utilized for an enhanced accomplishment in various areas such as sensing, catalysis and biomedical applications. Due to its various advantages MOs have proved as an excellent electrocatalytic material. They have extraordinary electrical, optical, and molecular properties. Moreover, it has an advantage of including more functional groups on the surface so that it can further immobilize other biological catalysts [12]. Comparatively, MOs have higher alkaline corrosion resistance than other materials in electrochemical environment due to the stabilization of the transition metals higher oxidation state. They also possess unique crystalline structures that inhibit the accumulation of metal oxide nanostructures [13].

Graphene (Gr), on the other hand, is widely used as sensor. It can be synthesized via mechanical cleavage, epitaxial growth, chemical vapor deposition (CVD), and organic synthesis methods as well [14, 15]. Gr has a single-atom-thick planar sheet of carbon atoms with honeycomb lattice structure. Gr has an excellent capability in the production of sensors and their development due to its excellent electrical, extraordinary electrochemical, high carrier mobility and optical properties [16-18]. It has very high electron transfer rate [19, 20] and has an excellent optical ability to extinguish fluorescence [21]. Its structure can overcome enormously high surface-to-volume ratio [22]. The eminent robustness and flexibility of graphene [23, 24] has found numerous applications in the field of electronics [25, 26] owing to its energy storage devices [27-29], transparent and conductive coatings and films [30, 31]. Recently, graphene has fascinated an enormous interest in scientific community owing to its capability in various technologies such as photonics [32], electronics of huge frequency [33], nano-electro-mechanical systems [34, 35] and gas sensor devices [36, 37].

Graphene and Graphene oxide (GO) are magnificent carbon particles having special properties such as excellent mechanical strength, electron transferring, lower density, and great heat conductance. Their good conducting, best electrocatalytic ability, and biocompatibility are in good terms to nanoparticles of metal and metal oxide thereby the GR and GO blend of metal nanoparticles found to be very useful in the production of electrochemical biosensors. The metal nanoparticles and carbon-based materials blend naturally exhibit collaborative properties in immobilizing enzymes and electrocatalytic applications.

The distinguished properties of MOs and the carbon nanostructures have great impact in the field of electrochemical sensors. Hence, this chapter deals with the various MOs decked Gr/rGO/ERGO modified electrodes for the efficacious investigation of/on the removal of prevailing hazardous pollutants both in the environmental and biological specimens as well.

2.   Electrochemical methods of determination of toxins

2.1   Heavy metal ions

The main environment pollution is imposed by heavy metals onto ecosystems with flora and fauna including aquatic lives. The water-soluble compounds which are highly stable can infiltrate everywhere thus ends up thousands of kilometres from the initial dribbling [38]. Therefore, critical evaluation on their threats in the ecosystems has to be undertaken. Correspondingly apart from intrinsic toxicity on living organisms the studies such as adsorption, distribution, metabolism, and elimination of a chemical compound should be investigated. The major threats under concern are physicochemical characteristics stability, potential ways of degradation and elimination, effects on microorganisms, and on higher organisms [39, 40].

The studies reported in the literature state that materials like polymers, NPs and GO were used in combination with metal oxides for augmenting the sensitivity and selectivity of the electrodes. The synthesis of graphene–TiO_2 composites was carried by hydrothermal or solvothermal method. Shen et al. [41] used GO as a precursor for the synthesis of RGO–TiO_2 composites in water under hydrothermal condition for RGO and tetrabutyl titanate as a single source precursor of TiO_2. Further, they are eco-friendly and this procedure has several benefits when compared to previous technologies due to their simplicity, high productivity, low cost and short processing times. The whole development is quite simple, scalable, and industrially compatible. The experimental parameters are controlled by the factors such as concentrations of precursor solutions and the reaction time. Gr–TiO_2 composites have controlled crystal facets and they can be easily attained by the hydrothermal method [42, 43]. Wen-Yi Zhou et al. [44] developed TiO_2 nanosheets by using the modified hydrothermal method in which they employed 98% concentrated H_2SO_4 solution as a solvent [45] followed by the heat treatment at different temperatures. The LODs of T-1-, T-2-, T-3 and T-4 based electrodes towards the detection of Hg(II) ions were reported to be 0.017, 0.024, 0.189, and 0.208 µM respectively and the sensitivity was 270.83 µA µM^{-1} cm^{-2}. The results of the electrodes met the requirement of the drinking water safety standard (0.03 µM) determined by the World Health Organization (WHO). Its performance also exceeds the previously reported electrodes.

Yan Wei et al., conducted various studies and reported the results of isolation of Cd(II), Pb(II), Cu(II), and Hg(II) ions by developing a glassy electrode using SnO_2 nanoparticles with graphene. The report revealed that SnO_2/reduced graphene oxide nanocomposite is a capable material which owns the advantages of the SnO_2 and graphene composite detecting heavy metal ions by electrochemical determination. Different voltammetric peaks were observed at different potentials with a separation of 212-480 mV between the stripping peaks for the stripping of Cd(II), Pb(II), Cu(II), and Hg(II) on the SnO$_2$/reduced graphene oxide nanocomposite electrode. This detection of four heavy metals can be done by the synthesis of the defined electrode and further it can be used concurrently with a high stability and a long-term usage. The LOD results obtained for Cd(II), Pb(II), Cu(II) and Hg(II) were reported to be 1.015 ×10^{-10} M, 1.839×10^{-10} M, 2.269×10^{-10} M and 2.789 ×10^{-10}M respectively. The LOD values attained were found to be below the guideline value given by the World Health Organization (WHO) [46]. The sensitivity results were found to be 18.4 μA μM^{-1}, 18.6 μA μM^{-1}, 14.98 μA μM^{-1}, 28.2 μA μM^{-1} respectively.

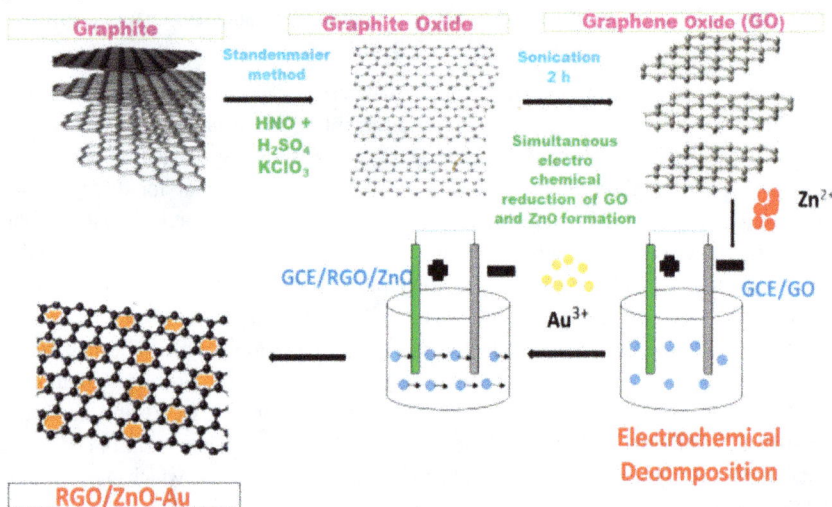

Fig. 1 Schematic illustration of the step wise production of ZnO/RGO/SPCE.

Wei Liu generated an effective nanocomposite electro catalyst by the blend of graphene with ZnO through *in situ* reduction of graphene oxide (ZnO/RGO). He prepared the GCE by hydrothermal one-pot method to construct a new ZnO/RGO nanocomposite. The ZnO/RGO composite was used to ascertain the detection of heavy metals such as Cu(II), Cd(II), Hg(II) and Pb(II) in aqueous solutions. Various potentials of separated stripping peaks were achieved for Hg(II), Pb(II), Cd(II) and Cu(II) on the nanocomposite of

ZnO/RGO. The limits of detection were reported to be 0.04 µM, 0.03 µM, 0.06 µM and 0.03 µM for Cu(II), Cd(II), Hg(II) and Pb(II) ions, respectively. The values are found to be extraordinarily lower than the guideline values fixed by the World Health Organization (WHO). Compared with the standards set by the World Health Organization (WHO), the resulting LODs are lower than the guideline values. The results indicated that this novel nanocomposite of ZnO/RGO, has greater advantages of ZnO as well as graphene to electrochemically detect heavy metal ions [47]. The hypothetical schematic illustration of the step wise production of ZnO-RGO/SPCE is represented in Fig1.

Sohee Lee *et al*, have devised a glassy electrode made up of an iron oxide (Fe_2O_3)/graphene (G) nanocomposite with in situ plated bismuth (Bi). This works as an electrochemical sensor for the detection of trace Zn^{2+}, Cd^{2+}, and Pb^{2+} ions. They have characterized the electrode by transmission electron microscopy, scanning electron microscopy, thermo-gravimetric analyzer, and X-ray diffraction for the synthesis of the Fe_2O_3/G/Bi nanocomposite electrode. The electrochemical properties of the composite modified electrode were investigated. In order to detect the heavy metal ions differential pulse anodic stripping voltammetry was applied. The synergetic effect between graphene and the Fe_2O_3 nanoparticles will depict the modified electrode's improved electrochemical catalytic activity high sensitivity toward trace heavy metal ions. Different parameters such as the preconcentration potential, bismuth concentration, preconcentration time, and pH were carefully optimized to detect the target metal ions. Further with these optimized conditions, the linear range of the electrode was reported as 1–100 µg L^{-1} for Zn^{2+}, Cd^{2+}, and Pb^{2+}, and the detection limits as 0.11 µg L^{-1}, 0.08 µg L^{-1}, and 0.07 µg L^{-1}, respectively (S/N = 3). Repeatability (% RSD) was found to be 1.68% for Zn^{2+}, 0.92% for Cd^{2+}, and 1.69% for Pb^{2+} for single sensor with 10 measurements and 0.89% for Zn^{2+}, 1.15% for Cd^{2+}, and 0.91% for Pb^{2+} for 5 different electrodes.

The characterisation of Fe_2O_3/G composite was carried out by SEM, TEM, TGA, and XRD studies to analyse the morphology and structure of the as-prepared Fe_2O_3/G composite. Fig. 2(a) shows the typical wrinkled and sheet-like character of RGO and Fig. 2(b) (c) shows the TEM and SEM images of the Fe_2O_3/G composite, respectively. The images of the Fe_2O_3/G composite depict the well-dispersed Fe_2O_3 nanoparticles that uniformly decorated the graphene sheets. The morphology and structure of the nanocomposite was further characterized by XRD. The characteristics of thermal decomposition of the Fe_2O_3/G composite were studied by TGA in air atmosphere. Fig. 3(a) depicts the XRD pattern of the RGO, Fe_2O_3/G nanocomposite and 3(b) represents the result of the of TGA Fe_2O_3/G nanocomposite.

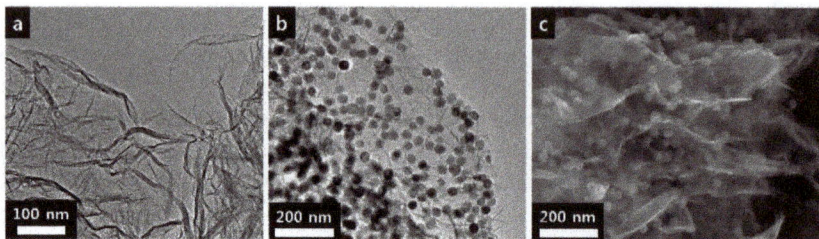

Fig. 2 (a) TEM image of RGO, (b) TEM image of Fe_2O_3/G nanocomposite and (c) SEM image of Fe_2O_3/G nanocomposite Reproduced from reference [49].

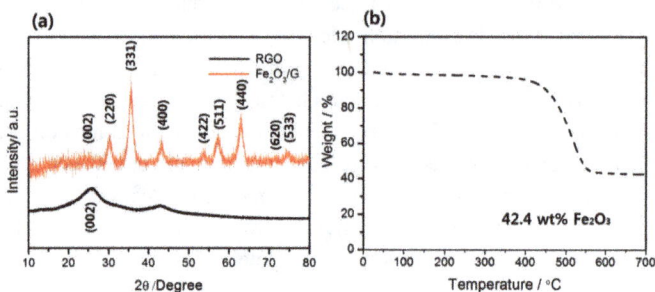

Fig. 3 (a) XRD pattern of the RGO, Fe_2O_3/G nanocomposite (b) and TGA result of the Fe_2O_3/G nanocomposite Reproduced from reference [49].

The Fe_2O_3/G/Bi composite electrode was successfully applied to the analysis of the trace metal ions in real samples. The solvent less thermal decomposition method was applied to the simple and easy synthesis of nanocomposite electrode materials. This can be further extended to the nanocomposites synthesis materials for the determination of heavy metal ions. They are attained through the promising electrode materials for high performance sensors [49]. The performance based on metal oxides with graphene oxide modified electrode is listed in the Table 1.

Table 1 *Analytical Performances of various metal ion sensors based on MO/Gr modified electrodes.*

S.No	Electrode Material	Analyte	LOD	Sensitivity	Ref.
1	RGO–TiO$_2$	Hg^{2+}	0.017 μM	270.83 μA μM^{-1} cm$^-_2$	[44]
2	RGO-SnO$_2$	Cd^{2+}	1.015×10^{-10} M	18.4 μA μM^{-1}	[46]
3	RGO-SnO$_2$	Pb^{2+}	1.839×10^{-10} M	18.6 μA μM^{-1}	[46]
4	RGO-SnO$_2$	Cu^{2+}	2.269×10^{-10}	14.98 μA μM^{-1}	[46]
5	RGO-SnO$_2$	Hg^{2+}	2.789×10^{-10} M	28.2 μA μM^{-1}	[46]
6	ZnO/RGO	Hg^{2+}	0.04 μM	-	[47]
7	ZnO/RGO	Pb^{2+}	0.03 μM	-	[47]
8	ZnO/RGO	Cd^{2+}	0.06 μM	-	[47]
9	ZnO/RGO	Cu^{2+}	0.03 μM	-	[47]
10	Fe$_2$O$_3$	Zn^{2+}	0.11 μg L^{-1}	-	[49]
11	Fe$_2$O$_3$	Cd^{2+}	0.08 μg L^{-1}	-	[49]
12	Fe$_2$O$_3$	Pb^{2+}	0.07 μg L^{-1}	-	[49]

2.2 Hydrogen peroxide

Zhi-Liang has synthesized graphene oxide nanoribbons (GONRs) via longitudinal unzipping of multi-walled carbon nanotubes (MWCNTs) nanoparticles with strong oxidants. Along with co-reduction of KMnO4 and GONRs, the MnO$_2$/reduced graphene oxide nanoribbons (MnO$_2$/rGONRs) composites were invented by the process of single-step hydrothermal method. The detection of hydrogen peroxide (H$_2$O$_2$) is carried out by the MnO$_2$/rGONRs composite electrode. This electrochemical sensor is utilized for investigating the electrochemical properties of the MnO$_2$/rGONRs. The (MnO2/rGONRs/GCE). The developed and modified glassy carbon electrode presented a distinct linear range of 0.25−2245 μM, and a detection limit of 0.071 μM (S/N= 3) (Table 2). The schematic presentation of the preparation of MnO$_2$/ rGO is given in Fig. 4.

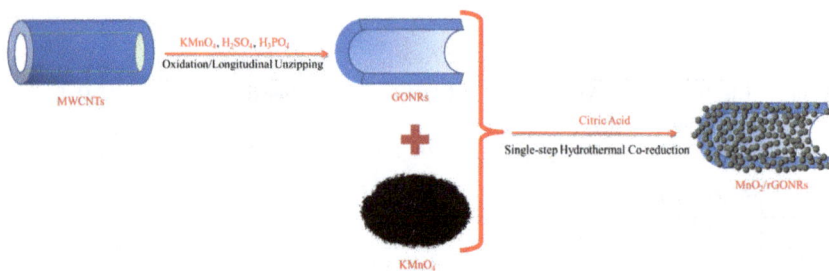

Fig. 4 Schematic representation for the preparation of MnO₂/rGONRs. Reproduced from the ref. [50].

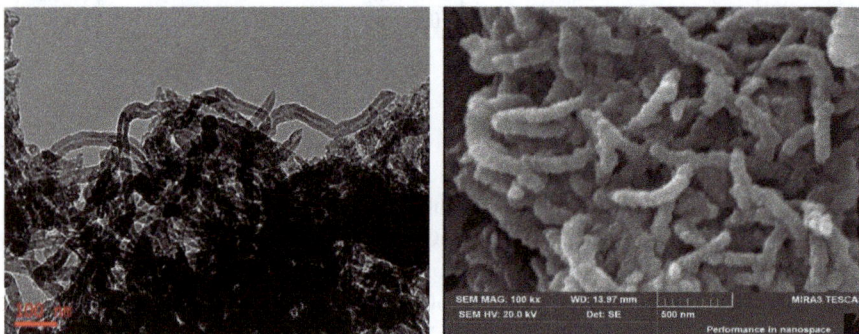

Fig. 5 (a) SEM image of MnO₂/rGONRs   *(b) TEM image of MnO₂/rGONRs*

Reproduced from reference [50].

The morphological and the structure of MnO_2 /rGO composite was characterised by the TEM and SEM images and it is represented in Fig 5(a), (b) respectively.

Further the hypothetical schematic presentation is also given in Fig 6. This electrochemical sensor also exhibited an exceptional analytical performance, reproducibility, precise, and great anti-interference ability [50].

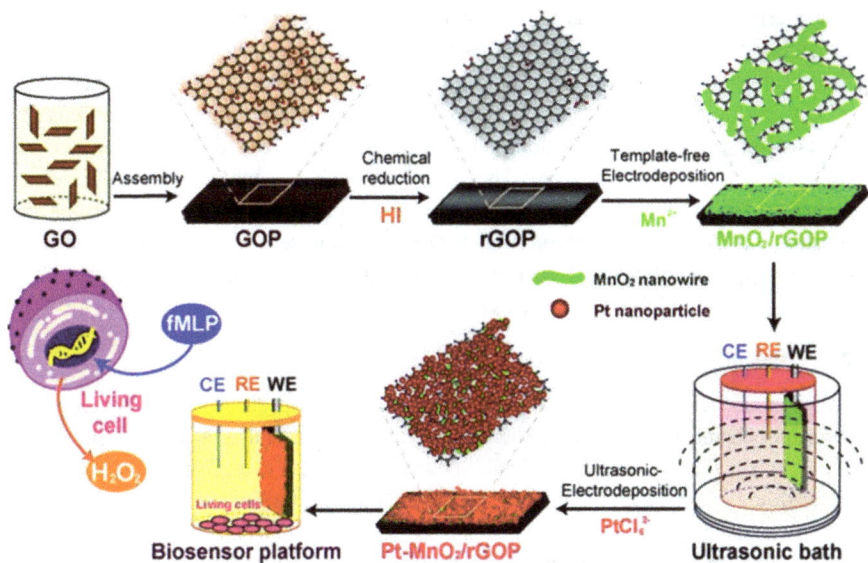

Fig. 6 Stepwise production of synthesis of MnO₂ nanoparticles with reduced graphene oxide. Reproduced from reference [51]. Advanced Catalytic Materials – Photo catalysis and other trends, DOI: 10.5772/61808

2.2   Cyanide

Cyanide is a very toxic industrial chemical but it is also widely used. Recent studies disclose that developing a sensor platform will exploit various enzymatic couplings for finding metabolites and species at extremely low concentrations. This can be a great threat to health and environmental concerns [52].

Hallaj and Haghighi, devised an improved glassy carbon electrode (GCE) for the amperometric determination of cyanide ions. Initially, aminopropyltriethoxysilane was coated on the surface of TiO_2 nanoparticles. And then, graphene oxide (GO) nanosheets were placed, on the surface of the GCE. The modification steps were followed by reductive deposition of reduced 4-nitrophenol (rNPh). The GO/TiO2-AS-rNPh, GCE electrode was used to design a photo electrochemical amperometric cyanide. They have a good characteristic limit of 0.1 μM detection limit and 165.5 $nA \cdot nM^{-1} cm^{-2}$ sensitivity, and an active linear range from 0.1 μM to 60 μM (Table 2). This modified GCE was tested for its response of potentially interfering anions and it has a good result. The characteristic features such as photocatalytic activity, stability, extensive linear analytical range, response time, lower detection limit and the best selectivity will make this GCE, a highly advantageous instrument [53]. They have also discussed the several factors which will affect the sensitivity, interference, durability, and of the performance enhancement of this sensor. Owing to its capability this cyanide sensor can detect cyanide at such extensive range of concentrations. They will have exclusive applications in the fields of homeland security, biomedical, and environmental monitoring. So this amazing sensitivity and range of detection limit fetches its use in several applications in the field of industry, biomedical, and security systems and further this can be developed by utilizing this ultra-high performance sensor

2.3   Phenolic compounds

Tanvir *et al.* devised an electrode comprised of nanocomposite GO–ZnO to investigate the electrochemical behaviour of phenol. The results revealed that the composite is highly stable, retained appropriate reproducibility and was precise to a great extent. The results showed that GO-ZnO is the most fitting composite for indicating the features of electrochemical sensors in analytical chemistry and environmental studies at an industrial level. From the analysis, the detection limit was observed to be 2.2 nM (S/N = 3) (Table 2). From the favourable results it indicates that the GO-ZnO composite can be considered as the capable electrode for investigating the electrochemical nature of phenol [54].

Yaling Tian *et al,* has developed a receptive and a potential consistent electrochemical sensor based on the Manganese oxide Nanowire (NW) MnO_2 NWs-rGO/GCE for the determination of BPA was fabricated. In addition, MnO_2 NWs-rGO/GCE displays the best reproducibility, exceptional selectivity, great sensitivity, and adequate stability. So on the basis of the augmented experimental conditions, quantitative analysis was conducted and the MnO_2 NWs-rGO/GCE exposed a better linear response to Bisphenol A(BPA) concentration ranging from 0.02–20 μM and 20–100 μM, and the detection limit was 6.0 nM (S/N = 3) (Table 2). This process has the best advantage such as sensitivity, good synthesis, quick response and economical. Therefore, an effective method for checking trace levels of BPA was established. This work widens the perspective of the application of GR-based electrochemical sensors and offers a varied outlook for repetitive sensing applications of BPA [55]. The hypothetical presentation of detection of Bisphenol A is repented in the Fig 7.

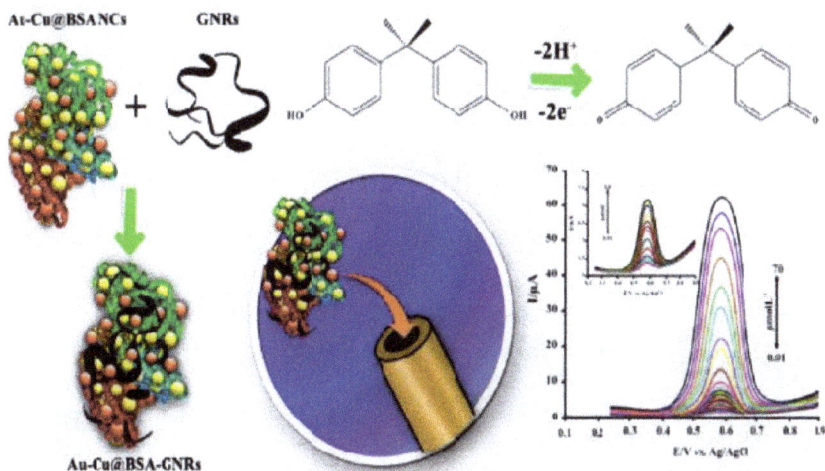

Fig. 7 Synthesis of the Au-Cu@BSA-GNRs/GCE sensor with the oxidation of BPA. Reproduced from Microchemical Journal (2018), doi:10.1016/j.microc.2018.10.044.

2.4   Hydrazine

Lei et al. synthesized MnO_2/graphene oxide nano composite by sonication method [56]. The synthesized material was coated onto the GC surface and GC/MnO_2/GO was obtained. This material was further used for detection [57] of hydrazine. Different techniques such as Scanning Electron Microscopy (SEM), Transmission Electron Microscopy (TEM) Fig 8(a), X-Ray Diffraction (XRD), Atomic Force Microscopy (AFM), X-ray Photo electron Spectroscopy (XPS), Cyclic Voltammetry (CV), UV-Vis, Raman, and FT-IR spectroscopy have been utilized for the characterization of graphene oxide nanocomposites. It clearly depicts a synthesized TEM image of graphene oxide nanocomposite, and it was found that graphene oxide has a smooth surface area after MnO_2 nanoparticle homogeneously dispersed on the graphene sheet. Thus the synthesized nanocomposite is identified to be more stable and has an electrocatalytic activity towards hydrazine. It has also explored that there was more interaction between graphene oxide and MnO_2 nanoparticle. The GO/MnO_2 nano composite, and the chemical composition of the GO/MnO_2 nanocomposite have been characterized by EDX. (Figure 8B) The high magnification of TEM image depicts more detail about the morphology and nanoparticle arrangement. The MnO_2/GO nanocomposite [39] is further confirmed by FT-IR Spectra and several characteristic peaks of GO and GO/MnO_2 nano composite was observed. The synthesized nanocomposite was characterized and also confirmed by XRD measurement. The reflection explained the face- centered-cubic phase structure for the MnO_2 nanoparticle formation. Fig 9, shows the amperometric experiments conducted at 0.6 V in pH 7 phosphate buffer solutions under continuous stirring [56]. The study state current has a linear relationship with the concentration of hydrazine range up to 1.12 mM with a coefficient of 0.999, the detection limit (LOD) 0.16µM and sensitivity of 1007 µAmM⁻¹cm⁻² respectively. The detection limit of MnO_2/GO/GCE is excellent and also it is a good candidate for hydrazine detection wit high performance and has depicted good selectivity and stability.

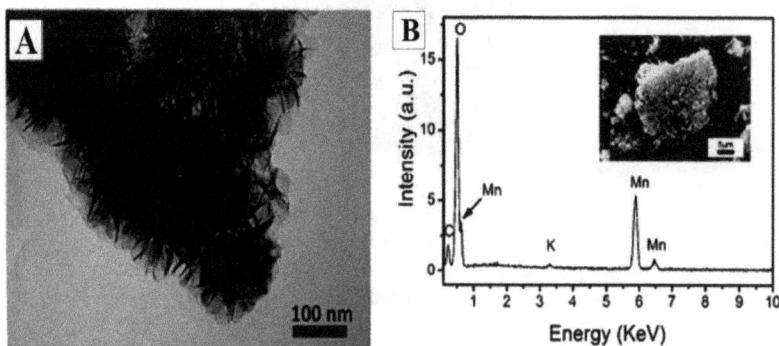

Fig. 8. *(a) TEM image of GO/MnO₂ nanocomposite.*   **Fig. 8 (b)** *EDX of GO/MnO₂ nano composite.*

Reproduced from reference [56]

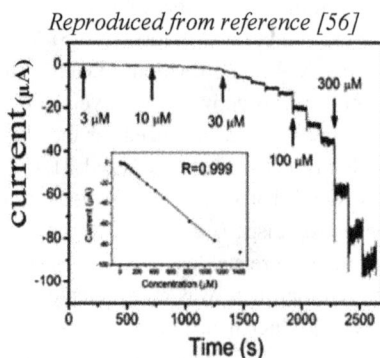

Fig. 9 Amperometric response of the GO/MnO₂ nanocomposite. Reproduced from reference, [56].

The overall oxidation and reduction of Mn species (57) is shown below.

$MnO_2(s) + 2e_- + 4H_3O_+ Mn_{2+} (aq) + 6H_2O$

The synthesized nanocomposite has greater sensitivity, selectivity, stability and linear range. A hypothetical schematic representation of detecting hydrazine is represented in the Figure 10.

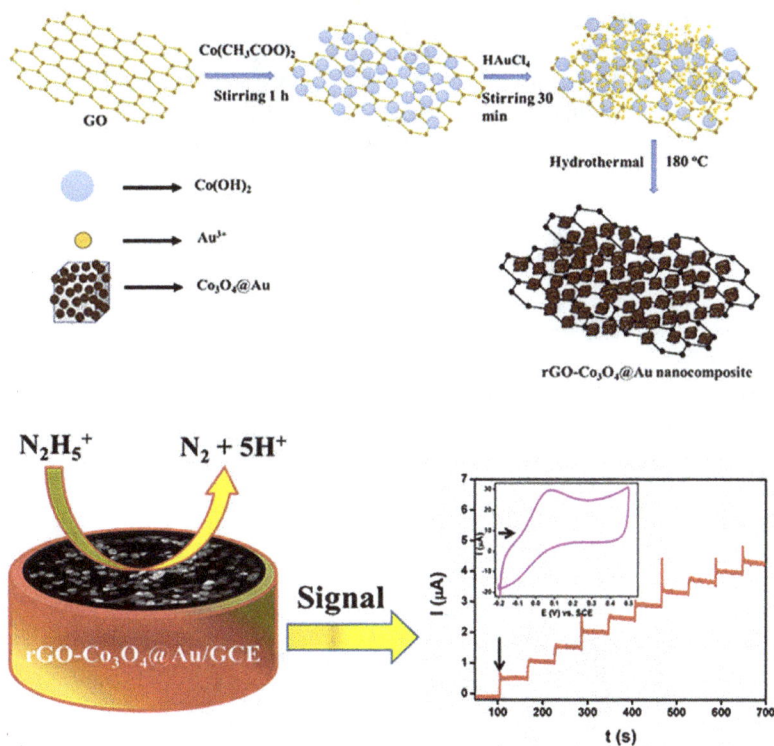

Fig. 10 Schematic representation of hydrazine sensing mechanism of cobalt oxide @gold nanocubes interleaved reduced graphene oxide nano composite to a glassy carbon electrode. Reproduced from Electrochimica Acta, Volume 259, 1 January 2018, Pages 606-616.

Yang *et al.* described the preparation of a nanocomposite consisting of Fe_3O_4 nanoparticles, polypyrrole and graphene oxide (Fe3O4/PPy/GO). The nanocomposite was prepared by combining chemical oxidative polymerization and co-precipitation. The Fe(III) ion is employed as both the oxidant for pyrrole and as a precursor of Fe_3O_4. Several techniques like Transmission Eelectron Microscopy (TEM), Energy-Dispersive X-ray spectroscopy (EDX), Fourier transform Infrared Spectroscopy (FTIR), X-ray Diffraction (XRD) and X-ray Photoelectron Spectroscopy (XPS) were applied for characterization of this nanocomposite. TEM observations exposed that huge numbers of Fe_3O_4 are homogeneously and densely distributed (Fig 11). This was further confirmed with EDX analysis (Fig 11). The Fe_3O_4/PPy/GO nanocomposite was developed into a glassy carbon electrode and has best operated at around 0.2 V (vs. SCE) exhibited excellent response to dissolved hydrazine over the 5.0 μM to 1.3 mM concentration range, a sensitivity of 449.7 μA mM−1 cm−2 and a low detection limit of 1.4 μM (at an S/N ratio of 3). Thus the manufactured electrode can detect the trace hydrazine in wastewater which will have high stability and its practical use as an electrochemical sensor [58].

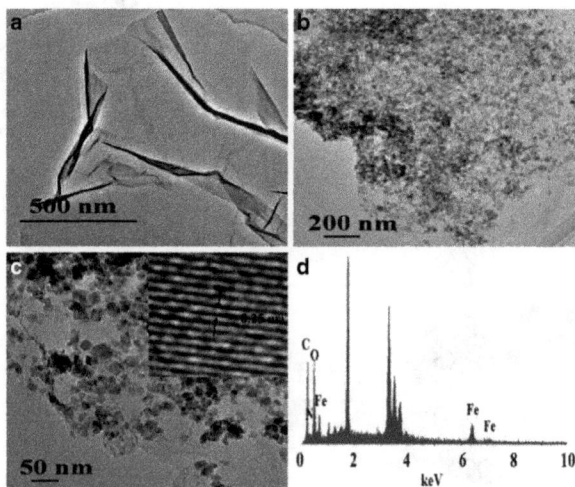

Fig. 11 TEM images of nanocomposites: (a) GO, (b, c) Fe_3O_4/PPy/GO and 11(d) EDX spectrum of Fe_3O_4/PPy/GO. [Reproduced from reference 58].

2.5   Agrochemicals

Dipa Dutta *et al.* stated that the majority of the urea sensors are biosensors and utilize urease, which will limit their use in harsh environments. Currently, because of their exceptional ability to endorse faster electron transfer, carbonaceous material composites and quantum dots are being used for fabrication of a sensitive transducer surface for urea biosensors. They also stated that an enzyme free ultrasensitive urea sensor fabricated using a SnO_2 quantum dots (QDs)/reduced graphene oxide (RGO) composite. Due to the collaborative effect of the elements, the SnO_2 QDs/RGO (SRGO) composite exhibited to be the best detector for electrochemical sensing. The various techniques were adopted to analyse the external and physiological structure f the composite and it was observed that SnO_2 QDs are decked on RGO layers. The electrochemical studies were conducted to assess the characteristics of the sensor in detecting the toxic urea.The SRGO/GCE electrode was subjected to amperometry studies and it is found to be sensitive to urea in the concentration range of $1.6 \times 10^{-14} -3.9 \times 10^{-12}$ M and with a detection limit of as low as 11.7 µM (Table 2). The different characteristics of SRGO such as analytical performance in the presence of interfering agents, economical, and easy synthesis methodology suggest that they can be the best and a promising electrochemical sensor for detecting the effective urea [58].

Ming *et al.* have devised a novel nonenzymatic sensor established on cobalt (II) oxide (CoO)-decked on reduced graphene oxide (rGO). Carbofuran (CBF) and Carbaryl (CBR) compounds were detected by using this novel electro chemical sensor. The two voltammetric peaks for CBF and CBR were obtained which was well-defined and separated with differential pulse with the CoO/rGO sensor in a mixed solution. Thus simultaneous detections of both the carbamate pesticides are possible. The sensor confirmed a linear relationship over a wide concentration range of 0.2–70 lM (R = 0.9996) for CBF and 0.5–200 lM (R = 0.9995) for CBR. The sensor displayed a lower detection limit of 4.2 lg/L for CBF and 7.5 lg/L for CBR (S/N = 3) (Table 2). This developed sensor yielded satisfactory results and detected CBF and CBR in fruit and vegetable samples [59].

2.6   Aromatic nitro compounds

Raja Nehru *et al.* devised an electrode for an accurate detection of toxic 4-nitrophenol (4-NP) which is considered to be the essential for the environment and human health. Several efficient and economical methods for environment monitoring have become necessary in the current scenario. This present research provides an excellent platform for discovering graphene oxide–TiO_2 composite materials with great potential for monitoring toxic contaminants. Hereby, a simple electrochemical sensor was devised to detect the toxic 4-Nitrophenol by a graphene oxide–TiO$_2$ composite. The hybrid of graphene oxide–TiO$_2$ composite material was prepared through an ultrasound assisted sonication method. The graphene oxide–TiO$_2$ composite shows superior performance for the detection of 4-NP with high sensitivity (3.9831 mA mM^{-1} cm^{-1}) coupled with a low limit of detection (LOD) and limit of quantification (LOQ), which were found to be 0.0039 mM and 0.0131 mM, respectively. The hybrid sensor exhibits satisfactory linear responses from 0.02 to 80.57 mM (Table 2). The hybrid GO–TiO$_2$ composite's performance was determined and its RSD was projected to be 2.35%, and the storage stability was 91.5%. The standard addition method is used for the detection of 4-NP in a river water samples by simple and sensitive methods. Given the excellent electrocatalytic activity for the detection of a hazardous environmental pollutant (4-NP), this hybrid GO–TiO$_2$ composite displays a good potential for checking the industrial wastewater samples [60].

Alam et al. synthesized polyethylene glycol mediated reduced graphene oxide/zinc oxide (r-GO/ZnO) nanocomposites by simple and economical chemical reduction method

using graphene oxide and zinc acetate as the precursors. The morphology and thermal decomposition of the nanocomposite material r-GO/ZnO nanocomposites were characterized by different techniques such as X-ray diffraction, transmission electron microscopy and thermogravimetric analysis, respectively. The energy dispersive spectra analysis depicted the results of elemental composition and mapping. The glassy carbon electrode (GCE) sensor for the detection of selective 2-nitrophenol (2-NP) was fabricated with a thin-layer of synthesized r-GO/ZnO composites. The electrochemical responses with enhanced performance depicted a high sensitivity large dynamic range and long-term stability towards the selective 2-NP. The 2-NP was attained using the fabricated r-GO/ZnO/GCE sensor. The calibration curve was found linear (r_2: 0.9916) over a wide range of 2-NP concentrations (10.0 nM ~10.0 mM). The detection limit and the sensitivity were calculated as 0.27 nM and 5.8 μA, mM^{-1}.cm^{-2} respectively based on 3N/S (Signal-to-Noise ratio) (Table 2). The present research demonstrated the detection of 2-NP by I-V method using r-GO/ZnO composites modified GCE electrode with very high sensitivity compared to various nanocomposites reported earlier. The synthesis of r-GO/ZnO composites using chemical reduction process is a very good way of launching sensor based r-GO/ZnO composites for toxic and carcinogenic chemicals.

Table 2 *Analytical Performances of electrochemical sensors based on MO/Gr modified electrodes towards the detection toxic chemicals.*

S.No	Electrode Material	Analyte	LOD	Linear Range	Sensitivity	Ref.
1	MnO$_2$/rGONRs / GCE	H$_2$O$_2$	0.071 µM	0.25–2245 µM	-	[50]
2	GO/TiO2-AS/GCE	CN	0.1 µM	0.1 µM to 60 µM.	165.5 nA·nM^{-1}cm^{-2}	[53]
3	GO–ZnO / GCE	Phenol	2.2 nM	3.3×10^{-8} mol L^{-1}	31 084 A m^{-2} M^{-1}	[54]
4	MnO$_2$ NWs-rGO/GCE	Bisphenol A	6.0 nM	0.02–20 µM and 20–100 µM	-	[55]
5	MnO2/graphene oxide	Hydrazine		1.12 mM		
6		Hydrazine			-	
7	SnO$_2$ QDs/RGO	Urea	11.7 µM	1.6×10^{-14} -3.9×10^{-12} M	1.38 µA/fM	[58]
8	CoO/rGO/GCE	Carbofuran	4.2 lg/L	0.2–70 lM	-	[59]
9	CoO/rGO/GCE	Carbaryl	7.5 lg/L	0.5–200 lM	-	[59]
10	TiO$_2$ /GO/GCE	4-Nitrophenol	0.0039 mM	0.02 to 80.57 mM	3.9831 mA mM^{-1} cm^{-1}	[60]
11	r-GO/ZnO	2-Nitro phenol	0.27 nM	10.0 nM ~10.0 mM	5.8 µA, mM^{-1}.cm$^{2.}$	[61]

Furthermore, the r-GO/ZnO modified electrode displays a good selectivity and has a good practical applicability in the waste water samples. In brief, the synthesized r-GO/ZnO modified GCE has been used for the sensitive detection of 2-NP. The r-GO/ZnO transformed electrode exhibits a high sensitivity for the detection of 2-NP compared to other electrodes. Thus the devised electrode displays many advantages, such as high sensitivity, low LOD, linear response range along with good selectivity for the detection of 2-NP. Ultimately the detection of 2-NP in waste water samples indicates the good practical application of the constructed electro chemical sensor. The manufactured electro chemical bio sensor can be further utilized for the detection of 2-NP in trace levels in the environmental samples. Thus from this novel approach an efficient route can be introduced for the development of efficient electro chemical sensor for health-care and environmental fields in broad ranges [61].

Conclusion and the upcoming perspective

In today's modern technology, due to the tremendous development in various fields such as agriculture, food, pharmaceutical and textile industries many toxic chemicals are released into the environment which is emerging as the major concern. Rapidly, this will affect the quality of life, human health and other living beings in the environment. Therefore, detection and assessment of trace level of contaminants in food, environment and biological samples are extremely crucial. Remarkably, electrochemical sensors are highly advantageous. An electrochemical sensing platform made of a reduced graphene oxide-metal oxide nanocomposites are used for detecting different toxic chemicals. Because of their low cost and the conductivity recently many new electrochemical sensors based on nanoparticles have been designed with excellent electrocatalytic properties using a simpler preparation for the detection of chemical pollutants.

This chapter highlights the recent developments in MOs with GOs as electrochemical sensors. There are different metal oxides which are capable for electro catalysis. But metal oxides alone are poorly conductive and also suffer interruption and accumulation during the electrochemical reactions. Indeed, some graphene-supported metal oxide nanocomposites have exhibited an amazing improvement in the electrocatalytic activity and stability towards electrochemical reactions.

In terms of electro catalysis, because of their potential advantages graphene-oxide supported metal oxides nanocomposites have exhibited promising applications. Mainly it is due to the large surface area and their flexibility, the graphene oxides can provide sufficient space to accommodate different nanomaterials and also check their accumulation. Owing, to their good characteristics of graphene, solid-air contact efficiency increases and concurrently adsorption of oxygen also increases. The electron transfer rate on the surface will be promoted by the electrical conductivity of graphene. Further, graphene will provide the structural defects for alteration with different functional groups to support of the selective electro catalysis.

To surmise, the inorganic nanomaterials incorporated onto the surface of graphene have fascinated a fabulous attention for the progress of new-generation electro-catalytic materials. These novel nano structures show superior electrocatalytic activity, selectivity, and long-term stability, which can serve as promising electrode material for different electrochemical reactions including detection and detoxification of chemical pollutants in the environment and biological samples.

References

[1] Jhumi Jain, Pammi Gauba, Heavy metal toxicity-implications on metabolism and health, Int J Pharma Bio Sci, 8(4) (2017) 452-460. https://doi.org/10.22376/ijpbs.2017.8.4.b452-460

[2] Ksenia S. Egorova and Valentine P. Ananikovm, Toxicity of Metal Compounds: Knowledge and Myths, Organometallics, 36 (2017) 4071-4090. https://doi.org/10.1021/acs.organomet.7b00605

[3] A.M. O'Mahony, J. Wang, Nanomaterial-based electrochemical detection of explosives: a review of recent developments, Anal. Methods, 5 (2013) 4296. https://doi.org/10.1039/c3ay40636a

[4] Maduraiveeran Govindhan, Bal-Ram Adhikari and Aicheng Chen, A. Chen, Nanomaterials-based electrochemical detection of chemical contaminants, RSC Adv. 4 2014) 63741. https://doi.org/10.1039/C4RA10399H

[5] JR Windmiller, J Wang, Wearable electrochemical sensors and biosensors: a review, Electroanalysis, 25 (1) (2013) 29-46. https://doi.org/10.1002/elan.201200349

[6] J Ma, D Yuan, K Lin, S Feng, T Zhou, Q Li, Applications of flow techniques in seawater analysis: A review., Trends Environ. Anal. Chem, 10 (2016) 1-10. https://doi.org/10.1016/j.teac.2016.02.003

[7] J.N. Stetter, W.R. Penrose, Y. Sheng, Sensors, chemical sensors, electrochemical sensors, and ECS. J. Electrochem. Soc, (2003)150: S11–S16. https://doi.org/10.1149/1.1539051

[8] E. Bakker, M. Telting-Diaz, Electrochemical sensors, Anal. Chem., 74 (2002) 2781-2800. https://doi.org/10.1021/ac0202278

[9] Iuliana Moldoveanua, Raluca-Ioana Stefan-van Stadena and Jacobus Frederick van Staden, Electrochemical Sensors Based on Nanostructured Materials, Handbook of Nano electrochemistry. 10.1007/978-3-319-15207-3_47-1.

[10] F. Vajedi, H. Dehghani, The characterization of TiO2-reduced graphene oxide nanocomposites and their performance in electrochemical determination for removing heavy metals ions of cadmium(II), lead(II) and copper(II), Mater. Sci. Eng B. 243 (2019) 189–198. https://doi.org/10.1016/j.mseb.2019.04.009

[11] J.M. George, A. Antony, B. Mathew, Metal oxide nanoparticles in electrochemical sensing and bio sensing: a review, Microchimica Acta, 185 (2018) 358. https://doi.org/10.1007/s00604-018-2894-3

[12] Arnab Halder, Minwei Zhang and Qijin Chi, Electrocatalytic Applications of Graphene–Metal Oxide Nano hybrid Materials, Advanced Catalytic Materials - Photocatalysis and Other Current Trends.

[13] A. Farmer and C. T. Campbell, Ceria Maintains, Smaller Metal Catalyst Particles by Strong Metal-Support Bonding, Science, 329, 5994 (2010) pp. 933–936. https://doi.org/10.1126/science.1191778

[14] Georgakilas, V., Otyepka, M., Bourlinos, A. B., Chandra, V., Kim, N., Kemp, K. C., Hobza, P., Zboril, R., Kim, K.S. Functionalization of Graphene: Covalent and Non-Covalent Approaches, Derivatives and Applications, Chem. Rev, 112 (2012) 6156−6214. https://doi.org/10.1021/cr3000412

[15] Georgakilas, V., Tiwari, J., Kemp, K.C., Perman, J., Bourlinos, A., Kim, K.S., Zboril, R. Non-Covalent Functionalization of Graphene and Graphene Oxide for Energy Materials, Bio sensing, Catalytic, and Biomedical Applications. Chem. Rev, 116 (2016) 5464−5519. https://doi.org/10.1021/acs.chemrev.5b00620

[16] Srikanth Ammu, Graphene based chemical sensors, Science Lettters, 4 (2015) 162.

[17] X. Du, I. Skachko, A. Barker, E.Y. Andrei, Approaching ballistic transport in suspended   grapheme, Nat. Nanotechnol., 3 (2008) 491. https://doi.org/10.1038/nnano.2008.199

[18] E. Pallecchi, F. Lafont, V. Cavaliere, F. Schopfer, D. Mailly, W. Poirier, A. Ouerghi, High Electron Mobility in Epitaxial Graphene on 4H-SiC (0001) via post-growth annealing under hydrogen, Sci. Rep., 4 (2014) 4558. https://doi.org/10.1038/srep04558

[19] C.L. Weaver, H. Li, X. Luo, X.T. Cui, A graphene oxide/conducting polymer nanocomposite for electrochemical dopamine detection: origin of improved sensitivity and specificity, J. Mater. Chem -B, 2 (2014) 5209. https://doi.org/10.1039/C4TB00789A

[20] K.P. Loh, Q. Bao, G. Eda, M. Chhowalla, Graphene oxide as a chemically tunable platform for optical applications, Nat. Chem., 2 (2010) 1015. https://doi.org/10.1038/nchem.907

[21] X. Sun, Z. Liu, K. Welsher, J.T. Robinson, A. Goodwin, S. Zaric, H. Dai, Nano-Graphene Oxide for Cellular Imaging and Drug Delivery, Nano Res., 1 (2008) 203. https://doi.org/10.1007/s12274-008-8021-8

[22] C.N.R. Rao, K.S. Subrahmanyam, H.S.S.R. Matte, and A. Govindaraj, Graphene: synthesis, functionalization and properties, MOD PHYS LETT B, 25(2011) 427-451. https://doi.org/10.1142/S0217984911025961

[23] V. Dua, S.P. Surwade, S. Ammu, S.R. Agnihotra, S. Jain, K.E. Roberts, S. Park, R.S. Ruoff S.K. Manohar, All-organic vapor sensor using inkjet-printed reduced grapheme oxide, Angew. Chem., 49 (2010) 2154. https://doi.org/10.1002/anie.200905089

[24] Y. Liu, B. Xie, Z. Zhang, Q, Zheng, Z. Xu, Mechanical properties of graphene papers, J MECH PHYS SOLIDS, 60 (4) (2012) 591. https://doi.org/10.1002/anie.200905089

[25] G. Eda, G. Fanchini, M. Chhowalla, Large-area ultrathin films of reduced graphene oxide as a transparent and flexible electronic material, Nat. Nanotechnol, 3 (2008) 270. https://doi.org/10.1038/nnano.2008.83

[26] M.F. El-Kady, V. Strong, S. Dubin, R.B. Kaner, Laser scribing of high-performance and flexible graphene-based electrochemical capacitor, Science, 335 (2012) 1326. https://doi.org/10.1126/science.1216744

[27] M.D. Stoller, S. Park, Y. Zhu, J. An, R.S. Ruoff, Graphene-Based Ultracapacitors, Nano Lett, 8 (2008) 3498. https://doi.org/10.1021/nl802558y

[28] Y. Wang, Z. Shi, Y. Huang, Y. Ma, C. Wang, M. Chen, Y. Chen, Supercapacitor devices based on Graphene materials, J. Phys, Chem.,113 (2009) 13103. https://doi.org/10.1021/jp902214f

[29] Dale A.C. Brownson, Dimitrios K. Kampouris, Craig E. Banks, An overview of graphene in energy production and storage applications, J. Power Sources, 196 (2011) 4873-4885. https://doi.org/10.1016/j.jpowsour.2011.02.022

[30] S. Watcharotone, D.A. Dikin, S. Stan,kovich, R. Piner, I. Jung, G.H.B. Dommett, G. Evmenenko, S.-E. Wu, S.-F. Chen, C.-P. Liu, S.T. Nguyen, R.S. Ruoff, Graphene-silica composite thin films as transparent conductors, Nano. Lett. 7 (2007) 1888. https://doi.org/10.1021/nl070477+

[31] V.C. Tung, L.-M. Chen, M.J. Allen, J.K. Wassei, K. Nelson, R.B. Kaner, Y. Yang, Low-temperature solution processing of graphene-carbon nanotube hybrid materials for high- performance transparent conductors, Nano Lett. 9 (2009) 1949. https://doi.org/10.1021/nl9001525

[32] F. Xia, T. Mueller, Y.-m. Lin, A. Valdes-Garcia, P. Avouris, Ultrafast graphene photodetector, Nat. Nanotechnol, 4 (2009) 839. https://doi.org/10.1038/nnano.2009.292

[33] F. Schwierz, Graphene transistors, Nat. Nanotechnol, 5 (2010) 487. https://doi.org/10.1038/nnano.2010.89

[34] S. Roy, Z. Gao, Nanostructure-based electrical, Nano Today, 4 (4) (2009) 318-334. https://doi.org/10.1016/j.nantod.2009.06.003

[35] C. Chen, J. Hone, Proc. Graphene nanoelectromechanical systems, IEEE 101 (2013) 1766. https://doi.org/10.1109/JPROC.2013.2253291

[36] Y. Liu, X. Dong, P. Chen, Biological and chemical sensors based on graphene materials, Chem. Soc. Rev, 41 (2012) 2283. https://doi.org/10.1039/C1CS15270J

[37] S. MacNaughton, S. Sonkusale, S. Surwade, S. Ammu, S. Manohar, S. MacNaughton, S. Sonkusale, S. Surwade, S. Ammu, S. Manohar, IEEE Sensors (2010) 894.

[38] Ksenia S. Egorova and Valentine P. Ananikov, Toxicity of Metal Compounds: Knowledge and Myths, Organometallics, 36 (2017) 4071-4090. https://doi.org/10.1021/acs.organomet.7b00605

[39] Woolley, A.A. Guide to Practical Toxicology: Evaluation, Prediction, and Risk, 2nd ed.; Informa Healthcare USA: New York, 2008. https://doi.org/10.1201/9781420043150

[40] A Textbook of Modern Toxicology; Hodgson, E., Ed.; Wiley: Hoboken, New Jersey, 2010.

[41] J.F. Shen, B. Yan, M. Shi, H.W. Ma, N. Li, M.X. Ye, One step hydrothermal synthesis of TiO_2-reduced graphene oxide sheets, J. Mater. Chem, 21(2011) 3415. https://doi.org/10.1039/c0jm03542d

[42] Wang ZY, Huang BB, Dai Y, Liu YY, Zhang XY, Qin XY, Wang JP, Zheng ZK, Cheng HF. Crystal facets controlled synthesis of graphene@TiO2 nanocomposites by a one-pot hydrothermal process, CRECF4, 14 (2012) 1687. https://doi.org/10.1039/C1CE06193C

[43] Changyuan Hu , Tiewen Lu , Fei Chen & Rongbin Zhang, A brief review of graphene–metal oxide composites synthesis and applications in photocatalysis, J. Chinese. Adv. Mater. Soc., 1910 (2013) 21-39. http://dx.doi.org/10.1080/22243682.2013.771917

[44] Wen-Yi Zhou, Jinyun Liu, Jieyao Song, Jinjin Li, Jinhuai Liu, Xing-Jiu Huang, Surface-Electronic-State-Modulated, Single-Crystalline (001) TiO2 Nanosheets for Sensitive Electrochemical Sensing of Heavy-Metal Ions, Anal. Chem, 89 (6) (2017) 3386–3394. https://doi.org/10.1021/acs.analchem.6b04023

[45] X. Han, Q. Kuang, M. Jin, Z. Xie, L. Zheng, Synthesis of titania nanosheets with a high percentage of exposed (001) facets and related photocatalytic properties, J. Am. Chem. Soc., 131 (2009) 3152-3153. https://doi.org/10.1021/ja8092373

[46] Yan Wei, Chao Gao, Fan-Li Meng, Hui-Hua Li, Lun Wang, Jin-Huai Liu, Xing-Jiu Huang, SnO2/Reduced Graphene Oxide Nanocomposite for the Simultaneous Electrochemical Detection of Cadmium(II), Lead(II), Copper(II) and Mercury(II): An Interesting Favorable Mutual Interference, J. Phys. Chem.C, 16 (2012) 1034-1041. https://doi.org/10.1021/jp209805c

[47] Wei Liu, Preparation of a zinc oxide-reduced graphene oxide nanocomposite for the determination of Cadmium(II), Lead(II), Copper(II), and Mercury(II) in water, Int. J. Electrochem. Sci., 2 (2017) 5392-5403. https://doi.org/10.20964/2017.06.06

[48] World Health Organization (WHO), Guidelines for drinking water quality, Sixty-first Meeting, Rome, June 10-19 2003

[49] Sohee Lee, Jiseop Oh, Dongwon Kim and Yuanzhe Piao, A sensitive electrochemical sensor using an iron oxide/graphene composite for the simultaneous detection of heavy metal ions, Talanta. https://doi.org/10.1016/j.talanta.2016.07.034

[50] Zhi-Liang Wu, Cheng-Kun Li, Jin-Gang Yu, Xiao-Qing Chen, MnO2/Reduced Graphene Oxide Nanoribbons: Facile hydrothermal preparation and their application in amperometric detection of hydrogen peroxide, SENSOR ACTUAT B-CHEM, 239 (2017) 544-552. https://doi.org/10.1016/j.snb.2016.08.062

[51] Solomon W. Leung, Maedeh Mozneb, James C.K. Lai, An ultra-sensitive Sol-Gel bio composite electrode sensor for cyanide detection, Sensors & Transducers, 191(8) 2015, 114-119.

[52] S. Mozneb, J.C.K. Lai, S.W. Leung, Cyanide detection by highly modified Sol-Gel bio composite sensor, in Proceeding of the NSTI, Nanotechnology Conference & Expo, Diagnostics & Imaging, Washington, D.C., 3 (3) (2015) 155-158.

[53] Hallaj, R., Haghighi, N. Photo electrochemical amperometric sensing of cyanide using a glassy carbon electrode modified with graphene oxide and titanium dioxide nanoparticles, Microchim Acta, 184, (2017) 3581-3590. https://doi.org/10.1007/s00604-017-2366-1

[54] Tanvir Arfin, Stephy N. Rangarim Graphene oxide–ZnO nanocomposite modified electrode for the detection of phenol, ANAL. METHODS. https://doi.org/10.1039/C7AY02650A

[55] Yaling Tian, eihong Deng, Yiyong Wu, Junhua Li, Jun Liu, Guangli Li, Quanguo H, MnO_2 nanowires-decorated reduced graphene oxide modified glassy carbon electrode for sensitive determination of bisphenol, J. Electrochem, 167 (2020) 046514. https://doi.org/10.1149/1945-7111/ab79a7

[56] Junyu Lei, Xiaofeng Lu , Wei Wang , Xiujie Bian , Yanpeng Xue , Ce Wang and Lijuan Li , Fabrication of MnO_2/graphene oxide composite nanosheets and their application in hydrazine detection RSC Adv., 2 (2012) 2541. https://doi.org/10.1039/c2ra01065h

[57] S. E. Baghbamidi, H. Beitollahi, S. Tajik. Graphene oxide nano-sheets/ferrocene derivative modified carbon paste electrode as an electrochemical sensor for determination of hydrazine. Anal. Bioanal. Electrochem.,6 (2014) 634.

[58] Z. Yang, Q. Sheng, S. Zhang, X. Zheng, J. Zheng, One-pot synthesis of Fe_3O_4/polypyrrole/graphene oxide nanocomposites for electrochemical sensing of hydrazine, Microchim. Acta. 184 (2017) 2219–2226. https://doi.org/10.1007/s00604-017-2197-0

[58] Dipa Dutta, Sudeshna Chandra, Akshaya K. Swain, Dhirendra Bahadu, SnO2 Quantum dots-reduced graphene oxide composite for enzyme-free ultrasensitive electrochemical detection of urea, ANA. CHEM., 86 (12) (2014), 5914-5921. https://doi.org/10.1021/ac5007365

[59] Ming Yan Wang, Jun Rao Huang, Meng Wang, Dong E Zhang, Jun Chen, Electrochemical nonenzymatic sensor based on CoO decorated reduced graphene oxide for the simultaneous determination of Carbofuran and Carbaryl in fruits and vegetables. Food Chem, 151 (2014) 191-197. https://doi.org/10.1016/j.foodchem.2013.11.046

[60] Raja Nehru, Praveen Kumar Gopi and Shen-Ming Chen, Enhanced sensing of hazardous 4-nitrophenol by a graphene oxide–TiO2 composite: Environmental pollutant monitoring applications, RSC. https://doi.org/10.1039/C9NJ06176B

[61] M.K. Alam, M.M. Rahman, M. Abbas, S.R. Torati, A.M. Asiri, D. Kim, C.G. Kim, Ultra-sensitive 2-nitrophenol detection based on reduced graphene oxide/ZnO nanocomposites, J. Electro anal. Chem. 788 (2017) 66–73. https://doi.org/10.1016/j.jelechem.2017.02.004

Chapter 6

Graphene-Metal Chalcogenide based Electrochemical Sensors for Toxic Chemicals

A. Arivarasan*

International Research Centre, Department of Physics, Kalasalingam Academy of Research and Education, Krishnankoil- 626 126, Tamilnadu, India

arivarasan.nanotech@gmail.com

Abstract

The developments in industrial and technological sectors result in severe environmental issues, such as environmental contamination and pollution. Among the various types of contaminations, toxic gases, heavy metal ions, pesticides and fertilizers, etc., play a major role in environmental issues. As an advancement of the sensors used to detect such toxic chemicals, electrochemical sensor attracts much attention in recent days. The development of graphene based nanocomposites for electrochemical sensors, paving a new path for the sensitive and efficient detection of such chemicals. Graphene/metal chalcogenide based nanocomposites have gained worldwide attention in recent decades and are being researched for use in different applications due to their unique physical and chemical properties. This chapter aims a comprehensive presentation of various toxic chemicals, their environmental effects and their conventional sensing mechanism. The role of graphene, metal chalcogenides and their nanocomposites on the sensing mechanism in electrochemical sensors were reviewed.

Keywords

Electrochemical Sensors, Graphene/Metal Chalcogenides, Nanocomposites, Toxic Chemicals

Contents

1.   Introduction

Accurate monitoring of toxic chemicals and detection of environmental pollutants has become a primary concern due to rapid progress in industrialization during the recent years. In this regard, designing of robust, low cost and portable sensors is an essential requirement and progressive research is being conducted in order to develop new ranges of chemical sensors with enhanced sensitivity. Over the last decades, instruments based on chemical reaction have been introduced to monitor the atmosphere and ecological conditions. Toxic chemical sensing, offer broad range of applications and play an important role in many areas, such as personal safety, medical diagnosis, pollutant detection and transportation industries [1]. For the quantitative analysis of toxic chemical/gases, various physical and chemical principles are involved. Different types of chemical sensors such as photo ionization sensors [2], IR sensors [3], fluorescent sensors [4-6], metal oxide semiconductor [7], catalytic gas sensors [8] and electrochemical gas sensors [9] are used for the detection of toxic chemicals. Several sensing platforms used in laboratory to monitor the toxic chemicals may include pellistores, optical sensors and semiconductor gas sensors. Pellistores change resistance in the presence of chemicals and consists of catalyst loaded ceramic pellets. Such sensing platform exhibits higher sensitivity however, faces zero drift at ppm concentration [10]. The problem of zero drift and cross sensitivity is observed in semiconductor gas sensors as well [11]. Optical sensors could achieve higher sensitivity, selectivity and stability than non-optical methods but due to miniaturization and relatively high cost, their applications on chemical sensors are seriously restricted [12]. In comparison to these sensing platforms, electrochemical sensors have significant advantages for quantifying and detecting hazardous chemicals/gases [13]. These sensors are relatively specific to individual chemical environment with sensitivity at ppm and ppb levels [14]. Few of the key attributes of the electrochemical sensors may include room temperature operating conditions, compact size, low cost, high portability and better selectivity [15]. Electrochemical sensors are extensively being employed in various applications, indicating the scope and potential of this area. In this context some toxic chemicals and pollutants play a major role in environmental issues and in the adverse effects on living organisms. The list of such chemicals and their role in environmental issues are discussed in the following section.

Table 1 List of applications of electrochemical sensors in industries.

Industry/Field	Compounds
Automotive	O_2, H_2, CO, NO_x, HCs,
IAQ	CO, CH_4, humidity, CO_2, VOCs
Food	Bactreria, biologicals, chemicals, fungal toxins, humidity, pH, CO_2,
Agriculture	NH_3, amines, humidity, CO_2, pesticides, herbicides
Medical	O_2, glucose, urea, CO_2, pH, Na^+, K^+, Ca^{2+}, C, bio-molecules, H_2S, Infectious disease, ketones, anesthesia gases
Water treatment	pH, Cl_2, CO_2, O_2, O_3, H_2S
Environmental	SO_x, CO_2, NO_x, HCs, NH_3, H_2S, pH, heavy metal ions
Industrial safety	Indoor air quality, toxic gases, combustible gases, O_2
Utilities (gas, electric)	O_2, CO, HCs, NO_x, SO_x, CO_2
Petrochemical	HC_x, conventional pollutants
Steel	Steel
Military	Agents, explosives, propellants
Aerospace	H_2, O_2, CO_2, humidity

2.   Environmental pollutants and toxic chemicals

Hydrazine

Hydrazine has emerged as highly promising material for various applications such as explosive, rocket propellant, fuel cells, pesticides, emulsifiers and catalysts [16]. The laboratory usage of hydrazine can cause the environmental hazards, because of its high toxicity, leaching to blood abnormalities as well as damage to the internal organs (liver and kidney) [17,18]. And also hydrazine plays a vital role in the construction of polymers, pharmaceuticals and pesticides [19]. Herein, the setting of hydrazine threshold limit value of 10 ppb and it can be classified as portable human carcinogen, neurotoxin and group B2 by United State Environmental Protection Agency (USEPA) and World Health Organisation (WHO) [20].

Nitrobenzene

Nitrobenzene (NB) is a hazardous organic material and finds its applications in various fields such as petroleum refineries, pesticides, herbicides, insecticides, dyes, shoe polishes, soap etc.[21] Its pollution in water, can affect human health and environment. The maximum concentration of the NB allowed in drinking water is 1 mg/L [22]. Therefore, the conversion methods (chemical and electrochemical reduction process) have been used for the reduction of nitrobenzene, which is demanded for environmental production and public health [23].

Nitrophenol

Nitrophenols are widely used in various industrial fields like pesticides, dye-stuffs, pharmaceuticals, explosive etc. [24]. The industrial effluents of phenolic compounds are inevitably released into the soil, leaching to serious environmental contaminations. The abundant 4-nitrophenol is harmful to human health because of its carcinogenicity and cause damage to kidney, liver and central nervous system. It also caused health issues such as, nausea, headaches, cyanosis and drowsiness [25].

Pesticides

Generally, pesticides are chemical compounds, which can be used to destroy or prevent the pests like mice, insects, weeds, microorganism and fungi etc. Most of the pesticides are hazardous in nature and it can kill the organisms by causing deadly diseases. Some of the toxic or hazardous pesticides that are used in the agricultural fields are organophosphate, parathion, carbamate, dichlorodiphenyltrichloroethane (DDT), mercury and arsenic derivatives [26,27]. Pesticides can exhibit harmful effects and may cause serious mishaps to the human being and cause infertility, neurological diseases, carcinogenetic and respiratory problems etc.

Heavy metals

Heavy metals enter the environment due to increasing industrial activities. Heavy metals such as Hg, As, Pb, and Cd are highly toxic and carcinogenic even at a trace level [28]. They are nonbiodegradable and can accumulate in the food chain, which poses a severe threat to the environment and human health. Heavy metal pollution becomes a concern for global sustainability. It is therefore essential to monitor heavy metals in the environment, drinking water, food, and biological fluids. Rapid development of nanotechnology has provided new opportunities for improving the performance of sensors in terms of sensitivity, limit of detection, selectivity, and reproducibility, and also enabled miniaturization with assistance of lab-on-chip (LOC) technology [29-31].

Compared to other sensors, the sensing signal of electrochemical sensors is collected through conducting wires instead of detectors. Hence, electrochemical sensors can be easily packed into a compact system. In addition, because heavy metals have the defined redox potential, the selectivity toward specific heavy metal ions can be achieved by bare electrodes without the need of a molecular recognition probe. Several techniques are employed in electrochemical sensing, including voltammetry, amperometry, potentiometry, impedemetry, and conductometry [32]. In particular, the anodic stripping voltammetry (ASV) method is readily amendable for determination of heavy metals. ASV analysis typically involves two steps [33,34]: (i) electrochemical deposition or accumulation of heavy metals at a constant potential to preconcentrate the analyte onto the electrode surface, and (ii) stripping or dissolution of the deposited analyte from the electrode surface.

Toxic gases

Nitrogen dioxide (NO_2)

It is one of the common toxic air pollutants, which is mostly found as a mixture of nitrogen oxides (NO_x) with different ratios (x). NO_2 is a reddish-brown, irritant, toxic gas having a characteristic sharp and biting odour. The LC_{50} (the lethal concentration for 50% of those exposed) for one hour of NO_2 exposure for humans has been estimated as 174 ppm. The noteworthy impacts of NO_2 include: respiratory inflammation of the airways, decreased lung function due to long term exposure, increased risk of respiratory conditions [35,36], increased responsiveness to allergens, contribution to the formation of fine particulate matter (PM) and ground level ozone which have adverse health effects, and contribution to acid rain causing damage to vegetation, buildings and acidification of lakes and streams [37,38].

Sulphur dioxide (SO$_2$)

It is the most common air pollutant, mostly found as a mixture of sulphur oxides (SO$_x$). It is an invisible gas with a nasty, sharp smell. The maximum concentration for SO$_2$ exposures of 30 min to 1 h has been estimated as 50 to 100 ppm. The main sources of SO$_2$ include burning of fossil fuels (fuel oil, coal) in power stations, oil refineries, other large industrial plants, motor vehicles and domestic boilers [39,40]. Excessive exposure of SO$_2$ causes harms on the eye, lung and throat [41,42].

Hydrogen sulfide (H$_2$S)

It is a highly toxic, malodorous, intensely irritating gas. The maximum concentration for H$_2$S exposure for one hour without grave after-effects has been estimated as 170 to 300 ppm. Minimal exposure to H$_2$S gas causes nose/eye irritation, olfactory nerve paralysis. Moderate amount may cause sore throat, cough, keratoconjunctivitis, chest tightness and pulmonary edema. Excessive exposure causes headaches, disorientation, loss of reasoning, coma, convulsions and even death [43,44].

3. Electrochemical sensors

3.1 Basic principles

The basic principle of an electrochemical sensor is that it detects the electron that is transferred during an electrochemical reaction [45]. Basically, the principal electrochemical sensing component is an electrochemical cell, where electrolyte is in contact with its surrounding through the electrodes. A typical electrochemical sensor consist of three electrodes namely, (i) sensing electrode (SE) or working electrode (WE), (ii) counter electrode (CE), and (iii) reference electrode (RE) In an electrochemical sensor, the analyte of interest diffuses through a membrane followed by its interaction at the surface of the sensing electrode resulting in either an oxidation or reduction mechanism. This results in a current flow between the sensing and counter electrode which produces an appreciable electrical signal.

3.2 Working mechanism

On the basis of electric parameters to be measured, electrochemical sensors are classified into potentiometric, amperometric, coulometric and conductometric sensors. Amperometric sensor is based on the measurement of redox current (amperes) when electroactive species get reduced or oxidized, while keeping potential constant. In potentiometric sensor, difference of electric potential (or voltage) is measured while maintaining a constant electric current (normally nearly zero) between the two electrodes.

Coulometric sensors generate the analytical signal corresponding to the charge (coulombs) consumed involving the analyte. Conductometric sensors measures the change in resistance or the conductivity of electrolyte, while a constant alternating-current (AC) potential is maintained between the two electrodes. Amperometric and potentiometric sensors are mostly used for the detection of toxic chemicals, while coulometric and conductometric sensors are rarely applied [46-48]. The working mechanism of a simple electrochemical sensor was shown in figure 6.1.

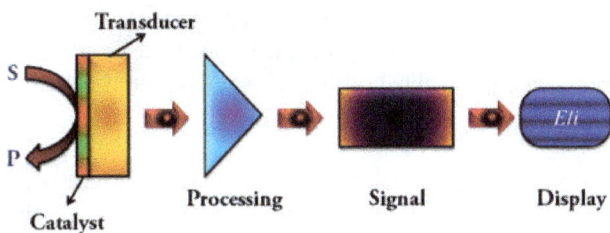

Figure 1 Working mechanism of an electrochemical sensor.

3.3   Advantages of electrochemical sensors

- Electrochemical sensor technology is an integral part of modern analytical chemistry and has captured vast attention. The potential applications of sensors can be found in different fields such as science, pharmaceuticals, medicine, analysis, chemistry, materials engineering, synthesis, molecular engineering, and biotechnology.

- Electrochemical sensors and detectors are fascinating to track down onsite pollution as well as to meet environmental requirements. This is especially important for biological samples, which are usually small in size and the injury of tissue should be reduced particularly when required for in vivo monitoring.

- Electrochemical sensors have many benefits over the conventional analytics that will lead to definite usage in the near future. They are a fascinating scientific tool because of the innate selectivity and sensitivity towards electroactive analytes, occasionally due to the accuracy and specificity, they are less time consuming, have flexibility and are easy to set up.

- They are lightweight and transferrable devices, easy to use and compact; resulting in swiftness in research data arrangement. The choice of appropriate functional principles, the design, and the material of a sensing toolbox depends on aspects such as portability, selectivity, sensitivity, the need for single or multivariate discovery, and any appropriate field applications.

- The choice of fabrication procedures and resources is important for the function of the sensors, and the performing ability of the sensors frequently depends on these important parameters. Therefore, future developments in device design should focus on developing innovative materials and novel technologies. The trends of using nanomaterials and porous particles in electrochemical sensing tools to improve selection and volatility have been progressively growing.

3.4   Working electrodes

Many solid electrodes made of gold, gold wire, platinum and certain chemically modified metals are in use to detect the above pollutants. Working electrode made of gold is observed to be superior due to its high affinity for metal ions, which can improve the pre-concentration effect. Gold-based electrodes are in limited use as sensor since undesirable gold amalgam may be formed which is known to destroy the surface features of the electrodes [49]. Gold nanoparticles (AuNPs) are richly used in analytic chemistry.

Modified electrodes are observed to be effective sensors. They serve as random arrays of microelectrodes and show distinct advantages over the conventional macro electrodes. They are characterised by increased mass transport, decreased resistance in solution, high sensitivity and better signal-to noise ratio [50]. In addition, the surface modified electrodes are miniaturized devices due to the small amount of analyte (typically less than 100 μL) [51]. Bare electrodes have been commonly used in the field of electrochemical sensors, biosensors, pesticide sensors and energy storage devices. Moreover, the unmodified electrode is chemically fouling nature, agglomeration, less electrode surface area and catalytic poisoning nature, low mechanical strength and poor chemical stability. Nanocomposites provide excellent opportunity and application in various technologically important fields such as energy generation and storage, small and medium electronics and low to high temperature applications.

4.   Nanomaterials based electrochemical sensors

The development of electrochemical sensors based on nanostructured materials has been an important area during the past few years, and it is still being growing up because of its essential importance in daily life. In the past few decades, a variety of materials including nanostructured metal oxides, nano-conducting polymers, carbon nanotubes, and their hybrid materials have been widely explored as an active material in the fabrication of chemical sensors. However, there has always been a surge for the discovery of novel nanomaterials for the development of electrochemical sensors with superior sensing performance such as high sensitivity and selectivity with fast response time [52]. The application of two-dimensional (2D) nanomaterials in chemical sensing gas gained attention in recent years because of their excellent semiconducting nature with layer-dependent physical and chemical properties and high surface-to-volume ratio [53]. Graphene, one of the first real 2D nanomaterials, has gained the attention of researchers in the scientific and industrial community since its invention in 2004.

5.   Graphene and graphene oxide

The term graphene was used for the first time in 1986. Graphene is derived from carbon and is usually formed in a two-dimensional (2D) sheet of carbon atoms in a hexagonal configuration. The carbon atoms in the structure of graphene are connected by sp^2 hybrid and each atom is bonded to three adjacent atoms with these bonds (single covalent) located at one page by 120 degree angle. According to this configuration, the carbon atoms are in weak position and forming a hexagonal page. Carbon bond length in graphene is about 0.142nm [54]. Graphene has a zero band gap and the electrons in it act like massless particles, and so it has an excellent electrical medium in 2D [55]. Over the years, research and study on graphene has shown using electron image microscopy that single layers can be observed in a larger structure between the layers of graphene [56,57]. The structures formed by graphene sheets covered with hydroxyl and epoxide functional groups are called "Graphene oxide" or "GO" [58]. These functional groups decrease the electrical and electronic transportation of carriers in GO. Therefore, it is more desirable to remove these functional groups, called "reduced graphene oxide" or "rGO". Figure 2 shows the reduction of GO from rGO lattice. In practice, when GO converts to rGO, not all of the functional groups are removed and some remain. As seen in optoelectronics applications, GO converted to rGO can be very important and so far, different research groups have presented several methods to synthesize GO and convert to rGO. Whichever method is most cost-effective, fast, large-scale, high yield, nontoxic and environmentally friendly will certainly be considered for use on an industrial scale.

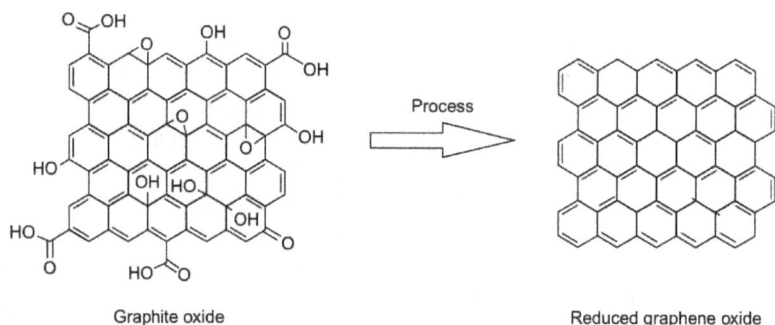

Graphite oxide   Reduced graphene oxide

Figure 2 Schematic representation of the reduction of rGO from GO.

5.1 Different methods to change GO to rGO

In recent years, many methods have been applied for the synthesis of graphene materials including epitaxial growth by chemical vapor deposition, mechanical exfoliation, liquid-phase exfoliation, chemical reduction of graphite oxide, arc discharge of graphite, in situ electron beam irradiation, epitaxial growth on SiC by a chemical vapor deposition (CVD) method, thermal fusion, laser reduction of polymers sheets, and unzipping of carbon nanotubes etc. Among these, the most common technique currently for synthesizing GO is to use graphite oxide, due to its cost-effectiveness and simplicity. [59]

Exfoliation method

Generally, exfoliation is a process in which materials expand by factors up to the hundreds along the special axis, resulting in puffed-up materials with low density and high temperature resistance. The exfoliation method for production of a high quality GO is a top-down process that is widely used in two ways, reversible and irreversible [60]. Reversible exfoliation of GO was first reported by Martin and Brocklehurst in 1964 and irreversible exfoliation of GO was first reported by Sylsworth in 1916 [61,62]. Given that exfoliation is a phase transition, which occurs at well-defined elevated temperatures, heating is always necessary. Accordingly, there are three different methods of heating for exfoliation: (1) external heating, (2) internal heating, and (3) excessive electrolytic intercalation [63]. The first two methods represent mechanical exfoliation, while the third method is called chemical exfoliation.

CVD method

In the CVD method, a bottom-up process, molecules are heated and converted to a gaseous state by chemical reaction. Depending on material quality, precursors, and the width and structure required, there are various types of CVD processes: thermal, plasma-enhanced (PECVD), cold wall, hot wall, reactive, and so on [64]. CVD growth of GO is consistently a two-step process. First, the precursor pyrolysis of materials form carbon, then, in the second step, the graphene oxide forms its carbon structure using the disassociated carbon atoms. The CVD process is reasonably straightforward, although some specialized equipment is necessary, and in order to create good quality graphene, it is important to strictly adhere to set guidelines concerning gas volumes, pressure, temperature, and time duration [65]. Reduction of GO is also an extremely vital process as it has a high impact on the quality of the rGO produced, and therefore will specify how close the rGO will come structurally to fresh graphene. The methods that reduce GO usually have a chemical, thermal or electrochemical basis. Some of these methods can produce very high quality rGO, similar to pristine graphene, but are too complex or time consuming to carry out frequently.

Thermal reduction

Thermal reduction of GO is generally done through thermal annealing and microwave and photo reduction. In the first strategy, thermal annealing can be done at both fast and slow rates. There are two drawbacks to thermal annealing. First, the temperature requires large energy consumption and critical treatment conditions. Second, if the reduction is performed to a collected GO structure, for example, a GO film, heating must be performed slowly enough to prevent the expansion of the structure; otherwise, quick heating may explode the structure as in the exfoliation of graphite oxide [66]. Microwave and photo reduction have the advantage of heating substances more uniformly and rapidly than conventional heating methods [67].

Chemical reduction

Chemical reduction of GO is often conducted at room temperature or with moderate heating and can be accomplished through chemical reagent reduction, photocatalyst reduction, electrochemical reduction, and solvothermal reduction methods. Chemical reduction methods are a cost-effective and widely available means for the mass production of rGO compared with thermal reduction. In chemical methods, a reduction agent in the form of gas or liquid is added to the GO for the elimination of functional groups [66].

Multistep reduction

Various groups have proposed multistep reduction to further optimize the reduction process for special purposes. Chemical reagents are usually not completely eliminated in the graphene after reduction, which can undermine the desired performance. Thus, additional actions like thermal annealing can be performed to achieve the desired properties.

Graphene is highly sensitive to changes in its chemical environment for several reasons: First, suspended graphene has extremely high electron mobility at room temperature, and the electron transport in graphene remains ballistic up to 0.3 μm at 300 K. Second, every carbon atom in graphene is a surface atom providing the greatest possible surface area per unit volume so that electron transport through graphene is highly sensitive to adsorbed molecular species. Third, graphene has inherently low electrical noise due to the quality of its crystal lattice and its very high electrical conductivity. These properties make graphene an ideal candidate for the ultrahigh sensitivity detection of different gases existing in various environments. High levels of sensitivity in detection processes are important for different industrial, environmental, public safety and military applications. The majority of graphene-based gas sensor work is based on changes in their electrical conductivity due to the adsorption of gas molecules on graphene's surface. These gas molecules act as donors or acceptors on graphene, similar to other solid-state sensors.

5.2   Future challenges

While pristine graphene and its derivatives such as rGO have proven to be remarkably effective in sensing trace amounts of various gas species in mixtures with air at room temperature and atmospheric pressure, there are a number of technical challenges that need to be overcome in order to enable practical applications:

- Specificity: Graphene is exquisitely sensitive to the chemical environment and hence is easily affected by a range of different gas species and mixtures. Therefore, definitive identification of contaminants is challenging. For example, exposure to NO_2 causes an increase in graphene conductance. However, a similar increase in conductance could also be generated by exposure to a different oxidizing agent such as O_2. Moreover it is possible to combine different gases in different concentrations to yield the same net change in conductance. For example, a mixture that contains an oxidizing agent such as NO_2 and a reducing agent such as NH_3 may not create any net change in the graphene's conductivity. Future research needs to focus on functionalizing the graphene with capture agents that will enable the specific binding of target gases to the graphene surface. Another approach may be to include multiple transduction mechanisms beyond simply electrical transduction. For example, optical, gravimetric (changes in graphene vibration frequencies), or ionization-based approaches (as have been successfully used with carbon nanotubes) may provide additional information that can collectively be used to pinpoint the identity of the gas specie being detected and lower the rate of false alarms.

- Reversibility: The thermal energy at room temperature is typically not enough to overcome the activation energy needed for molecular desorption for gases such as NO_2 and NH_3 on graphene surfaces. This necessitates high temperature desorption in an inert Ar or vacuum environment to clean the graphene surface and expel the chemisorbed molecules. Such treatment precludes the reversible operation of graphene devices in the field where such high-temperature annealing treatment may not be feasible. Overcoming this challenge will require innovative methods such as functionalizing graphene surfaces to control the binding energy of target molecules to the graphene surface.

- Reliability: Electrical conductivity of graphene is exquisitely sensitive to changes in the environmental conditions such as moisture, temperature, residual charge build-up, or contamination, and this creates additional difficulties for reliable and repeatable sensing. One possible solution to this issue is to use arrays/films of graphene sheets or macro-GFs that may be less susceptible to extraneous factors.

- Cost: To be competitive with commercial sensors, graphene-based devices must be mass producible at low cost. Unlike carbon nanotubes, which do not exist in nature, graphene sheets are already present in graphite. Therefore top-down methods such as exfoliation of GO could be used to mass produce graphene nanosheets at low cost. Such top-down options do not exist for most other categories of nanofillers. Moreover CVD synthesis of macro-GFs and roll-to-roll deposition of graphene on large area substrates by CVD could substantially lower the cost of graphene-based chemical sensors.

6. Metal chalcogenides

Metal chalcogenide nanomaterials are found to have various applications in photovoltaics, display devices, batteries and sensor devices as a result of their good electrical, optical and chemical properties. A chalcogenide is a chemical compound consisting of at least one chalcogen anion, such as sulfur (S), selenium (Se), and tellurium (Te), and at least one more electropositive element. Among metal chalcogenide materials, sulfur-based compounds are being extensively studied because of their high carrier mobility, large bandgap, and good photovoltaic properties [68, 69]. In addition, ternary, quaternary, or penternary chalcogenides are a new class of alternative catalytic materials because they can offer a high flexibility for tuning the bandgap without relying on toxic elements. However, metal chalcogenides would undergo photocorrosion when irradiated in the absence of sacrificial electron donors. Therefore, it processes a suitable band engineering to prohibit rapid recombination of e^-/h^+ pairs and backward reactions. There are some ways to improve the shortage of metal chalcogenide materials as listed:

- Metal ion doping: Dopants could form traps between VB and CB of the photocatalyst. They act not only as light absorption centers but also as recombination-avoiding sites by trapping electrons or holes. Therefore, dopants could enhance the charge separation required of the photocatalytic reaction.

- Nonmetal ion doping: The impurity states of nonmetal ion dopants are close to the VB. Thus, they would shift the VB edge upward, which results in a narrowing of bandgap energy.

- Dye sensitization: Dye sensitization extends the range of excitation energies of the photocatalyst into visible region, making more complete use of solar energy.

- Composite semiconductors: Combining two different semiconductors with suitable CB and VB of the photocatalyst could induce better collection of the photogenerated electrons and holes on the different semiconductor surfaces and enhance the redox reactions of the electrons and holes, respectively.

- Surface defects: Surface defects can serve as adsorption sites where a charge transfer to the adsorbed species can prevent the recombination of the e^-/h^+ pairs.

7. Graphene/metal chalcogenide nanocomposites

Graphene-based composites can be employed in both phases of the sensors demanding on the requirement. This unique property of graphene has arisen from its high electrical conductivity, better electrochemical (high electron transfer rate), mechanical and optical properties, and a very good biocompatible nature. At present, various graphene-based composites materials (metal nanoparticles, metal oxides, carbon hybrids) have been utilized to develop several kinds of electrochemical and biosensors. In this chapter, we focus specifically on the enormous potential of graphene–metal chalcogenide composites for the detection of chemicals, biomolecules, and biomacromolecules. Graphene–metal chalcogenides composites have numerous advantages in utilization as an electrochemical sensor when compared to other composite materials. The advantage arises from the hybridization of graphene and metal chalcogenides (sulfides, selenides, tellurides, and polonides but not oxides), where the high electrical conductivity and surface area of the graphene counterpart may contribute to high current; redox centers imparted by the metal and chalcogenide sites are highly beneficial for electrocatalytic sensing of variety of molecules. The schematic representation of ex-situ synthesis CdS nanosheet loaded graphene sheets was shown in figure 3.

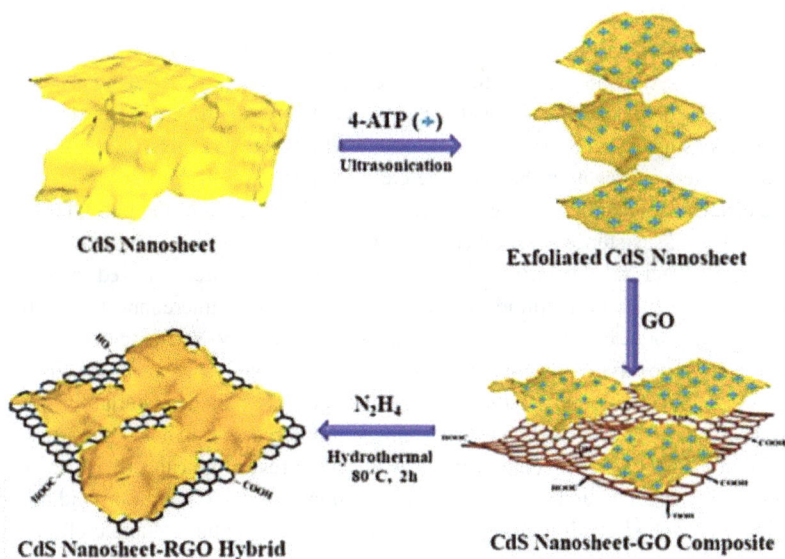

Figure 3 Ex-situ synthesis of CdS nanosheet loaded reduced graphene oxide.

Recently various types of graphene/metal chalcogenide composites were reported for various electrochemical devices. Zhang et al. synthesized noble metal-free rGO-$Zn_xCd_{1-x}S$ nanocomposite by a facile co-precipitation-hydrothermal reduction strategy and they reported the high solar photocatalytic H_2-production activity [70]. These results demonstrate that the unique features of rGO make it an excellent supporting material for $Zn_{0.8}Cd_{0.2}S$ NPs as well as a good electron collector and transporter. Another attempt was reported by Zhang et al. for copper sulfide (CuS) nanostructures [71]. Under visible light illumination, the CuS/rGO nanocomposites showed enhanced photocurrent response and improved photocatalytic activity in the degradation of methylene blue (MB) compared to the pristine CuS. In addition to CuS/ rGO nanocomposites, CuSe/rGO nanocomposites are also the best materials for photocatalytic activity. Recently, the photocatalytic performance of Cu_3Se_2 NPs and Cu_3Se_2/rGO nanocomposites via a simple co-precipitation method was reported by M. Nouri et al. [72]. They observed that rGO caused significantly enhanced photocatalytic performance. In fact, the photocatalytic activity of Cu_3Se_2/rGO nanocomposites was comparable with P-25 NPs, known as highly efficient photocatalytic materials.

Tayyebi et al. decorated PbS QDs onto a graphene surface in a semi coreshell structure using supercritical ethanol for photovoltaic cell application [73]. They reported that integrating the PbS QDs with graphene brings about two significant outcomes. First, graphene acts as a substrate for nucleation and growth of the PbS QDs, and at the same time hinders their aggregation. Second, the electron-hole pairs in the excited PbS QDs could be efficiently separated through the transport of electrons from the PbS QDs to the graphene. Overall, existence of graphene improves the optoelectronic characteristics of the PbS QDs, and the prepared PbS/graphene nanocomposites showed promising photovoltaic properties. Furthermore, Zhao et al. synthesized 3D interconnected spherical graphene framework-decorated SnS NPs (3D SnS@SG) by self-assembly of GO nanosheets and positively charged polystyrene/ SnO_2 nanospheres, followed by a controllable in situ sulfidation reaction during calcination as an anode material for lithium battery [74] and studied their electrochemical performances. Their measurements revealed that the 3D SnS@SG demonstrates an excellent reversible capacity ($800mAhg^{-1}$ after 100 cycles at 0.1°C and $527.1mAhg^{-1}$ after 300 cycles at 1°C) and outstanding rate capability ($380mAhg^{-1}$ at 5°C). They suggested that the high initial columbic efficiency, cycling stability, and rate capability of the 3D SnS@SG could be attributable to the unique structural features of the 3D interconnected spherical graphene framework and the synergistic effects of the subtle SnS NPs and 3D SG. Kumar et al. presented a sonication method to metal-free ultra-thin MoS_2 (UM) on rGO and used this composite to enhance photocatalytic hydrogen production of CdS nanorods [75]. They proposed that the excited electron could be transferred to MoS_2 nanosheets via rGO nanosheets, effectively separating the photogenerated charge carriers, and enhancing the surface shuttling properties followed by reduction of protons to molecular H_2. In-situ synthesis of CuS decorated graphene sheet was shown in figure 4.

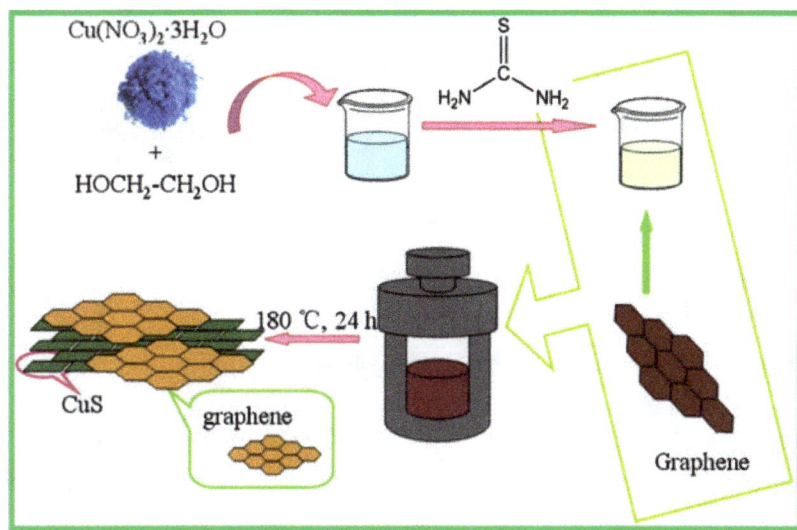

Figure 4 Schematic representation of synthesis of Graphene/CuS nanocomposites.

8. Graphene/metal chalcogenide based electrochemical sensors for toxic chemicals

8.1 MoS₂-RGO hybrids based NO₂ sensor

Wang et al. reported a novel NO_2 sensor using molybdenum disulphide nanoparticles (MoS_2 NPs) decorated RGO (MoS_2-RGO) hybrids as sensing materials, where MoS_2-RGO hybrids were prepared by a two-step wet-chemical method [76]. MoS_2 NPs prepared by modified liquid exfoliation method from bulky MoS_2 powder. Then, MoS_2-RGO hybrids were prepared by self-assembly of MoS_2 NPs and GO nanosheets, followed by a facile hydrothermal treatment progress. The schematic representation of the synthesis procedure was shown in figure 5.

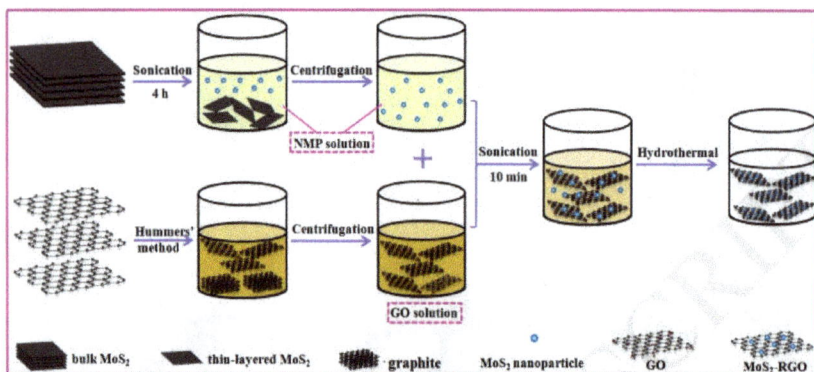

Figure 5. Schematic representation of the synthesis of MoS$_2$ decorated reduced graphene oxide.

For fabrication of gas sensor, 0.3 µL of MoS$_2$-RGO aqueous dispersion was dropped on the surface of electrode, followed by dryness at room temperature. The gas sensing properties of MoS$_2$-RGO hybrids were determined under specific condition (25% RH). The response of a sensor was defined as the ratio (response: $S = R_a/R_g$) of the sensor resistance in air (R_a) to that in the target gas (R_g). The time taken by the sensor to achieve 90% of the total resistance change was defined as the response time in the case of adsorption and recovery time in the case of desorption. The response time and recovery time of MoS$_2$-RGO-based NO$_2$ sensor are 360 s and 1155 s, respectively. The good response and recovery property of MoS$_2$-RGO-based NO$_2$ sensor is different from the previously reported sensors based on MoS$_2$-RGO hybrids, where those sensors could be not recovered [77-79]. However, the present NO$_2$ sensors show slow response and recovery rate, because of the strong adsorption of NO$_2$ on MoS$_2$ surface [80]. The strong adsorption of NO$_2$ by RGO is also the major reason for slow response and recovery rate of RGO-based room-temperature NO$_2$ sensors [81]. It is well known that increasing operating temperature could improve the response and recovery rate. But there is no tremendous change of response to NO$_2$ is observed at different operating temperature, the response time and recovery time are greatly reduced with increasing operating temperature. The response time and recovery time of MoS$_2$-RGO based sensor are 8 s and 20 s at 160 °C, which are much shorter than that of the sensors operated at 120 °C (36 s and 165 s) and 140 °C (27 s and 78 s). By further increasing operating temperature, unstable resistance value of the sensor is observed, which could be attributed to poor stability of RGO [82]. Compared to the sensing performance (1.37 KΩ) of pure RGO surface upon exposure to NO$_2$, MoS$_2$-RGO hybrids exhibit higher response. It indicates that the sensing performances of RGO-based sensor have been strongly enhanced by introduction of MoS$_2$ NPs.

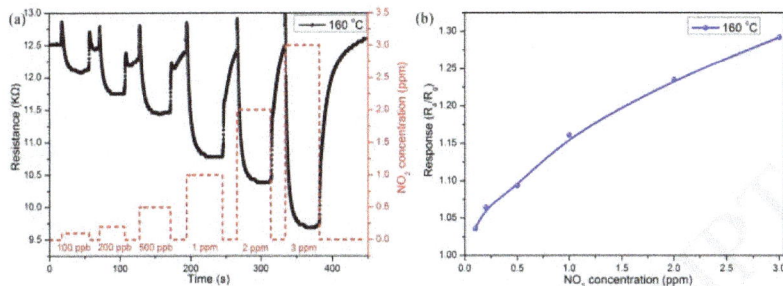

Figure 6 (a) The response and recovery curve of MoS$_2$-RGO based sensor to various NO$_2$concentrations of 100 ppb, 200 ppb, 500 ppb, 1 ppm, 2 ppm and 3 ppm, (b) the relationship line of NO$_2$ response and NO$_2$ concentration.

The decoration of MoS$_2$ can introduce more vacancies and structural defects, which can also act as adsorption sites for NO$_2$ molecules [79,80]. When MoS$_2$ content exceeds the optimal proportion, the aggregation of MoS$_2$ NPs may be formed on RGO nanosheets. As a result, the active sites could be covered by each other. Figure 6a shows a typical response curve toward different NO$_2$ concentrations from 100 ppb to 3 ppm at 160 °C, where the responses are 1.04, 1.06, 1.09, 1.16, 1.24 and 1.29, respectively, indicating that NO$_2$ sensor thus constructed can be used for detection of NO$_2$. Upon exposure to NO$_2$, the sensor resistance exhibits a pronounced decrease because NO$_2$ is a known electron acceptor due to the unpaired electron on the N atom. Upon NO$_2$ adsorption, since the electron extraction from MoS$_2$-RGO is causing a decrease in sensor resistance, indicating that MoS$_2$-RGO hybrids exhibit a p-type characteristic. The low gas response may be related to the relatively low content of defects in pure RGO and MoS$_2$-RGO hybrids, instead of the inferior crystalline quality. Figure 6b shows the relationship line between NO$_2$ responses to corresponding concentrations from 100 ppb to 3 ppm, revealing that the response of sensor increases with increasing NO$_2$ concentrations, attributed to increased absorption amount of NO$_2$ by introducing high concentration of NO$_2$. It is seen that the responses to NO$_2$ increase from 1.04 (100 ppb) to 1.29 (3 ppm) with increasing NO$_2$ concentration. Additionally, the sensors show relatively fast response rate about 8-10 s for detection of NO_2 with various concentration. In contrast, the recovery times gradually increase from 1 s (100 ppb) to 32 s (3 ppm), indicating good response and recovery rate.

The sensing property of the MoS_2-RGO- hybrid is different for different gases due to the chemical properties and reactivity of gas molecules. It is seen that the response of the sensor to NO_2 is much higher than that of other gases, including Cl_2, and NO Additionally, the response of MoS_2-RGO hybrids towards 3 ppm NO_2 is also larger than 100 ppm VOCs, including acetone, ethanol, formaldehyde, toluene and ammonia. All these observations indicate that MoS_2-RGO hybrids exhibit high selectivity for NO_2 sensing. The above results suggest that MoS_2-RGO-based sensor can steadily detect NO_2 at room temperature.

The excellent NO_2 sensing performances of MoS_2-RGO-based sensor are attributed to the synergic effect of RGO with MoS_2, and the reason for enhanced NO_2 sensing performances is possibly concluded as follow.

- At first, as a p-type semiconductor, RGO is conductive and has chemically active defect sites, making it a promising candidate as active materials in gas sensors. RGO could interact with gas molecules from weak van der Waals forces to strong covalent bonding. Both of them can alter the local carrier density of RGO, changing the electrical resistance. Besides, RGO has low electrical noise because of its 2D atomic layer structure and high-quality hexagonal lattice. Therefore, very little charge transfer between RGO and gas can induce an obvious change of its resistance. In addition, MoS_2 itself has been known to be effective in NO_2 sensing due to change in the electronic structure and charge transfer induced by NO_2 adsorption.

- It should be noted that the formation of p-n junction between p-type RGO and n-type MoS_2 plays an important role in improving the sensing performance for MoS_2-RGO-based sensor.

- The incorporation of MoS_2 NPs onto the RGO significantly decreases the conductivity of RGO, which is agreement with the resistance of MoS_2-RGO hybrids increases due to the information of p-n junction.

Figure 7a shown the schematic illustration of electron transfer from MoS_2 to RGO, which is attributed to the conduction band of MoS_2 (larger than 4.6 eV) is higher than that of work function of RGO (4.75 eV). In this work, n-type MoS_2 NPs are deposited onto the surface of p-type RGO, resulting in formation p-n heterojunction. Thus, the electrons will transfer from MoS_2 to RGO until the equilibrium of the Fermi level was achieved, as shown in Figure 7b. When the MoS_2 RGO hybrids are exposed to NO_2, the electrons will transfer from MoS_2 to NO_2, leading to a decrease of the MoS_2 carrier concentration. The NO_2 molecules converse into NO_3^- , after obtain electrons from MoS_2-RGO hybrids. By putting the sensors into air again, NO_3^- will leave from the MoS_2-RGO hybrids (Figure 7c), and NO_2 molecules are obtained after give the electrons back to hybrids (Figure 7d). Thus, the resistance of the sensors could recover back to the initial value. Thus, MoS_2-RGO hybrids can form an excellent charge transfer pathway. The charge transport quickly through the sp^2 carbon structures of the RGO, resulting in a very large and fast variation of the resistance.

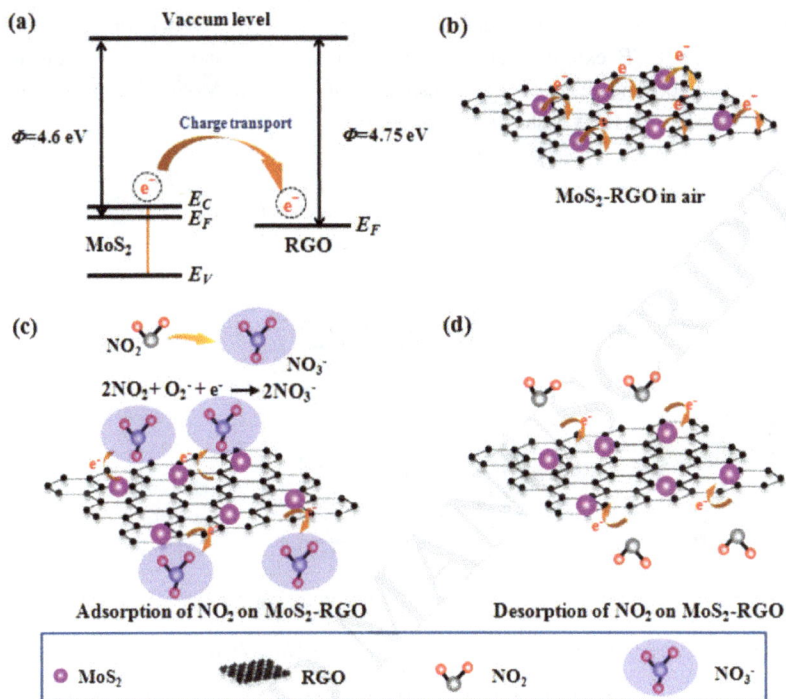

Figure 7 Schematic illustration of electron transfer process and the NO_2 sensing mechanism of MoS_2 decorated reduced graphene oxide modified working electrode.

8.2   Co doped MoSe₂ decorated graphene based sensor

The electrochemical sensor performances of Co-doped $MoSe_2$ with different dopant concentrations was studied by Ramaraj et al. using a hydrothermal method and encapsulation with GO [83]. They identified that the low concentration of Co doping does not affect the layered structural nature of $MoSe_2$. On the other hand, the higher concentration of doping partially changes the layered structure of $MoSe_2$ into a bulk rod-like structure. In the following electrochemical characterizations, the layered structure with defect/distortion due to Co doping ($CoMoSe_2$) exhibits excellent electrochemical properties in both electrochemical sensor and supercapacitor applications. GO@$CoMoSe_2$/GCE exhibited an ultralow detection limit and sensitivity toward the sensing of Metol. The electrochemical responses of Co doped $MoSe_2$ decorated graphene based sonsors for metol was shown in figure 8.

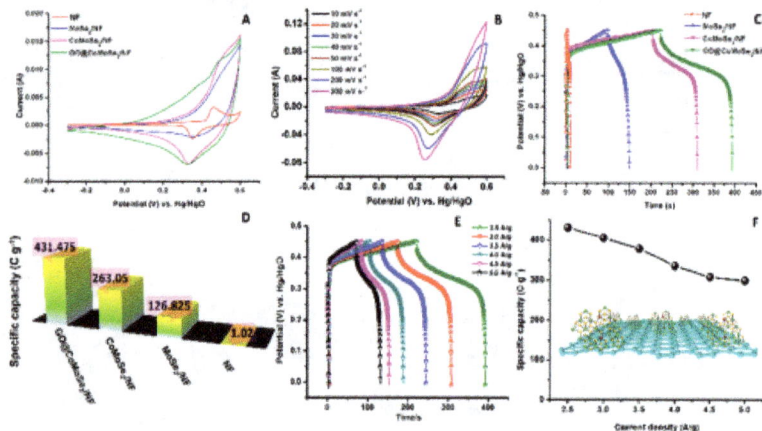

Figure 8   Electrochemical performances of GO@CoMoSe₂ modified electrodes against Metol.

According to the results, both the heterogeneous metal doping and the hybridization of carbon conductive material can effectively enhance the electrocatalytic activity and electronic conductivity of layered $MoSe_2$ for electrochemical sensor and energy storage applications. The charge transfer mechanism between the active components of the electrochemical cell was lineated in figure 9.

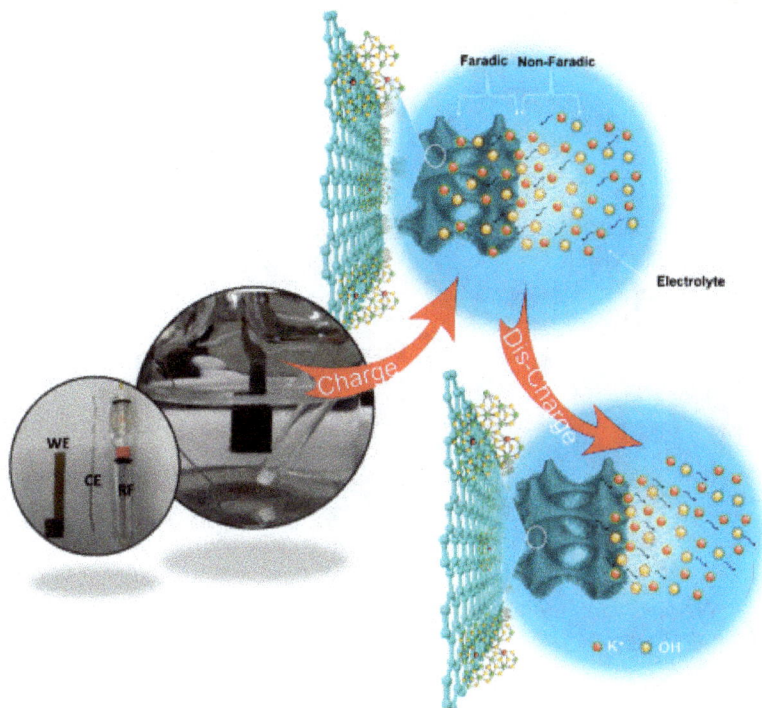

Figure 9 Charge transfer mechanism of GO@CoMoSe$_2$ electrodes.

An et al. developed an electrochemical sensor for the detection of H_2O_2 based on CdS nanorods/RGO composites by a simple one-pot hydrothermal reaction [84]. The inhibited recombination of electron–hole pairs and the promoted separation and transport of charge carriers were detected in the time-resolved fluorescence decays and electrochemical measurements. The promising application as an enzyme-free biosensor for H_2O_2 was also selected as a preliminary test of the electrocatalytic behavior of CdS nanorods/RGO composites. The CdS nanorods/RGO composite sensors exhibited superior analytical performance compared to blank CdS nanorods and CdS nanoparticles/RGO composites. These sensors are low cost and show a rapid and highly sensitive response. This method opens up a facile route to preparing graphene-based nanocomposites with excellent photocatalytic and electrocatalytic properties.

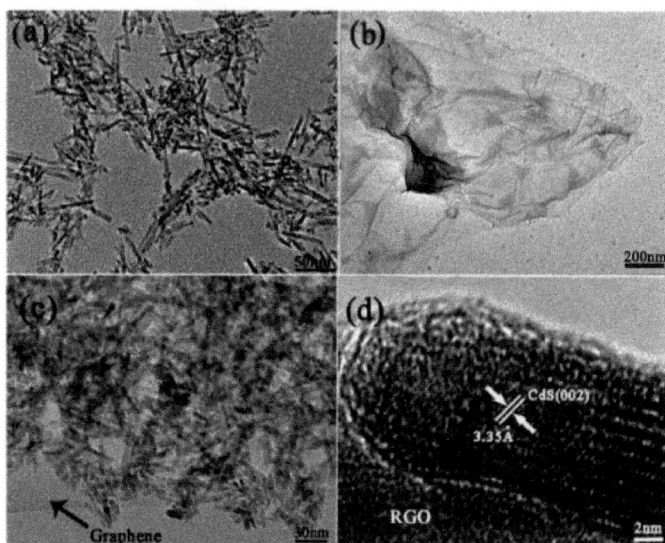

Figure 10 TEM images of (a)- CdS nanorods, (b)- Graphene, (c)- CdS nanorods loaded graphene sheets and (d)- HRTEM images of CdS nanorods loaded graphene sheets.

The gas (NO_2) sensing property of graphene oxide (rGO)/molybdenum disulfide (MoS_2) composites was revealed by Niu et al. [85]. They have developed a facile method to synthesize a fiber-like reduced graphene oxided (rGO)/molybdenum disulfide (MoS_2) composites through a wet-spinning and a hydrothermal method. In the presence of Dodecyl dimethyl benzyl ammonium chloride, graphene provides a substrate for nucleation of MoS_2. In this composite, MoS_2 nanosheets can be anchored onto the surface of graphene through both physical adsorption and electron transfer by hydrothermal method. This process inhibits both graphene and MoS_2 layers from stacking together. The gas sensing properties are evaluated in an intelligent gas sensing analysis system. It is demonstrated that the obtained composite fiber devices show an excellent sensitivity and selectivity to NO_2 and NH_3 than the individual components in different light illumination conditions and the proper proportion of components was detected. Various types of graphene/metal chalcogenide based nanocomposites were proposed as the electrode materials for the electrochemical sensing of various toxic chemicals and heavy metals.

Table 2 summarises the recently reported electrochemical sensors based on graphene/metal chalcogenides for the detection of toxic chemicals.

Table 2: Graphene/metal chalcogenides based electrochemical sensors for the detection of toxic chemicals and heavy metal ions.

S.No.	Electrode material	Analyte	Technique	Linear range (μM)	LoD (nM)	Ref.
1.	Au NPs@ MoS_2-rGO/GCE	Hydroquinone Catechol resorcinol	voltammetric	0.1–950 3–560 40–960	40 95 146	[86]
2.	CdS/RGO	CO_2	Potentiometry	200-1000 (ppm)	-	[87]
3.	CuS/RGO	H_2O_2	Amperometric	5-1500	-	[88]
4.	CdS/RGO	Cu^{2+}	voltammetric	0.5–120	16	[89]
5.	$Cd_xZn_{1-x}S$/ rGO	Cu^{2+}	voltammetric	0.02–20	67	[90]
6.	CuS| rGO	Hydrazine	Amperometric	1-1000	300	[91]
7.	CuS-frGO	Cr (IV)	voltammetric	0-200	26.60	[92]
8.	Gr/MoS_2/MWCNT	H_2O_2	amperometric	5-145	830	[93]
9.	MoS_2/GO/myoglobin hybrid	NO_2	amperometric	1-3600	360	[94]
10.	rGO-MoS_2/GCE	NO	amperometric	0.2–4800	170	[95]
11.	MoS_2/Gr	Methyl parathion	amperometric	0.01-1900	3.2	[96]
12.	rGO-MoS_2-PEDOT	NO	amperometric	1-1000	590	[97]
13.	CuS/GO/MWCNTs	H_2O_2	amperometric	450-6000	600	[98]
14.	rGO-MoS_2	NO_2	amperometric	1–104 Ppm	0.147 ppm	[99]

Conclusion

This chapter has discussed and highlighted the role of Graphene/metal chalcogenides in electrochemical sensors for the detection of toxic chemicals and environmental pollutants. Electrochemical sensors based on graphene/metal chalcogenides provide a crucial analytical tool, as the demand for the ultrasensitive, rapid, and selective determination of toxic chemical increases. They may be easily adapted to selective and multiple sensing a range of chemical contaminants inexpensively, in contrast to spectroscopic and chromatographic instrumentation. Some unique physicochemical and electrochemical properties of graphene and metal chalcogenides may facilitate the fabrication of potentially robust electrochemical sensor systems, with attributes such as sensitivity and efficient. The architectures of graphene/metal chalcogenides interfaces also afford a suitable environment for detection and improvements in sensitivity and long-term stability. Additionally, graphene/metal chalcogenides enabled sensors are capable of being embedded into/integrated with portable or miniaturized devices for specific applications, due to the selective catalytic activity, stability and biocompatibility of the sensor design.

References

[1] L. P. Martin, A.-Q. Pham, R. S. Glass, Electrochemical hydrogen sensor for safety monitoring, Solid State Ionics, 175 (2004) 527-530. https://doi.org/10.1016/j.ssi.2004.04.042

[2] M. Suresh, N. J. Vasa, V. Agarwal, J. Chandapillai, UV photo-ionization based asymmetric field differential ion mobility sensor for trace gas detection, Sens. Actuator B-Chem., 195 (2014) 44-51. https://doi.org/10.1016/j.snb.2014.01.008

[3] C. Di Natale, V. Ferrari, A. Ponzoni, G. Sberveglieri, and M. Ferrari, Sensors and Microsystems: Proceedings of the 17th National Conference, Brescia, Italy, 5-7 February 2013, 268 (2013) Springer Science & Business Media. https://doi.org/10.1007/978-3-319-00684-0

[4] C. Li and G. Shi, Carbon nanotube-based fluorescence sensors J. Photochem. Photobiol., C: Photochem. Rev., 19 (2014) 20-34. https://doi.org/10.1016/j.jphotochemrev.2013.10.005

[5] Q. Zhu and R. C. Aller, Planar fluorescence sensors for two-dimensional measurements of H_2S distributions and dynamics in sedimentary deposits, Mar. Chem., 157 (2013) 49-58. https://doi.org/10.1016/j.marchem.2013.08.001

[6] R. Jackson, R. P. Oda, R. K. Bhandari, S. B. Mahon, M. Brenner, G. A. Rockwood, and B.A. Logue, Development of a Fluorescence-Based Sensor for Rapid Diagnosis of Cyanide Exposure, Anal. Chem., 86 (2014) 1845-1852. https://doi.org/10.1021/ac403846s

[7] K. Zakrzewska, Mixed oxides as gas sensors, Thin solid films, 391 (2001) 229-238. https://doi.org/10.1016/S0040-6090(01)00987-7

[8] Z. Darmastuti, C. Bur, P. Möller, R. Rahlin, N. Lindqvist, M. Andersson, A. Schütze, A. Lloyd Spetz, SiC–FET based SO_2 sensor for power plant emission applications Sens. Actuator B-Chem., 194 (2014) 511-520. https://doi.org/10.1016/j.snb.2013.11.089

[9] N. S. Lawrence, L. Jiang, T. G. Jones, R. G. Compton, A Thin-Layer Amperometric Sensor for Hydrogen Sulfide: The Use of Microelectrodes To Achieve a Membrane-Independent Response for Clark-Type Sensors, Anal. Chem., 75 (2003) 2499-2503. https://doi.org/10.1021/ac0206465

[10]   E. Jones, Techniques and mechanism in gas sensing., (1987) 17.

[11]   J. W. Fergus, Perovskite oxides for semiconductor-based gas sensors, Sens. Actuator B-Chem., 123 (2007) 1169-1179. https://doi.org/10.1016/j.snb.2006.10.051

[12]   X. Liu, S. Cheng, H. Liu, S. Hu, D. Zhang, and H. Ning, A Survey on Gas Sensing Technology, Sensors, 12 (2012) 9635-9665. https://doi.org/10.3390/s120709635

[13]   A. W. E. Hodgson, P. Jacquinot, L. R. Jordan, and P. C. Hauser, Amperometric gas sensors with detection limits in the low ppb range, Anal Chim Acta, 393 (1999) 43-48. https://doi.org/10.1016/S0003-2670(99)00189-0

[14]   E. Bakker and M. Telting-Diaz, Electrochemical Sensors Anal. Chem., 74 (2002) 2781-2800. https://doi.org/10.1021/ac0202278

[15]   G. Alberti, F. Cherubini, R. Palombari, Amperometric solid-state sensor for NO and NO_2 based on protonic conduction, Sens. Actuator B-Chem., 37 (1996) 131-134. https://doi.org/10.1016/S0925-4005(97)80127-X

[16]   K.J. Huang, L. Wang, Y.J. Liu, T. Gan, Y.M. Liu, L.L. Wang, Y. Fan, Synthesis and electrochemical performances of layered tungsten sulfide-graphene nanocomposite as a sensing platform for catechol, resorcinol and hydroquinone, Electrochim. Acta, 107 (2013) 379-387. https://doi.org/10.1016/j.electacta.2013.06.060

[17]   R.V. Burg, T. Stout, Hydrazine, J. Appl. Toxicol., 11 (1991) 447-450. https://doi.org/10.1002/jat.2550110612

[18]   D.P. Elder, D. Snodin, A. Teasdale, Control and analysis of hydrazine, hydrazides and hydrazones-Genotoxic impurities in active pharmaceutical ingredients (APIs) and drug products J. Pharm. Biomed. Anal., 54 (2011) 900-910. https://doi.org/10.1016/j.jpba.2010.11.007

[19]   U. Ragnarsson, Synthetic methodology for alkyl substituted hydrazines, Chem. Soci. Review, 30 (2001) 205-213. https://doi.org/10.1039/b010091a

[20]   U.S. Environmental Protection Agency (EPA), Integrated Risk Information System (IRIS) on hydrazine, Hydrazine sulfate: National centre for environmental assessment, office of research and development: Washington, DOC, 1999. and development: Washington, DOC, 1999.

[21]   A. Agarwal, P.G. Tratnyek, Reduction of Nitro Aromatic Compounds by Zero-Valent Iron Metal, Environ. Sci. Technol., 30 (1996) 153-160. https://doi.org/10.1021/es950211h

[22]   Ambient water quality criteria for nitrobenzene (EP1.29/3:N63/3) in: U.S.E.P, Agency (Ed.), Washington, DC: (1980).

[23]   Y. Mu, R.A. Rozendal and K. Rabaey, J. Keller, Nitrobenzene Removal in Bioelectrochemical Systems, J. Environ. Sci., 43 (2009) 8690-8695. https://doi.org/10.1021/es9020266

[24]   C. Schummer, C. Groff, J.A. Chami, F. Jaber and M. Millat, Analysis of phenols and nitrophenols in rainwater collected simultaneously on an urban and rural site in east of France, Sci. Total Environ., 407 (2009) 5637-5643. https://doi.org/10.1016/j.scitotenv.2009.06.051

[25]   X.M. Xu, Z. Liu, X. Zhang, S. Duan, S. Xu, C.L. Zhou, β-Cyclodextrin functionalized mesoporous silica for electrochemical selective sensor: Simultaneous determination of nitrophenol isomers, Electrochim. Acta, 58 (2011) 142- 149. https://doi.org/10.1016/j.electacta.2011.09.015

[26]   S. Wu, F. Huang, X. Lan, X. Wang, J. Wang and C. Meng, Electrochemically reduced graphene oxide and Nafion nanocomposite for ultralow potential detection of organophosphate pesticide, Sens. Actuators, B, 2013, 177 (2013) 724-729. https://doi.org/10.1016/j.snb.2012.11.069

[27]   A.H.A. Hassan, S.L. Moura, F.H.M. Ali, W.A. Moselhy, M.I. Pividori, Electrochemical sensing of methyl parathion on magnetic molecularly imprinted polymer, Biosens. Bielectronics, 118 (2018) 181-187. https://doi.org/10.1016/j.bios.2018.06.052

[28]   J.H. Duffus, Heavy metals: A meaningless term? Pure Appl. Chem. 74 (2002) 793-807. https://doi.org/10.1351/pac200274050793

[29]   G. Aragay, F. Pino, A. Merkoci, Nanomaterials for sensing and destroying pesticides. Chem. Rev. 112 (2012) 5317-5338. https://doi.org/10.1021/cr300020c

[30]   H.J. Huang, Y.K. Choi, Chemical sensors based on nanostructured materials. Sens. Actuators, B 122 (2017) 659-671. https://doi.org/10.1016/j.snb.2006.06.022

[31]   M. Li, H. Gou, I. Al-Ogaidi, N. Wu, Nanostructured Sensors for Detection of Heavy Metals: A Review, ACS Sustainable Chem. Eng., 1 (2013) 713-723. https://doi.org/10.1021/sc400019a

[32]   D.W. Kimmel, G. LeBlanc, M.E. Meschievitz, D.E. Cliffel, Electrochemical sensors and biosensors. Anal. Chem. 84 (2012) 685- 707. https://doi.org/10.1021/ac202878q

[33]   L. Rassaei, F. Marken, M. Sillanpää, M. Amiri, C.M. Cirtiu, M. Sillanpää, Nanoparticles in electrochemical sensors for environmental monitoring. Trends Anal. Chem. 30 (2011) 1704-1714. https://doi.org/10.1016/j.trac.2011.05.009

[34]   T. Kuila, S. Bose, P. Khanra, A.K. Mishra, N.H. Kim, J.H. Lee, Recent advances in graphene-based biosensors. Biosens. Bioelectron. 26 (2011) 4637-4648. https://doi.org/10.1016/j.bios.2011.05.039

[35]   M.A. Bauer, M.J. Utell, P.E. Morrow, D.M. Speers, F.R. Gibb, Inhalation of 0.30 ppm nitrogen dioxide potentiates exercise-induced bronchospasm in asthmatics. Am. Rev. Respir. Dis., 134 (1986)1203–1208.

[36]   Ehrlich, R. Effect of nitrogen dioxide on resistance to respiratory infection. Bacteriol. Rev., 30 (1966) 604–614. https://doi.org/10.1128/MMBR.30.3.604-614.1966

[37]   S. Genc, Z. Zadeoglulari, S.H. Fuss, K. Genc, The adverse effects of air pollution on the nervous system. J. Toxicol. 2012, 2012. https://doi.org/10.1155/2012/782462

[38]   S.W. Lee, W. Lee, Y. Hong, G. Lee, D.S. Yoon, Recent advances in carbon material-based NO_2 gas sensors. Sens. Actuators B Chem., 255 (2018) 1788–1804. https://doi.org/10.1016/j.snb.2017.08.203

[39]   Y.Y. Guo, Y.R. Li, T.Y. Zhu, M. Ye, Investigation of SO_2 and NO adsorption species on activated carbon and the mechanism of NO promotion effect on SO_2, Fuel 143 (2015) 536–542. https://doi.org/10.1016/j.fuel.2014.11.084

[40]   C. Chatterjee, A. Sen, Sensitive colorimetric sensors for visual detection of carbon dioxide and sulfur dioxide, J. Mater. Chem. A 3 (2015) 5642–5647. https://doi.org/10.1039/C4TA06321J

[41] R.R. Khan, M.J.A. Siddiqui, Review on effects of Particulates; Sulfur Dioxide and Nitrogen Dioxide on Human Health. Int. Res. J. Environment Sci. 3 (2014) 70–73.

[42] J. Nisar, Z. Topalian, A.D. Sarkar, L. Osterlund, R. Ahuja, TiO_2-based gas sensor: A possible application to SO_2. ACS Appl. Mater. Interfaces 2013, 5 (2013) 8516–8522. https://doi.org/10.1021/am4018835

[43] C. Chou, Hydrogen Sulfide: Human Health Aspects: Concise International Chemical Assessment Document 53; World Health Organization: Geneva, Switzerland, 2003.

[44] A.D. Wiheeb, I.K. Shamsudin, M.A. Ahmad, M.N. Murat, J. Kim, M.R. Othman, Present Technologies for Hydrogen Sulfide Removal from Gaseous Mixtures. Rev. Chem. Eng. 29 (2013) 449–470. https://doi.org/10.1515/revce-2013-0017

[45] L. Xiong and R. G. Compton, Amperometric Gas detection: A Review, Int. J. Electrochem. Sci., 9 (2014) 7152-7181.

[46] F. Opekar and K. Štulík, Encyclopedia of Analytical Chemistry., (2001).

[47] D. S. Silvester, Recent advances in the use of ionic liquids for electrochemical sensing, Analyst, 136 (2011) 4871-4882. https://doi.org/10.1039/c1an15699c

[48] M. Ahmad Khan, F. Qazi, Z. Hussain, M.U. Idrees, S. Shahid, S. Saeeda, Review Recent Trends in Electrochemical Detection of NH_3, H_2S and NO_x Gases, Int. J. Electrochem. Sci., 12 (2017) 1711 – 1733. https://doi.org/10.20964/2017.03.76

[49] C. Welch, O. Nekrassova, X. Dai, M. Hyde, R. G. Compton, Fabrication, Characterisation and Voltammetric Studies of Gold Amalgam Nanoparticle Modified Electrodes Chem. Phys. Chem. 5 (2004) 1405–1410. https://doi.org/10.1002/cphc.200400263

[50] J. Cassidy, J. Ghoroghchian, F. Sarfarazi, J.J. Smith,. Pons, Simulation of edge effects in electroanalytical experiments by orthogonal collocation—VI. Cyclic voltammetry at ultramicroelectrode ensembles, Electrochim. Acta, 31 (1986) 629-636. https://doi.org/10.1016/0013-4686(86)87029-3

[51] L. Shen, G.R. Zhang, W. Li, M. Biesalski, B.J.M. Etzold, Modifier-Free Microfluidic Electrochemical Sensor for Heavy-Metal Detection, ACS Omega, 2 (2017)593-4603. https://doi.org/10.1021/acsomega.7b00611

[52] R. Ramachandran T-W Chen, S-M Chen, T. Baskar, R. Kannan, P. Elumalai, P. Raja, T. Jeyapragasam, K. Dinakaran, G. Gnanakumar, Review of Advance Developments on Electrochemical Sensors for Detection of Toxic and Bioactive

Molecules, Inorganic Chemistry Frontiers, 2019.
https://doi.org/10.1039/C9QI00602H

[53]   S. S. Varghese, S. H. Varghese, S. Swaminathan, K. K. Singh, and V. Mittal, Two-Dimensional Materials for Sensing: Graphene and BeyondElectronics 4 (2015) 651. https://doi.org/10.3390/electronics4030651

[54]   D.V. Kosynkin, et al., Longitudinal unzipping of carbon nanotubes to form graphene nanoribbons, Nature 458 (7240) (2009) 872–876. https://doi.org/10.1038/nature07872

[55]   N. Yang, et al., Two-dimensional graphene bridges enhanced photoinduced charge transport in dye-sensitized solar cells, ACS Nano 4 (2) (2010) 887–894. https://doi.org/10.1021/nn901660v

[56]   R. Heyrovska, Atomic structures of graphene, benzene and methane with bond lengths as sums of the single, double and resonance bond radii of carbon, arXiv (2008). Preprint arXiv:0804.4086.

[57]   Wikipedia, Available from: Wikipedia.org.

[58]   B.C. Brodie, On the atomic weight of graphite, Philos. Trans. R. Soc. Lond. 149 (1859) 249–259. https://doi.org/10.1098/rstl.1859.0013

[59]   S.R. Kumar, et al., Graphene oxide: strategies for synthesis, reduction and frontier applications, RSC Adv. 6 (2016) 64993–65011. https://doi.org/10.1039/C6RA07626B

[60]   D.D.L. Chung, Exfoliation of graphite, J. Mater. Sci. 22 (12) (1987) 4190–4198. https://doi.org/10.1007/BF01132008

[61]   W.H. Martin, J.E. Brocklehurst, The thermal expansion behaviour of pyrolytic graphitebromine residue compounds, Carbon 1 (2) (1964) 133–134. https://doi.org/10.1016/0008-6223(64)90067-3

[62]   S.H. Anderson, D.D.L. Chung, Exfoliation of single crystal graphite and graphite fibers intercalated with halogens, Synth. Met. 8 (3–4) (1983) 343–349. https://doi.org/10.1016/0379-6779(83)90118-2

[63]   D.D.L. Chung, L.W. Wong, Electromechanical behavior of graphite intercalated with bromine, Carbon 24 (5) (1986) 639–647. https://doi.org/10.1016/0008-6223(86)90154-5

[64]   M.S.A. Bhuyan, et al., Synthesis of graphene, Int. Nano Lett. 6 (2) (2016) 65–83. https://doi.org/10.1007/s40089-015-0176-1

[65]   Graphenea, Available from: www.graphenea.com.

[66]   S. Pei, H.-M. Cheng, The reduction of graphene oxide, Carbon 50 (9) (2012) 3210–3228. https://doi.org/10.1016/j.carbon.2011.11.010

[67]   Y. Zhu, et al., Microwave assisted exfoliation and reduction of graphite oxide for ultracapacitors, Carbon 48 (7) (2010) 2118–2122. https://doi.org/10.1016/j.carbon.2010.02.001

[68]   D. Aldakov, A. Lefranc,ois, P. Reiss, Ternary and quaternary metal chalcogenide nanocrystals: synthesis, properties and applications, J. Mater. Chem. C 1 (2013) 3756-3776. https://doi.org/10.1039/c3tc30273c

[69]   J. Theerthagiri, K. Karuppasamy, G. Durai, A.H.S. Rana, P. Arunachalam, K. Sangeetha, P. Kuppusami, H-S. Kim, Recent Advances in Metal Chalcogenides (MX; X = S, Se) Nanostructures for Electrochemical Supercapacitor Applications: A Brief Review, Nanomaterials, 2018, 8 (2018) 256. https://doi.org/10.3390/nano8040256

[70]   J. Zhang, et al., Noble metal-free reduced graphene oxide-$Zn_xCd_{1-x}S$ nanocomposite with enhanced solar photocatalytic H2-production performance, Nano Lett. 12 (9) (2012) 4584–4589. https://doi.org/10.1021/nl301831h

[71]   Y. Zhang, et al., Biomolecule-assisted, environmentally friendly, one-pot synthesis of CuS/ reduced graphene oxide nanocomposites with enhanced photocatalytic performance, Langmuir 28 (35) (2012) 12893–12900. https://doi.org/10.1021/la303049w

[72]   M. Nouri, et al., High solar-light photocatalytic activity of using Cu_3Se_2/rGO nanocomposites synthesized by a green co-precipitation method, Solid State Sci. 73 (2017) 7–12. https://doi.org/10.1016/j.solidstatesciences.2017.09.001

[73]   A. Tayyebi, et al., Supercritical synthesis and characterization of graphene–PbS quantum dots composite with enhanced photovoltaic properties, Ind. Eng. Chem. Res. 54 (30) (2015) 7382–7392. https://doi.org/10.1021/acs.iecr.5b00008

[74]   B. Zhao, B. Zhao, Three-dimensional interconnected spherical graphene framework/SnS nanocomposite for anode material with superior lithium storage performance: complete reversibility of Li_2S, ACS Appl. Mater. Interfaces 9 (2) (2017) 1407–1415. https://doi.org/10.1021/acsami.6b10708

[75]   D.P. Kumar, et al., Ultrathin MoS_2 layers anchored exfoliated reduced graphene oxide nanosheet hybrid as a highly efficient cocatalyst for CdS nanorods towards

enhanced photocatalytic hydrogen production, Appl. Catal. B Environ. 212 (2017) 7–14. https://doi.org/10.1016/j.apcatb.2017.04.065

[76]   Z. Wang, T. Zhang, C. Zhao, T. Han, T. Fei, S. Liu, G. Lu, Rational synthesis of molybdenum disulfide nanoparticles decorated reduced graphene oxide hybrids and their application for high-performance NO_2 sensing, Sensors and Actuators B: Chemical, 2017, Accepted Manuscript. https://doi.org/10.1016 /j.snb.2017.12.181

[77]   Y. Niu, R. Wang, W. Jiao, G. Ding, L. Hao, F. Yang, X. He, MoS2 graphene fiber based gas sensing devices, Carbon 95 (2015) 34-41. https://doi.org/10.1016/j.carbon.2015.08.002

[78]   B. Cho, J. Yoon, S.K. Lim, A.R. Kim, D.-H. Kim, S.-G. Park, J.-D. Kwon, Y.-J. Lee, K.-H. Lee, B.H. Lee, H.C. Ko, M.G. Hahm, Chemical sensing of 2D Graphene/MoS2 heterostructure device, ACS Appl. Mater. Interfaces 7 (2015) 16775-16780. https://doi.org/10.1021/acsami.5b04541

[79]   Y. Niu, W. Jiao, R. Wang, G. Ding, Y. Huang, Hybrid nanostructures combining graphene-MoS2 quantum dots for gas sensing, J. Mater. Chem. A 4 (2016) 8198-8203. https://doi.org/10.1039/C6TA03267B

[80]   H. Long, A. Harley-Trochimczyk, T. Pham, Z. Tang, T. Shi, A. Zettl, C. Carraro, M. A. Worsley, R. Maboudian, High surface area MoS_2/graphene hybrid aerogel for ultrasensitive NO2 detection, Adv. Funct. Mater. 26 (2016) 5158-5165. https://doi.org/10.1002/adfm.201601562

[81]   L.T. Duy, D.-J. Kim, T.Q. Trung, V.Q. Dang, B.-Y. Kim, H.K. Moon, N.-E. Lee, High performance three-dimensional chemical sensor platform using reduced graphene oxide formed on high aspect-ratio micro-pillars, Adv. Funct. Mater. 25 (2015) 883-890. https://doi.org/10.1002/adfm.201401992

[82]   J. Wu, K. Tao, Y.Y. Guo, Z. Li, X.T. Wang, Z.Z. Luo, S.L. Feng, C.L. Du, D. Chen, J.M. Miao, L.K. Norford, A 3D chemically modified graphene hydrogel for fast, highly sensitive, and selective gas sensor, Adv. Sci. 4 (2017) 1600319. https://doi.org/10.1002/advs.201600319

[83]   S. Ramaraj, M. Sakthivel, S-M Chen, K-C. Ho, Active-Site-Rich 1T-Phase $CoMoSe_2$ Integrated Graphene Oxide Nanocomposite as an Efficient Electrocatalyst for Electrochemical Sensor and Energy Storage Applications, Anal. Chem., 91 (2019) 8358-8365. https://doi.org/10.1021/acs.analchem.9b01152

[84]   X. An, X. Yu, J. C. Yu, G. Zhang, CdS nanorods/reduced graphene oxide nanocomposites for photocatalysis and electrochemical sensing, J. Mater. Chem. A, 1 (2013) 5158-5164. https://doi.org/10.1039/c3ta00029j

[85]   Y. Niu, R. Wang, W. Jiao, G. Ding, L. Hao, F. Yang, X. He, MoS_2 graphene fiber based gas sensing devices, Carbon 95 (2015) 34-41. https://doi.org/10.1016/j.carbon.2015.08.002

[86]   G. Ma, H. Xu, M. Wu, L. Wang, J. Wu, F. Xu, A hybrid composed of MoS_2, reduced graphene oxide and gold nanoparticles for voltammetric determination of hydroquinone, catechol, and resorcinol, Microchimica Acta, 186 (2019), 689. https://doi.org/10.1007/s00604-019-3771-4

[87]   A. Hasani, H. S. Dehsari, A. A. Zarandi, A. Salehi, F. A. Taromi, H. Kazeroni, Visible Light-Assisted Photoreduction of Graphene Oxide Using CdS Nanoparticles and Gas Sensing Properties, Journal of Nanomaterials , 2015 (2015) 930306-11. https://doi.org/10.1155/2015/930306

[88]   Jing Bai, and Xiue Jiang, A Facile One-pot Synthesis of Copper Sulfide Decorated Reduced Graphene Oxide Composites for Enhanced Detecting of H_2O_2 in Biological Environments, Analytical Chemistry, 2013. https://doi.org/10.1021/ac400659u

[89]   I Ibrahim, H. N Lim, N. M Huang, A Pandikumar, Cadmium Sulphide-Reduced Graphene Oxide-Modified Photoelectrode-Based Photo-electrochemical Sensing Platform for Copper(II) Ions, PLoS ONE , 11(2016) e0154557. https://doi.org/10.1371/journal.pone.0154557

[90]   Yan J, Wang K, Liu Q, Qian J, Dong X, Liu W, et al. One-pot synthesis of $Cd_xZn_{1-x}S$–reduced graphene oxide nanocomposites with improved photoelectrochemical performance for selective determination of Cu^{2+}, RSC Advances, 3(2013) 14451–7. https://doi.org/10.1039/c3ra41118d

[91]   Y.J. Yang, W. Li, X. Wu, Copper sulphide/reduced graphene oxide nanocomposite for detection of hydrazine and hydrogen peroxide at low potential in neutral medium, Electrochimica Acta, 123 (2014) 260–267. https://doi.org/10.1016/j.electacta.2014.01.046

[92]   P. Borthakur, M. R. Das, S. Szunerits, R Boukherroub, CuS Decorated Functionalized Reduced Graphene Oxide: A Dual Responsive Nanozyme for Selective Detection and Photoreduction of Cr(VI) in an Aqueous Medium, ACS Sustainable Chem. Eng., 7 (2019) 16131−16143. https://doi.org/10.1021/acssuschemeng.9b03043

[93] M. Govindasamy, V. Mani, S.M. Chen, R. Karthik, K. Manibalan, R. Umamaheswari, MoS_2 Flowers Grown on Graphene/Carbon Nanotubes: A Versatile Substrate for Electrochemical Determination of Hydrogen Peroxide. Int. J. Electrochem. Sci., 11 (2016) 2954–2961. https://doi.org/10.20964/110402954

[94] J. Yoon, J.W. Shin, J. Lim, M. Mohammadniaei, G.B. Bapurao, T. Lee, J.W. Choi, Electrochemical nitric oxide biosensor based on amine-modified MoS_2/graphene oxide/ myoglobin hybrid. Colloid Surf. B-Biointerfaces, 159 (2017) 729–736. https://doi.org/10.1016/j.colsurfb.2017.08.033

[95] J. Hu, J. Zhang, Z.T. Zhao, J. Liu, J.F. Shi, G. Li, P.W. Li, W.D. Zhang, K. Lian, S. Zhuiykov, S. Synthesis and electrochemical properties of rGO-MoS_2 heterostructures for highly sensitive nitrite detection. Ionics, 24 (2018) 577–587. https://doi.org/10.1007/s11581-017-2202-y

[96] M. Govindasamy, S.M. Chen, V. Mani, M. Akilarasan, S. Kogularasu, B. Subramani, Nanocomposites composed of layered molybdenum disulfide and graphene for highly sensitive amperometric determination of methyl parathion. Microchim. Acta, 184 (2017) 725–733. https://doi.org/10.1007/s00604-016-2062-6

[97] R. Madhuvilakku, S. Alagar, R. Mariappan, S. Piraman, Glassy Carbon Electrodes Modified with Reduced Graphene Oxide-MoS_2-Poly (3, 4- Ethylene Dioxythiophene) Nanocomposites for the Non-Enzymatic Detection of Nitrite in Water and Milk, Analytica Chimica Acta, 2019, Accepted Manuscript. https://doi.org/10.1016/j.aca.2019.09.043

[98] W. Jin, Y. Fu, W. Cai, In-situ growth of CuS decorated graphene oxide-multiwalled carbon nanotubes for the ultrasensitive H_2O_2 detection in alkaline solution, New Journal of Chemistry, 2019, Accepted Manuscript. https://doi.org/10.1039/C8NJ06134C

[99] A. Mukherjee, L. R. Jaidev, K. Chatterjee, A. Misra, Nanoscale heterojunctions of rGO-MoS_2 composites for nitrogen dioxide sensing at room temperature, Nano Express 1 (2020) 010003. https://doi.org/10.1088/2632-959X/ab7491

Chapter 7

Graphene-Polymer based Nanocomposites for Electrochemical Sensing of Toxic Chemicals

A.R. Marlinda[1]*, M.R. Johan[1]

[1]Nanotechnology and Catalysis Research Centre, University of Malaya, 50603, Kuala Lumpur, Malaysia

marlinda@um.edu.my

Abstract

This chapter (with 94 refs.) gives an overview of the progress in the past few years on the development of graphene-polymer based for use in sensors and analytical tools for the determination of toxic chemicals. Sensing of toxic molecules is critical to environmental monitoring, control of chemical processes, agricultural, and medical applications. In particular, the detection of heavy metal ions such as mercury, cadmium, arsenic, chromium, thallium and lead which are extremely harmful pollutants in the biosphere due to their toxicity and even trace amounts of them pose a detrimental risk to human health. Graphene and their polymer composites are synthesized by using various synthesis techniques. Following details is an overview of the significant synthesis techniques of graphene-polymer based nanocomposites that have been reported in the past few years. We also discussed the various analytical electrochemical detection methods for toxic chemicals such as potentiometric, voltammetric and electrochemical impedance spectroscopy methods. Subsections cover electrochemical sensors on graphene-polymer based nanocomposites with different kind of polymers used, and finally their detection sensors on toxic chemical containing heavy metal ions.

Keywords

Polymerization, Graphene Derivatives, Electroanalysis, Nanohybrid, Biosensor, Potentiometric Technique, Voltammetric Technique

Contents

1.   Introduction

A toxic chemical is a substance or a particular mixture that can be poisonous or cause damage to the organism. Many chemicals can be obtained persistently from organic pollutants such as flame retardants (PBDEs), dioxins/furans (PCDD/Fs), polycyclic aromatic hydrocarbons (PAHs), polychlorinated biphenyls (PCBs) [1-4] that can be found on waste sites, heavy metals/metalloid concentrations and even industrial waste. Any chemical can be toxic because they can harm us when it contacts or enters the body. The toxicity level of any chemical is described by the types of effects it causes and its potency under a certain condition. Different chemical causes different type of effects. For example, ingesting gasoline can cause burns, vomiting, diarrhea and, in very large amounts, drowsiness or death, but not cancer. However, there are certain chemical such as asbestos and benzene which may have no noticeable effects at the beginning of exposure but can cause cancer after years [5]. The health effects depend on the amount of chemical exposure, for example, two aspirin tablets can help to relieve a headache, but taking an entire bottle of aspirin can cause stomach pain, nausea, vomiting, headache, convulsions and even death.

Meanwhile, heavy metals include mercury (Hg) [6,7], cadmium (Cd) [8], arsenic (As) [9], chromium (Cr) [10], thallium (Tl) [11], and lead (Pb) [12] are extremely harmful pollutants in the biosphere due to their toxicity and even trace amounts of them pose a detrimental risk to human health. Although, heavy metals are naturally abundant elements found from earth's crust, but most environmental contamination can occur from anthropogenic human activities such as mining and smelting operations. Human exposure through industrial activities include metal processing in refineries, coal burning in power plants, petroleum combustion, nuclear power stations and high tension lines, plastics, textiles, microelectronics, wood preservation and paper processing plants can risk harmful health effects [13]. Among the harmful health effect of heavy metals, they can induce multiple organ damage, even at low degree of exposure. They are also classified as human carcinogens (known or probable) according to the U.S. Environmental Protection Agency, and the International Agency for Research on Cancer [14-16].

Since these chemicals can be toxic, it is crucial to know the amount of a substance that enters or exposes a person. Thus, it is necessary to have a rapid, sensitive, and simple analytical method for the detection and monitoring of these environmental pollutants. The development of a sensor for the precise and selective measurement for metal ions in the presence of potential biomolecules interference at the abnormal levels characteristic of living systems can make a great contribution to disease diagnosis. Although many sensor materials reported showed good potential in detecting toxic chemicals, this chapter will emphasize on graphene and their hybrids with different kind polymer composites. These graphene-polymer based composites are material mainly used for detecting heavy metals and hazardous chemicals in respective detection techniques. The graphene-polymer nanocomposites have recently received a great attention among worldwide researchers in the field of sensors due to its novel chemical, optical and physical properties. Some of its unique properties of graphene-polymer composites are related to their high simplicity and flexibility [17,18]. Polymer–carbon composite is another sensing material that received significant attention since it was introduced many centuries ago. As a filler component, it can improve mechanical properties, chemical inertness and stability, versatile processing techniques and low cost [19]. The graphene-polymer also performed an important role in the sensitivity and selectivity especially in the sensing platform [20,21]. Likewise, the functionalized graphene acts as an efficient nanofiller in polymer composites to improve its engineering properties. Moreover, even with a small quantity of functionalized graphene can improve the mechanical, electronic, optical, thermal and magnetic properties significantly [22-25]. The mechanical properties of a nanocomposite material are evaluated based on the enhancement of the performance as characterized by the elastic modulus, tensile strength, elongation, and toughness [26,27]. Therefore, interfacial interactions play key roles to achieve these mechanical properties. Even more, graphene-polymer nanocomposites possess an impressive functional properties, such as electrical (semi-) conductivity, unique photonic/optical transportation, anisotropic transport, low permeability, and fluorescence quenching [28,29]. Some interesting properties of graphene-polymer based nanocomposites are summarized in Figure 1.

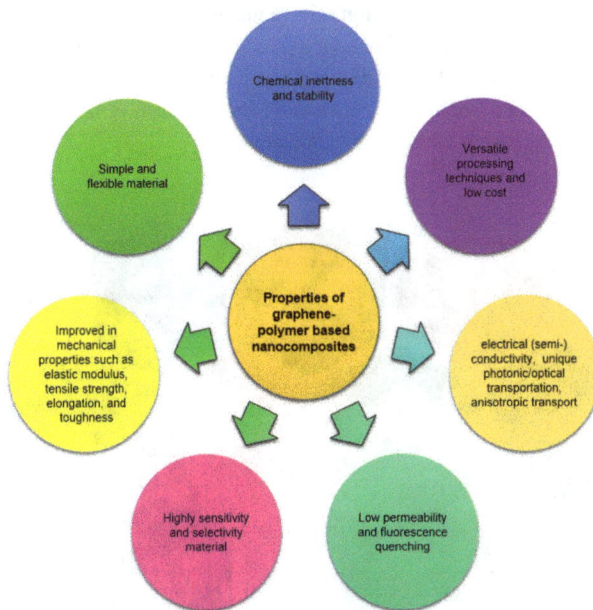

Figure 1 Properties of graphene-polymer based nanocomposites.

Frequently, the graphene-polymer based nanocomposites have been used in a wide range of practical demanding applications nowadays. Figure 2 shows multifunction applications of graphene-polymer based nanocomposites. Some conjugated polymer functionalized graphene like polythiophenes or its derivative composites are widely used in photovoltaic devices [30], light-emitting diodes [31,32], transparent conducting electrodes [33], gas barrier membranes [34,35], and biosensors [36-38]. Meanwhile, conducting polymers such as polyaniline and polypyrrole functionalized graphene composites show a great potential application in supercapacitors [39,40] and lithium-ion battery electrode [41]. As for supercapacitor devices based on graphene/PANI composite film, it exhibits large electrochemical capacitance value of 210 F g^{-1} at a discharge rate of 0.3 A g^{-1} [39]. Another example of a composite that has been used for supercapacitor devices is based on the rGO/polypyrrole composite which shows a higher specific capacitance value of 424 F g^{-1} [36]. Besides that, non-covalently polymer functionalized graphene composites contribute a variety of sensor applications [42,43], e.g. GO/methylcellulose and GO/poly(vinyl alcohol) hybrid system acts as a good sensor for the detection of nitroaromatics [38] and Au (III) ions in aqueous media, respectively [44]. The poly(ethylene glycol) functionalized nanographene is used as new vectors for the delivery of cancer drug into the cells [45]. Another branch of biomedical applications by using graphene-polymer based including controlled drug release, enzyme immobilization, sensors and actuators, and as tissue culture substrates.

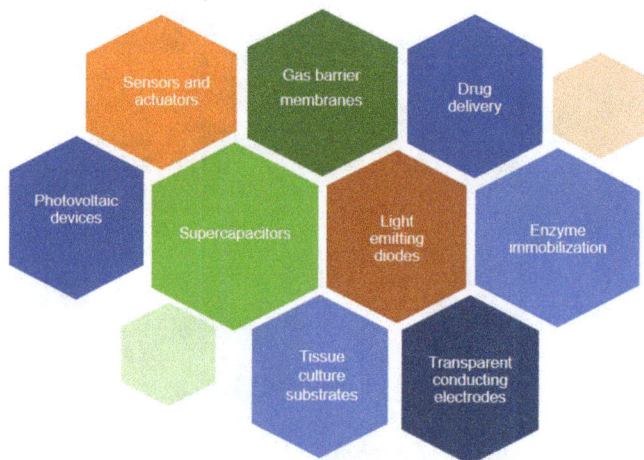

Figure 2 Multifunction of graphene-polymer based nanocomposites applications.

2. Methods of detection

Various techniques for detecting heavy metal ions have been developed as a result of the rapid development of nanotechnology, and each technique has its advantages and disadvantages. Along with the production of a variety of techniques for the detection. To date, there are various analytical techniques, such as atomic absorption spectrometry [46], inductively coupled plasma mass spectrometry [47], inductively coupled plasma atomic emission spectrometry [48], capillary electrophoresis [49], X-ray fluorescence spectrometry [50], Neutron activation analysis [51], Ion chromatography [52] and UV-VIS spectrometry (UV-VIS) [53], have been performed for heavy metals quantification. However, these spectrometric methods require expensive instruments and not suitable for in situ analysis due to the ponderous and also needing professional operators to handle complicated instruments. Thus, a simple and inexpensive method that not only detects but also quantities heavy metal ions is desirable for the real-time monitoring of environmental, biological, and industrial samples. On the contrary, the electrochemical method considered as an alternative to these expensive spectroscopic techniques has been acknowledged as an efficient method to detect heavy metal ions [54-58]. Advantages of electrochemical analysis, including has high sensitivity, high accuracy, wide measuring range, simple device, economical, user-friendly, reliable and suitable for in-field applications [59]. For metal analysis, the principal electrochemical techniques used are potentiometric, voltammetric and electrochemical impedance spectroscopy methods [60]. The brief principal aspects of electrochemical operation are explained as below.

2.1   Potentiometric technique [61]

It is a classical technique in which information about the composition of the sample is provided through the potential appearing between two electrodes. Ion-selective electrodes (ISEs) provide rapid selective potentiometric techniques for the determination of the major cations (metal samples). The detection limit is in the nanomolar range (or even lower).

2.2   Voltammetric technique

It is a method in which information about the composition of electrolytic solutions is provided by measuring the current as a function of applied potential [62]. For voltammetry, stripping analysis is one of the most widely used in metal ion analysis due to highly sensitive voltammetric method, which makes possible to determine minimal concentrations of analyte (sub-nanograms) [63,64]. In contrast to conventional analytical techniques, these techniques can provide the best limit of detection (LOD) but low cost, environmentally friendly, quicker and easy to apply for analysis of heavy metal ions in a variety of real samples. There are several voltmmetric techniques currently use for metal ions detection including cyclic voltammetry (CV), differential pulse voltammetry(DPV), linear scan voltammetry (LSV), chronoamperometry (CA), square wave voltammetry (SWV), differential pulse anodic stripping voltammetry (DPASV) and square wave anodic stripping voltammetry (SWASV). The comparison of different voltammetry with graphene-polymer modified electrodes have comprehensively summarized in Table 1.

Table 1 Analytical performances of various methods of modified electrodes for metal ions determination.

Modified electrode[a]	Methods	Ion/compound	Detection limit	Ref.
-(IL-rGO/AuNDs/Nafion/GCE)	SWV	Iron	35 nM	[65]
PGMGPE	CV	Hg^{2+}	6.6 µM	[66]
		Pb^{2+}	0.8 µM	
Nafion-G	DPASV	Pb^{2+} and Cd^{2+}	0.02 µg L^{-1}	[67]
3D rGO@PANI	CV	Pb^{2+}	0.035 nM	[6]
PPy–RGO	SWASV	Hg^{2+}	15 nM	[68]
G/PANI/PS	SWASV	Pb^{2+}	3.30 µg L^{-1}	[69]
		Cd^{2+}	4.43 µg L^{-1}	
sGO/PPy	DPASV	Pb^{2+}	0.07 ppb	[70]
RGO-CS/PLL	DPASV	Cd^{2+}	0.01 µg L^{-1}	[71]
		Pb^{2+}	0.02 µg L^{-1}	
		Cu^{2+}	0.02 µg L^{-1}	
PEI-RGO	DPASV	Cu^{2+}	0.3 µM L^{-1}	[72]
GR-CD/PPy	DPV	Hg^{2+}	0.47 nM L^{-1}	[73]
PA/PPy/GO	DPV	Cd^{2+}	2.13 µg L^{-1}	[74]
		Pb^{2+}	0.41 µg L^{-1}	

[a]IL-rGO/AuNDs=ionic liquid-reduced graphene oxide (IL-rGO) supported gold nanodendrites; PGMGPE=polyglycine-modified graphene paste electrode; G=graphene; 3D rGO@PANI=three-dimensional reduced graphene oxide and polyaniline; PPy–RGO=polypyrrole/reduced graphene oxide; G/PANI/PS= graphene/polyaniline/polystyrene; sGO/PPy=cysteine-functionalized graphene oxide/polypyrrole; RGO-CS/PLL=reduced graphene oxide-chitosan/poly-l-lysine; PEI-RGO=Polyethyleneimine-reduced graphene oxide; GR-CD/PPy=polypyrrole decorated graphene/β-cyclodextrin; PA/PPy/GO=phytic acid functionalized polypyrrole/graphene oxide

2.3 Electrochemical impedance spectroscopy (EIS)

EIS is a non-destructive method used for detecting and determining the target species concentrations based on electrical impedance changes the sensor analyte interface [75]. It is a highly sensitive characterization technique used to study a variety of electrochemical systems since EIS can provide accurate error-free kinetic and mechanistic information using a variety of techniques and output formats. The EIS is a modulation technique in which impedance measured by applying a small amplitude sinusoidal excitation signal to the system under study (electrochemical system) and measuring the response signal [76]. Impedance measurement provides information such as processes occurring on electrode surfaces, the conductivity of the solution and ability numerical expression of an aqueous solution to carry electric current [77]. One-step fabrication of DNA-modified three-dimensional reduced graphene oxide and chitosan nanocomposite (CS@3D-rGO@DNA) was reported for highly sensitive detection of Hg^{2+} in drinking water [78]. The direct adsorption of Hg^{2+} onto the nanocomposite detection was investigated by EIS with a detection limit of 0.016 nM. The authors measured the sensitivity of CS@3D-rGO@DNA-modified electrode based on EIS Nyquist plots toward with different concentrations of Hg^{2+} from 0.1 nM to 10 nM are shown in Figure 3a. Whereas, the linear plot and the dependence of Rct on the concentration of Hg2+ (0.1 to 10 nM) were calculated by using software shown in Figure 3b.

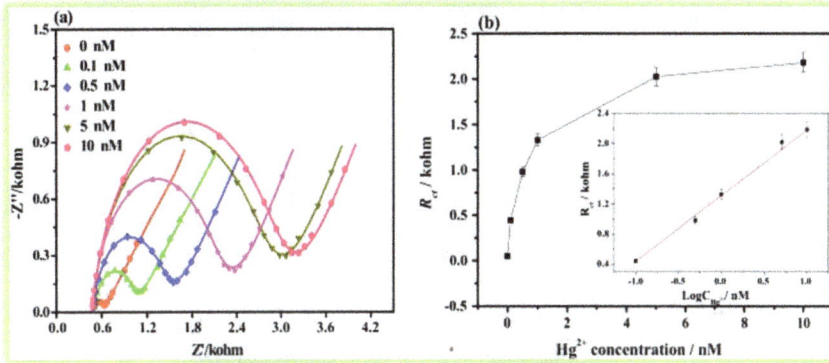

Figure 3 (a) EIS Nyquist plots for the detection of different concentrations of Hg^{2+} ions with concentrations are of 0 to 10 nM. (b) The linear fit plots R_{ct} as function of the logarithm of Hg^{2+} concentration [78].

3.   Electrochemical sensors using graphene-polymer nanocomposites

Many approaches are used for the modification of the working electrode surface by using graphene incorporated polymer based composites. Generally, traditional fabrication routines include solution-based processing [79-82] and melt-based processing [83,84]. The most widely used for the fabrication of graphene-polymer based nanocomposites approaches are in situ polymerization, chemical grafting, latex emulsion blending, layer-by-layer (LbL) assembly, and directed assembly [85-87]. For the in situ polymerization method, intercalated monomers within expanded graphite clusters can promote their efficient exfoliation into single sheets throughout the polymer matrix caused by catalysis reactions [88].

3.1 Sensors using graphene modified conducting polymer

3.1.1 Graphene-polyaniline (PANI) nanocomposites

Yang et al. synthesized nanorod-like nanocomposite of three-dimensional reduced graphene oxide and polyaniline (3D-rGO@PANI) by an in situ chemical oxidative polymerization method for detecting Hg^{2+} in aqueous solution [6]. The synthesis schematic illustration is shown in Figure 4. The results demonstrated that the electrochemical biosensor based on 3D-rGO@PANI nanocomposite showed high sensitivity and selectivity toward Hg^{2+} within a concentration range from 0.1 nM to 100 nM with a low detection limit of 0.035 nM. The authors proved that the presence of 3D-rGO within the nanocomposite further improves the specific surface area and electrochemical performance.

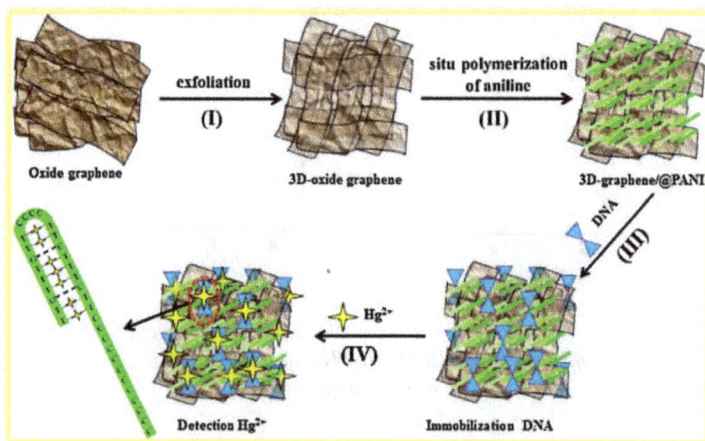

Figure 4 Schematic diagram of the detection of Hg^{2+} nanorod-like nanocomposite of three-dimensional reduced graphene oxide and polyaniline (3D-rGO@PANI) by an in situ chemical oxidative polymerization method [6].

Muralikrishna et al. prepared PANI/GO hydrogels for highly sensitive electrochemical determination of Pb^{2+} [89]. They synthesized hydrogels via in situ polymerization of aniline in the presence of GO nanosheets followed by hydrogel formation at an elevated temperature as shown in figure 5. The synthesized nanomaterial exhibits significant properties for the highly sensitive electrochemical determination as well as removal of environmentally harmful lead (Pb^{2+}) ions. The detection limit obtained for this electrode is 0.04 nM with the longer linear concentration range. The sensor showed excellent repeatability and reproducibility and successfully tested by using real water samples.

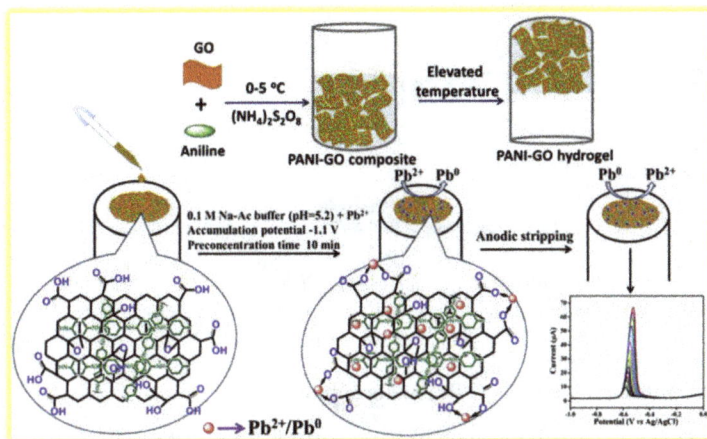

Figure 5 Schematic diagram of the detection of Pb^{2+} electrochemical sensor based PANI/GO hydrogels synthesized via in situ polymerization method [89].

3.1.2 Graphene/polyaniline/polystyrene nanoporous fibre

Graphene/polyaniline/polystyrene (G/PANI/PS) nanoporous fibres modified electrode was developed by using electrospinning fabrication for highly sensitive and simultaneous determination of lead (Pb^{2+}) and cadmium (Cd^{2+}) in real water samples [69]. Promphet et al. prepared nanoporous structure on Graphene/polyaniline/polystyrene fibre which increases an electroactive surface area, that leading to enhance electrochemical sensitivity of carbon electrode as shown in Figure 6. The authors employed square-wave anodic stripping voltammetry technique (SWASV) for detecting Pb^{2+} and Cd^{2+} in the presence of bismuth (Bi^{3+}) on G/PANI/PS nanoporous fibre-modified SPCE with detection limit of 3.30 μg L^{-1} and 4.43 μg L^{-1} respectively.

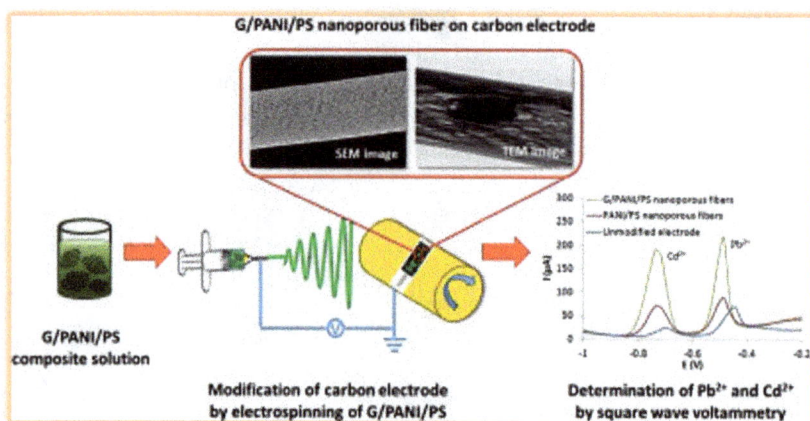

Figure 6 Schematic diagram of the detection of Pb^{2+} and Cd^{2+} electrochemical sensors based (G/PANI/PS) nanoporous fibers modified electrode synthesized via electrospinning fabrication method [69].

3.1.3 Graphene-polypyrrole (PPy) nanocomposites

Palanisamy et al. reported the synthesis of polypyrrole decorated graphene/β-cyclodextrin (GR-CD/PPy) composite for low-level electrochemical detection of Hg^{2+} in various water samples [73]. The GR-CD/PPy composite was synthesized by chemical oxidation of PPy monomer in GR-CD solution using $FeCl_3$. The $FeCl_3$ solution was added into the GR-CD/pyrrole suspension, the mixture was stirred for 1 h for the polymerization of pyrrole to PPy. The result of GR-CD/PPy composite was obtained after a day treated in an air oven at 50 °C. The authors employed the differential pulse voltammetry (DPV) technique for the detection of Hg^{2+}, and the results showed the limit of detection (LOD) of $0.47\ nM\ L^{-1}$ that below the guideline level of Hg^{2+} set by the World's Health Organization (WHO) and U.S. Environmental Protection Agency (EPA). Figure 7 illustrates the schematic of GR-CD/PPy composite for detecting Hg^{2+}.

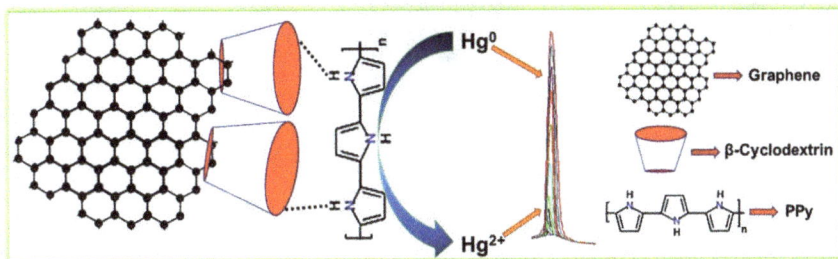

Figure 7 Schematic diagram of the detection of Hg^{2+} based on GR-CD/PPy composite modified electrode synthesized via polymerization method [73].

Another example for the synthesis of electrochemical sensor based on graphene-polypyrrole nanocomposites is reported by Dai et al. [74]. Dai synthesized polypyrrole/graphene oxide (PPy/GO) nanocomposites via in situ chemical oxidation polymerization and followed by phytic acid (PA) molecules functionalization with nanocomposites through electrostatic attraction. According to the authors, a purified GO nanosheets were obtained by exfoliation of natural graphite are dispersed in water, giving a yellow-brown dispersion with a concentration of 0.5 mg/mL. Then a pyrrole solution is added by magnetic stirring for more than 1 h. Subsequently, the oxidation agent containing $FeCl_3$-$6H_2O$ is added into the mixture before continued stirring for 4 h at temperature of 0 to 4 °C. The final product of PA/PPy/GO nanocomposites is obtained after sonication PPy/GO nanocomposites into a PA solution for 2 h. The synthesis steps of PA/PPy/GO nanocomposite are shown in Figure 8. The synthesized of phytic acid-functionalized polypyrrole/graphene oxide nanocomposites were then developed for determination of Cd^{2+} and Pb^{2+} simultaneously. The electrochemical determination of Cd^{2+} and Pb^{2+} metal ions were traced by differential pulse voltammetry (DPV) technique revealed the detection limit of 2.13 μg L^{-1} for Cd^{2+} and 0.41 μg L^{-1} for Pb^{2+} respectively.

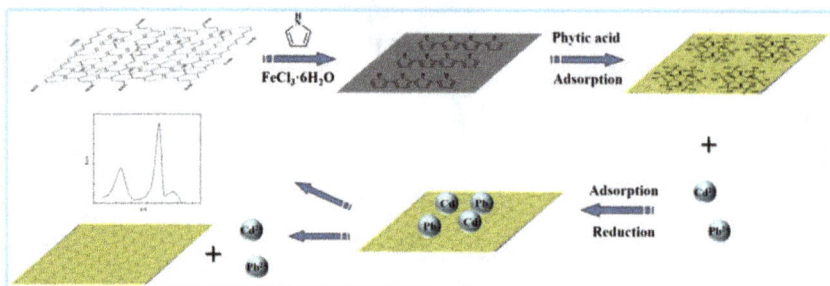

Figure 8 Schematic diagram of the detection of Cd^{2+} and Pb^{2+} elctrochemical sensors based on PA/PPy/GO nanocomposites modified electrode synthesized via in situ chemical oxidation polymerization method [74].

3.1.4 Graphene-poly (3,4 - ethylenedioxythiophene) (PEDOT) nanocomposites

Zuo et al. reported an electrochemical determination of Hg^{2+} at trace level based on poly(3,4-ethylenedioxythiophene) nanorods/graphene oxide nanocomposite modified glassy carbon electrode (PEDOT/GO/GCE) by using a simple liquid–liquid interfacial polymerization approach [90]. The differential pulse stripping voltammetry (DPSV) was applied to determine low concentrations of Hg^{2+} on PEDOT/GO/GCE. The electrochemical sensor exhibited a good linear relationship of peak currents and the concentration of Hg^{2+} in range of 10.0 nM to 3.0 μM with detection limit was estimated to be 2.78 nM. The electrochemical sensor showed good selectivity for Hg^{2+} detection and applicable in real water samples testing.

Yasri et al. also reported a partially oxidized graphene flakes (po-Gr) modified with poly(3,4-ethylenedioxythiophene)/poly(styrenesulfonate) (PEDOT:PSS) by sonication method [91]. First, the po-Gr flakes were obtained by electrochemical exfoliation of graphite sheet. Then the po-Gr dispersion was mixed with PEDOT:PSS solution by a probe sonicator during sonication process. By using differential pulse stripping voltammetry (DPSV) technique, the final composite of the po-Gr/PEDOT:PSS conducting film exhibited satisfactorily for sensitive and selective detection of Hg^{2+} in real samples with a limit of detection of 0.19 μM.

3.2   Sensors using graphene modified electroactive polymer

3.2.1   Graphene-Nafion nanocomposites

Besides fabrication with conducting polymers, other electroactive polymers including Nafion, poly(dimethylsiloxane) (PDMS), polydopamine, poly-L-lysine (PLL), polyallylamine, and polyethyleneimine have also been widely used in heavy metal ion sensing [92]. Another sensing platform to detect trace levels of toxic elements, such as lead and cadmium by differential pulse anodic stripping voltammetry (DPASV) technique was earlier introduced by Li at al. [67]. Li dispersed graphene nanosheets into Nafion-isopropyl-alcohol solution by using in situ plated bismuth film electrode (BFE) fabrication. Then, the final Nafion-G composite film was obtained after the coated glassy carbon electrode was evaporated under an infrared lamp. The Nafion-G composite film modified GC electrode exhibited enhance electrochemical sensing with the limit of detection around $0.02\ \mu g\ L^{-1}$ for Pb^{2+} and Cd^{2+} as shown in Figure 9.

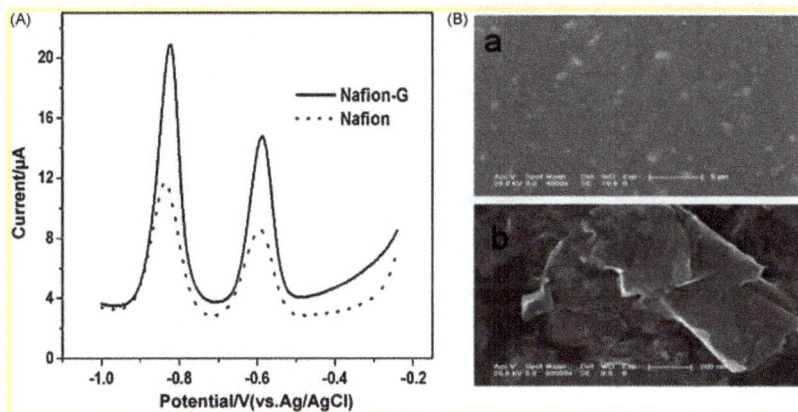

Figure 9 (A) DPASVs for 20 μg L^{-1} each of Cd^{2+} and Pb^{2+} on an in situ plated Nafion-BFE, Nafion-G-BFE in presence of Bi^{3+} solution, (B) SEM image of Bi film deposited on the Nafion (a) and Nafion-G (b) modified GCE [67].

Another example of Nafion composite fabrication in removal of heavy metal ions is employed when the Nafion cation is introduced into ionic liquid-reduced graphene oxide (IL-rGO) supported gold nanodendrites (AuNDs) [65]. The ionic liquids (ILs) have received great attention because of their tunable structures and unique physicochemical properties such as can provide a wide electrochemical window, high ionic conductivity, superior thermal stability, good solubility, and biocompatibility [93,94]. Furthermore, the introduction of IL moieties into functional graphene composites can increase their solubility and improve electrochemical performance [65]. The IL-rGO/AuNDs/Nafion modified electrode is applied for iron removal in coastal waters by square wave voltammetry (SWV) technique. The IL-rGO/AuNDs/Nafion modified electrode showed good responses for iron ions with a linear relation with its concentrations ranging from 0.30 to 100 $\mu mol\,L^{-1}$ and the detection limit of 35 $nmol\,L^{-1}$. The schematic diagram of the stepwise self-assembly procedures of IL-rGO/AuNDs/Nafion modified electrode is shown as figure 10.

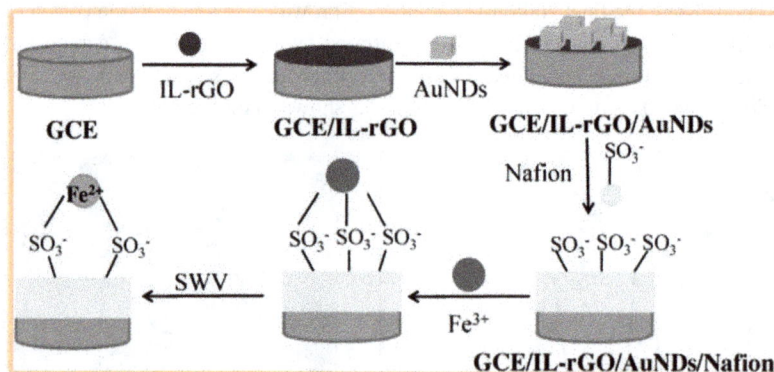

Figure 10 Schematic diagram of the stepwise self-assembly procedures IL-rGO/AuNDs/Nafion modified electrode for electrochemical determination of iron [65].

References

[1] M.H. Wong, S.C. Wu, W.J. Deng, X.Z. Yu, Q. Luo, A.O.W. Leung, C.S.C. Wong, W.J. Luksemburg, A.S. Wong, Export of toxic chemicals – A review of the case of uncontrolled electronic-waste recycling, Environ. Pollut. 149 (2007) 131-140. https://doi.org/10.1016/j.envpol.2007.01.044

[2] ATSDR, Toxicological Profile for Chlorinated Dibenzo-p-dioxins (CDDs) (US Department of Health and Human Services, Public Health Service, Atlanta, GA, 1998)

[3] N. Negash, H. Alemu, M. Tessema, Determination of phenol and chlorophenols at single-wall carbon nanotubes/poly (3,4-ethylenedioxythiophene) modified glassy

carbon electrode using flow injection amperometry, Am. J. Anal. Chem. 5 (2014) 188–198. https://doi.org/10.4236/ajac.2014.53023

[4] J.D. Berset, R. Holzer. Organic micropollutants in Swiss agriculture: distribution of polynuclear aromatic hydrocarbons (PAH) and polychlorinated biphenyls (PCB) in soil, liquid manure, sewage sludge, and compost samples: a comparative study, Int. J. Environ. Anal. Chem. 59 (1995)145-165. https://doi.org/10.1080/03067319508041324

[5] Website :https://www.health.ny.gov/environmental/chemicals/toxic_substances.htm

[6] Y. Yang, M. Kang, S. Fang, M. Wang, L. He, J. Zhao, H. Zhang, Z. Zhang, Electrochemical biosensor based on three-dimensional reduced graphene oxide and polyaniline nanocomposite for selective detection of mercury ions, Sens. Actuator B-Chem. 214 (2015) 63-69. https://doi.org/10.1016/j.snb.2015.02.127

[7] K. Zangeneh Kamali, A. Pandikumar, S. Jayabal, R. Ramaraj, H.N. Lim, B.H. Ong, C.S.D. Bien, Y.Y. Kee, N.M. Huang, Amalgamation based optical and colorimetric sensing of mercury(II) ions with silver@graphene oxide nanocomposite materials, Microchim. Acta 183 (2016) 369-377. https://doi.org/10.1007/s00604-015-1658-6

[8] Tayyaba Kokab, Afzal Shah, Faiza Jan Iftikhar, Jan Nisar, Mohammad Salim Akhter, Sher Bahadur Khan, Amino acid-fabricated glassy carbon electrode for efficient simultaneous sensing of zinc(II), cadmium(II), copper(II), and mercury(II) ions, ACS Omega 4 (2019) 22057-22068. https://doi.org/10.1021/acsomega.9b03189

[9] S. Kumar, G. Bhanjana, N. Dilbaghi, R. Kumar, A. Umar, Fabrication and characterization of highly sensitive and selective arsenic sensor based on ultra-thin graphene oxide nanosheets, Sens. Actuator B-Chem. 227 (2016) 29-34. https://doi.org/10.1016/j.snb.2015.11.101

[10] A. Prakash, S. Chandra, D. Bahadur, Structural, magnetic, and textural properties of iron oxide-reduced graphene oxide hybrids and their use for the electrochemical detection of chromium, Carbon 50 (2012) 4209-4219. https://doi.org/10.1016/j.carbon.2012.05.002

[11] M. Nasiri-Majd, M.A. Taher, H. Fazelrad, Synthesis and application of nano-sized ionic imprinted polymer for the selective voltammetric determination of thallium, Talanta, 144 (2015) 204-209. https://doi.org/10.1016/j.talanta.2015.05.058

[12] S.-Y. Kuo, H.-H. Li, P.-J. Wu, C.-P. Chen, Y.-C. Huang, Y.-H. Chan, Dual Colorimetric and Fluorescent Sensor Based On Semiconducting Polymer Dots for

Ratiometric Detection of Lead Ions in Living Cells, Anal. Chem., 87 (2015) 4765-4771. https://doi.org/10.1021/ac504845t

[13] P.B. Tchounwou, C.G. Yedjou, A.K. Patlolla, D.J. Sutton, Heavy metal toxicity and the environment, Exp. Suppl. 101 (2012) 133-164. https://doi.org/10.1007/978-3-7643-8340-4_6

[14] World Health Organization. Biological Monitoring of Metals, 1994. http://apps.who.int/iris/bitstream/10665/62052/1/WHO_EHG_94.2.pdf.

[15] L. Jarup, Hazards of heavy metal contamination, Br. Med. Bull. 68 (2003) 167–182. https://doi.org/10.1093/bmb/ldg032

[16] G. Aragay, J. Pons, A. Merkoci, Recent trends in macro-, micro-, and nanomaterial-based tools and strategies for heavy-metal detection, Chem. Rev. 111 (2011) 3433–3458. https://doi.org/10.1021/cr100383r

[17] J.K. Abraham, B. Philip, A. Witchurch, V.K. Varadan, C. Reddy, A compact wireless gas sensor using a carbon nanotube/PMMA thin fifilm chemiresistor, Smart Mater. Struct. 13 (2004) 1045–1049. https://doi.org/10.1088/0964-1726/13/5/010

[18] T. Alizadeh, Chemiresistor sensors array optimization by using the method of coupled statistical techniques and its application as an electronic nose for some organic vapors recognition, Sens. Actuators B 143 (2010) 740–749. https://doi.org/10.1016/j.snb.2009.10.018

[19] L.H. Baekeland, The synthesis constitution and uses of Bakelite, J. Ind. Eng. Chem. 1 (1909) 149–161. https://doi.org/10.1021/ie50003a004

[20] T. Kuila, S. Bose, A. Mishra, P. Khanra, N. Kim, J. Lee, Chemical functionalization of graphene and its applications, Prog. Mater. Sci. 57 (2012) 1061–1105. https://doi.org/10.1016/j.pmatsci.2012.03.002

[21] M. Craciun, S. Russo, M. Yamamoto, S. Tarucha, Tuneable electronic properties in graphene, Nano Today 6 (2011) 42–60. https://doi.org/10.1016/j.nantod.2010.12.001

[22] D.G. Papageorgiou, I.A. Kinloch, R.J. Young, Mechanical properties of graphene and graphene-based nanocomposites, Prog. Mater Sci. 90 (2017) 75-127. https://doi.org/10.1016/j.pmatsci.2017.07.004

[23] M. Yoonessi, Y. Shi, D.A. Scheiman, M. Lebron-Colon, D.M. Tigelaar, R. Weiss, et al., Graphene polyimide nanocomposites; thermal, mechanical, and high temperature shape memory effects. ACS Nano 6 (2012) 644–655. https://doi.org/10.1021/nn302871y

[24] X. Yang, X. Wang, J. Yang, J. Li, L. Wan, Functionalization of graphene using trimethoxysilanes and its reinforcement on polypropylene nanocomposites, Chem. Phys. Lett. 570 (201) 125–131.

[25] N.H. Kamaruddin, A.R. Marlinda, M. Said, F. Abd Wahab, G.B. Tong, N.A. Hamizi, Z.Z. Chowdhury, S. Sagadevan, M.R. Johan, Synergistic effects of rubber band infused graphene nanocomposite on morphology, spectral, and dynamic mechanical properties, Polym. Compos. 41 (2020) 1475-1480. https://doi.org/10.1002/pc.25470

[26] H. Fischer, Polymer nanocomposites: from fundamental research to specific applications, Mater. Sci. Eng. C 23 (2003) 763–772. https://doi.org/10.1016/j.msec.2003.09.148

[27] M.A. Rahman, G.B. Tong, N.H. Kamaruddin, F.A. Wahab, N.A. Hamizi, Z.Z. Chowdhury, S. Sagadevan, N. Chanlek, M.R. Johan, Effect of graphene infusion on morphology and performance of natural rubber latex/graphene composites, J. Mater. Sci.-Mater. Electron. 30 (2019) 12888-12894. https://doi.org/10.1007/s10854-019-01650-0

[28] K. Hu, D.D. Kulkarni, I. Choi, V.V. Tsukruk, Graphene-polymer nanocomposites for structural and functional applications, Prog. Polym. Sci. 39 (2014) 1934-1972. https://doi.org/10.1016/j.progpolymsci.2014.03.001

[29] D. Cai, M. Song, Recent advance in functionalized graphene/polymer nanocomposites, J. Mater. Chem. 20 (2010) 7906-7915. https://doi.org/10.1039/c0jm00530d

[30] D. Yu, Y. Yang, M. Durstock, J.-B. Baek, L. Dai, Soluble P3HT-grafted graphene for efficient bilayer–heterojunction photovoltaic devices, ACS Nano 4 (2010) 5633-5640. https://doi.org/10.1021/nn101671t

[31] B.R. Lee, J.-w. Kim, D. Kang, D.W. Lee, S.-J. Ko, H.J. Lee, C.-L. Lee, J.Y. Kim, H.S. Shin, M.H. Song, Highly efficient polymer light-emitting diodes using graphene oxide as a hole transport layer, ACS Nano 6 (2012) 2984-2991. https://doi.org/10.1021/nn300280q

[32] J.P. Singh, U. Saha, R. Jaiswal, R.S. Anand, A. Srivastava, T.H. Goswami, Enhanced polymer light-emitting diode property using fluorescent conducting polymer-reduced graphene oxide nanocomposite as active emissive layer, J. Nanopart. Res. 16 (2014) 2693. https://doi.org/10.1007/s11051-014-2693-7

[33] A.M. Díez-Pascual, J.A. Luceño Sánchez, R. Peña Capilla, P. García Díaz, Recent Developments in Graphene/Polymer Nanocomposites for Application in Polymer Solar Cells, Polymers 10 (2018) 217. https://doi.org/10.3390/polym10020217

[34] Y. Cui, S.I. Kundalwal, S. Kumar, Gas barrier performance of graphene/polymer nanocomposites, Carbon 98 (2016) 313-333. https://doi.org/10.1016/j.carbon.2015.11.018

[35] B.M. Yoo, H.J. Shin, H.W. Yoon, H.B. Park, Graphene and graphene oxide and their uses in barrier polymers, J. Appl. Polym. Sci. 131 (2014) 39628. https://doi.org/10.1002/app.39628

[36] H. Kim, Y. Miura, C.W. Macosko, Graphene/polyurethane nanocomposites for improved gas barrier and electrical conductivity. Chem. Mater. 22 (2010) 3441–3450. https://doi.org/10.1021/cm100477v

[37] D.A.C. Brownson, D.K. Kampouris, C.E. Banks, An overview of graphene in energy production and storage applications, J. Power Sources 196 (2011) 4873–4885. https://doi.org/10.1016/j.jpowsour.2011.02.022

[38] X. Huang, X. Qi, F. Boey, H. Zhan, Graphene-based composites, Chem Soc..Rev. 41 (2012) 666–686. https://doi.org/10.1039/C1CS15078B

[39] Q. Wu, Y. Xu, Z. Yao, A. Liu, G. Shi, Supercapacitors based on flexible graphene/polyaniline nanofiber composite films, ACS Nano 4 (2010) 1963-1970. https://doi.org/10.1021/nn1000035

[40] H.-H. Chang, C.-K. Chang, Y.-C. Tsai, C.-S. Liao, Electrochemically synthesized graphene/polypyrrole composites and their use in supercapacitor, Carbon 50 (2012) 2331-2336. https://doi.org/10.1016/j.carbon.2012.01.056

[41] T Cassagneau, J.H. Fendler, High density rechargeable lithium-ion batteries self-assembled from graphite oxide nanoplatelets and polyelectrolytes. Adv Mater. 10 (1998) 877–881. https://doi.org/10.1002/(SICI)1521-4095(199808)10:11<877::AID-ADMA877>3.0.CO;2-1

[42] A. Kundu, R.K. Layek, A.K. Nandi, Enhanced fluorescent intensity of graphene oxide–methyl cellulose hybrid in acidic medium: Sensing of nitro-aromatics, J. Mater. Chem. 22 (2012) 8139-8144. https://doi.org/10.1039/c2jm30402c

[43] R.K. Layek, A.K. Nandi, A review on synthesis and properties of polymer functionalized graphene, Polymer 54 (2013) 5087-5103. https://doi.org/10.1016/j.polymer.2013.06.027

[44] A. Kundu, R.K. Layek, A. Kuila, A.K. Nandi, Highly fluorescent graphene oxide-poly(vinyl alcohol) hybrid: An effective material for specific Au^{3+} ion sensors, ACS Appl. Mater. Interfaces 4 (2012) 5576-5582. https://doi.org/10.1021/am301467z

[45] Z. Liu, J.T. Robinson, X. Sun, H. Dai, PEGylated nanographene oxide for delivery of water-insoluble cancer drugs, J. Am. Chem. Soc. 130 (2008) 10876-10877. https://doi.org/10.1021/ja803688x

[46] Z. Fang, J. Ru° ziˇ cka, ˇ E. Hansen, An efficient flow-injection system with online ion-exchange preconcentration for the determination of trace amounts of heavy metals by atomic absorption spectrometry, Anal. Chim. Acta 164 (1984) 23–29. https://doi.org/10.1016/S0003-2670(00)85614-7

[47] D.E. Nixon, K.R. Neubauer, S.J. Eckdahl, J.A. Butz, M.F. Burritt, Comparison of tunable bandpass reaction cell inductively coupled plasma mass spectrometry with conventional inductively coupled plasma mass spectrometry for the determination of heavy metals in whole blood and urine, Spectrochim. Acta B 59 (2004) 1377–1387. https://doi.org/10.1016/j.sab.2004.05.013

[48] O.V.S. Raju, P.M.N. Prasad, V.Varalakshmi, Y.V.R. Reddy, Determination of heavy metals in ground water by Icp-Oes in selected coastal area of Spsr Nellore district , Andhra Pradesh, India, Int. J. Innov. Res. Sci. Eng. Technol. 3 (2014) 9743–9749.

[49] Y. Li, Y. Jiang, X.P. Yan, Probing mercury species-DNA interactions by capillary electrophoresis with on-line electrothermal atomic absorption spectrometric detection, Anal. Chem. 78 (2006) 6115–6120. https://doi.org/10.1021/ac060644a

[50] O.W. Lau, S.Y. Ho, Simultaneous determination of traces of iron, cobalt, nickel, copper, mercury and lead in water by energy-dispersive x-ray fluorescence spectrometry after preconcentration as their piperazino-1,4-bis(dithiocarbamate) complexes, Anal. Chim. Acta 280 (1993) 269–277. https://doi.org/10.1016/0003-2670(93)85131-3

[51] V.P. Guinn, C.D. Wagner, Instrumental Neutron Activation Analysis, Anal. Chem. 32 (1960) 317-323. https://doi.org/10.1021/ac60159a005

[52] C. Sarzanini, M.C. Bruzzoniti, Metal species determination by ion chromatography, TrAC, Trends Anal. Chem. 20 (2001) 304-310. https://doi.org/10.1016/S0165-9936(01)00071-1

[53] C.O.B. Okoye, A.M. Chukwuneke, N.R. Ekere, J.N. Ihedioha, Simultaneous ultraviolet-visible (UV-VIS) spectrophotometric quantitative determination of Pb, Hg,

Cd, As and Ni ions in aqueous solutions using cyanidin as a chromogenic reagent, Int. J. Phys. Sci. 8 (2013) 98-102. https://doi.org/10.5897/IJPS12.670

[54] J.M. Liu, L.P. Lin, X.X. Wang, S.Q. Lin,W.L. Cai, L.H. Zhang, Z.Y. Zheng, Highly selective and sensitive detection of Cu (II) with lysine enhancing bovine serum albumin modified-carbon dots fluorescent probe. Analyst 137 (2012) 637–642. https://doi.org/10.1039/c2an35130g

[55] C.C. Huang, J.C. He, Electrosorptive removal of copper ions from waste water by using ordered mesoporous carbon electrodes, Chem. Eng. J. 221 (2013) 469–475. https://doi.org/10.1016/j.cej.2013.02.028

[56] J. Lu, X. Zhang, N. Liu, H. Li, Z. Yu, X. Yan, Electrochemical sensor for mercuric chloride based on graphene-MnO_2 composite as recognition element, Electrochim. Acta 174 (2015) 221–229. https://doi.org/10.1016/j.electacta.2015.05.181

[57] C.V. Gherasim , J. Krivcik , P. Mikulasek, Investigation of batch electrodialysis process for removal of lead ions from aqueous solutions, Chem. Eng. J. 256 (2014) 324–334. https://doi.org/10.1016/j.cej.2014.06.094

[58] X. Huakun, X. Jingkun, Z. Xiaofei, D. Xuemin, L. Limin, Z. Yinxiu, Z. Youshan, W. Wenmin, A new electrochemical sensor based on carboimidazole grafted reduced graphene oxide for simultaneous detection of Hg^{2+} and Pb^{2+}. J. Electroanal. Chem. 782 (2016)250–255. https://doi.org/10.1016/j.jelechem.2016.10.043

[59] L. Eddaif, A. Shaban, J. Telegdi, Sensitive detection of heavy metals ions based on the calixarene derivatives-modified piezoelectric resonators: a review, Int. J. Environ. Anal. Chem. 99 (2019) 824-853. https://doi.org/10.1080/03067319.2019.1616708

[60] O.A. Farghaly, R.S.A. Hameed, Analytical Application using modern electrochemical techniques: a review, Int. J. Electrochem. Sci. 9 (2014) 3287.

[61] C.M.A. Brett, Electrochemical sensors for environmental monitoring . Strategy and examples *, Pure Appl. Chem. 73 (2001) 1969–1977. https://doi.org/10.1351/pac200173121969

[62] J. Kudr, L. Richtera, L. Nejdl, K. Xhaxhiu, P. Vitek, B. Rutkay-nedecky, D. Hynek, P. Kopel, V. Adam, and R. Kizek, Improved electrochemical detection of zinc ions using electrode modified with electrochemically reduced graphene oxide, Materials 9 (2016) 1–12. https://doi.org/10.3390/ma9010031

[63] K. Keawkim, S. Chuanuwatanakul, O. Chailapakul, S. Motomizu, Determination of lead and cadmium in rice samples by sequential injection/anodic stripping voltammetry using a bismuth film/crown ether/nafion modified screen-printed carbon

electrode, Food Control 31 (2013) 14-21.
https://doi.org/10.1016/j.foodcont.2012.09.025

[64] S. Dal Borgo, V. Jovanovski, B. Pihlar, S.B. Hocevar, Operation of bismuth film electrode in more acidic medium, Electrochim. Acta 155 (2015) 196-200.
https://doi.org/10.1016/j.electacta.2014.12.086

[65] F. Li, D. Pan, M. Lin, H. Han, X. Hu, Q. Kang, Electrochemical determination of iron in coastal waters based on ionic liquid-reduced graphene oxide supported gold nanodendrites, Electrochim. Acta 176 (2015) 548-554.
https://doi.org/10.1016/j.electacta.2015.07.011

[66] C. Raril, J.G. Manjunatha, Fabrication of novel polymer-modified graphene-based electrochemical sensor for the determination of mercury and lead ions in water and biological samples, J. Anal. Sci. Technol. 11 (2020) 3.
https://doi.org/10.1186/s40543-019-0194-0

[67] J. Li, S. Guo, Y. Zhai, E. Wang, High-sensitivity determination of lead and cadmium based on the Nafion-graphene composite film, Anal. Chim. Acta, 649 (2009) 196-201.
https://doi.org/10.1016/j.aca.2009.07.030

[68] Z.-Q. Zhao, X. Chen, Q. Yang, J.-H. Liu, X.-J. Huang, Selective adsorption toward toxic metal ions results in selective response: electrochemical studies on a polypyrrole/reduced graphene oxide nanocomposite, Chem. Commun. 48 (2012) 2180-2182. https://doi.org/10.1039/C1CC16735A

[69] N. Promphet, P. Rattanarat, R. Rangkupan, O. Chailapakul, N. Rodthongkum, An electrochemical sensor based on graphene/polyaniline/polystyrene nanoporous fibers modified electrode for simultaneous determination of lead and cadmium, Sens. Actuator B-Chem. 207 (2015) 526-534. https://doi.org/10.1016/j.snb.2014.10.126

[70] R. Seenivasan, W.-J. Chang, S. Gunasekaran, Highly sensitive detection and removal of lead ions in water using cysteine-functionalized graphene oxide/polypyrrole nanocomposite film electrode, ACS Appl. Mater. Interfaces 7 (2015) 15935-15943. https://doi.org/10.1021/acsami.5b03904

[71] Z. Guo, D.-D. Li, X.-K. Luo, Y.-H. Li, Q.-N. Zhao, M.-M. Li, Y.-T. Zhao, T.-S. Sun, C. Ma, Simultaneous determination of trace Cd(II), Pb(II) and Cu(II) by differential pulse anodic stripping voltammetry using a reduced graphene oxide-chitosan/poly-l-lysine nanocomposite modified glassy carbon electrode, J. Colloid Interface Sci. 490 (2017) 11-22. https://doi.org/10.1016/j.jcis.2016.11.006

[72] R. Hu, H. Gou, Z. Mo, X. Wei, Y. Wang, Highly selective detection of trace Cu^{2+} based on polyethyleneimine-reduced graphene oxide nanocomposite modified glassy carbon electrode, Ionics 21 (2015) 3125-3133. https://doi.org/10.1007/s11581-015-1499-7

[73] S. Palanisamy, K. Thangavelu, S.-M. Chen, V. Velusamy, M.-H. Chang, T.-W. Chen, F.M.A. Al-Hemaid, M.A. Ali, S.K. Ramaraj, Synthesis and characterization of polypyrrole decorated graphene/β-cyclodextrin composite for low level electrochemical detection of mercury (II) in water, Sens. Actuator B-Chem. 243 (2017) 888-894. https://doi.org/10.1016/j.snb.2016.12.068

[74] H. Dai, N. Wang, D. Wang, H. Ma, M. Lin, An electrochemical sensor based on phytic acid functionalized polypyrrole/graphene oxide nanocomposites for simultaneous determination of Cd(II) and Pb(II), Chem. Eng. J. 299 (2016) 150-155. https://doi.org/10.1016/j.cej.2016.04.083

[75] S. Guruva, R. Avuthu, B.B. Narakathu, A. Eshkeiti, and S. Emamian, Detection of heavy metals using fully printed three electrode electrochemical, Sensors *IEEE* (2014) 1–4. https://doi.org/10.1016/j.snb.2010.05.053

[76] A.P. Bhondekar, M. Dhiman, A. Sharma, and A. Bhakta, A novel iTongue for Indian black tea discrimination, Sens. Actuator B-Chem. 148 (2010) 601-609.

[77] B. Bansod, T. Kumar, R. Thakur, S. Rana, I. Singh, A review on various electrochemical techniques for heavy metal ions detection with different sensing platforms, Biosens. Bioelectron. 94 (2017) 443-455. https://doi.org/10.1016/j.bios.2017.03.031

[78] Z. Zhang, X. Fu, K. Li, R. Liu, D. Peng, L. He, M. Wang, H. Zhang, L. Zhou, One-step fabrication of electrochemical biosensor based on DNA-modified three-dimensional reduced graphene oxide and chitosan nanocomposite for highly sensitive detection of Hg(II), Sens. Actuator B-Chem. 225 (2016) 453-462. https://doi.org/10.1016/j.snb.2015.11.091

[79] T. Ramanathan, S. Stankovich, D.A. Dikin, H. Liu, H. Shen, S.T. Nguyen, L.C. Brinson, Graphitic nanofillers in PMMA nanocomposites—An investigation of particle size and dispersion and their influence on nanocomposite properties, J. Polym. Sci., Part B: Polym. Phys. 45 (2007) 2097-2112. https://doi.org/10.1002/polb.21187

[80] S. Kim, I. Do, L.T. Drzal, Multifunctional xGnP/LLDPE nanocomposites prepared by solution compounding using various screw rotating systems, Macromol. Mater. Eng. 294 (2009) 196-205. https://doi.org/10.1002/mame.200800319

[81] J. Liang, Y. Huang, L. Zhang, Y. Wang, Y. Ma, T. Guo, Y. Chen, Molecular-level dispersion of graphene into poly(vinyl alcohol) and effective reinforcement of their nanocomposites, Adv. Funct. Mater. 19 (2009) 2297-2302. https://doi.org/10.1002/adfm.200801776

[82] X. Zhao, Q. Zhang, D. Chen, P. Lu, Enhanced mechanical properties of graphene-based Poly(vinyl alcohol) Composites, Macromolecules, 44 (2011) 2392-2392. https://doi.org/10.1021/ma200335d

[83] K. Kalaitzidou, H. Fukushima, L.T. Drzal, A new compounding method for exfoliated graphite–polypropylene nanocomposites with enhanced flexural properties and lower percolation threshold, Compos. Sci. Technol. 67 (2007) 2045-2051. https://doi.org/10.1016/j.compscitech.2006.11.014

[84] Y.F. Zhao, M. Xiao, S.J. Wang, X.C. Ge, Y.Z. Meng, Preparation and properties of electrically conductive PPS/expanded graphite nanocomposites, Compos. Sci. Technol. 67 (2007) 2528-2534. https://doi.org/10.1016/j.compscitech.2006.12.009

[85] H. Kim, Y. Miura, C.W. Macosko, Graphene/polyurethane nanocomposites for improved gas barrier and electrical conductivity, Chem. Mat. 22 (2010) 3441-3450. https://doi.org/10.1021/cm100477v

[86] H. Li, S. Pang, S. Wu, X. Feng, K. Müllen, C. Bubeck, Layer-by-Layer assembly and UV photoreduction of graphene–polyoxometalate composite films for electronics, J. Am. Chem. Soc. 133 (2011) 9423-9429. https://doi.org/10.1021/ja201594k

[87] D. Cho, S. Lee, G. Yang, H. Fukushima, L.T. Drzal, Dynamic mechanical and thermal properties of phenylethynyl-terminated polyimide composites reinforced with expanded graphite nanoplatelets, Macromol. Mater. Eng. 290 (2005) 179-187. https://doi.org/10.1002/mame.200400281

[88] H. Hu, X. Wang, J. Wang, L. Wan, F. Liu, H. Zheng, R. Chen, C. Xu, Preparation and properties of graphene nanosheets–polystyrene nanocomposites via in situ emulsion polymerization, Chem. Phys. Lett. 484 (2010) 247-253. https://doi.org/10.1016/j.cplett.2009.11.024

[89] S. Muralikrishna, D.H. Nagaraju, R.G. Balakrishna, W. Surareungchai, T. Ramakrishnappa, A.B. Shivanandareddy, Hydrogels of polyaniline with graphene oxide for highly sensitive electrochemical determination of lead ions, Anal. Chim. Acta 990 (2017) 67-77. https://doi.org/10.1016/j.aca.2017.09.008

[90] Y. Zuo, J. Xu, X. Zhu, X. Duan, L. Lu, Y. Gao, H. Xing, T. Yang, G. Ye, Y. Yu, Poly(3,4-ethylenedioxythiophene) nanorods/graphene oxide nanocomposite as a new

electrode material for the selective electrochemical detection of mercury (II), Synth. Met. 220 (2016) 14-19. https://doi.org/10.1016/j.synthmet.2016.05.022

[91] N.G. Yasri, A.K. Sundramoorthy, W.-J. Chang, S. Gunasekaran, Highly selective mercury detection at partially oxidized graphene/poly(3,4-Ethylenedioxythiophene):poly(Styrenesulfonate) nanocomposite film-modified electrode, Front. Mater. 1 (2014) 1-10. https://doi.org/10.3389/fmats.2014.00033

[92] Y. Zuo, J. Xu, X. Zhu, X. Duan, L. Lu, Y. Yu, Graphene-derived nanomaterials as recognition elements for electrochemical determination of heavy metal ions: a review, Microchim. Acta, 186 (2019) 171. https://doi.org/10.1007/s00604-019-3248-5

[93] L. Timperman, A. Vigeant, M. Anouti, Eutectic mixture of protic ionic liquids as an electrolyte for activated carbon-based supercapacitors, Electrochim. Acta, 155 (2015) 164-173. https://doi.org/10.1016/j.electacta.2014.12.130

[94] S.J. Yoo, L.-J. Li, C.-C. Zeng, R.D. Little, Polymeric ionic liquid and carbon black composite as a reusable supporting electrolyte: Modification of the electrode surface, Angew. Chem. Int. Edit. 54 (2015) 3744-3747. https://doi.org/10.1002/anie.201410207

Materials Research Forum LLC
https://doi.org/10.21741/9781644900956-8

Chapter 8

Graphene-Carbon Nanotubes Nanocomposite Modified Electrochemical Sensors for Toxic Chemicals

*A. Sivakami[a], *S. Bagyalakshmi[b], *K.S Balamurugan[c], Nurul Izrini Ikshan[d,e]

[a]Department of Physics, School of Science, Malla Reddy University, Medchal, Hyderabad-500100, Telangana, India.

[b]Department of Physics, Sri Ramakrishna Institute of Technology, Coimbatore-641010,

Tamil Nadu, India.

[c]Department of Electronics and Communication Engineering, Bharat Institute of Engineering and Technology, Hyderabad-501510, Telangana, India.

[d]Faculty of Applied Sciences, UniversitiTeknologi MARA (UiTM), 40450 Shah Alam, Selangor, Malaysia

[e]Ionic Materials and Devices (iMade), UniversitiTeknologi MARA (UiTM), 40450 Shah Alam, Selangor, Malaysia

*sivakamitce@gmail.com

Abstract

The unique mechanical, electrical, physical and chemical properties of carbon nanotubes (CNTs) and graphene is showing excellent detection of toxic chemicals. The graphene (GR)-Carbon nanotube (CNT) nanocomposite showed large active surface area, high porosity and electrical conductivity than graphene based or CNT based ones. The electrochemical performance of GR-CNT nanocomposite can be enhanced due to synergistic effects operating between GR and CNT. The electrochemically modified GR-CNT nanocomposites has been used in different applications such as biomedical, pharmaceutical, environmental, energy harvesting, food sector applications. This chapter summarizes the electrochemical sensing of GR-CNTs nanocomposites for detection of heavy metal ions, phenolic compounds, nitrite, nitrate, hydrogen peroxide and etc. GR-CNTs nanocomposite based electrochemical sensors showed the great selectivity, sensitivity and reproducibility for detection of environmental pollutants.

Keywords

Graphene-CNT, Modified Electrode, GR-MWCNTs, Composite, Toxic, Metal Ions, Detection

Contents

1.   Introduction

Sensors play an important role in therapeutic fields such as molecular imaging, gas sensing, environmental monitoring, diabetics and water remediation [1, 2]. Electrochemical biosensors are commonly used for detecting the environmental pollutants because of their low cost, fast response times, simple to utilize, and due to their size of small structure. In the medical field, food protection, and to monitor the environment these electrochemical sensors have been used widely [3, 4]. Because of tremendous needs to produce highly cost-effective, large selective and sensitive sensors, carbon nanotubes and graphene based materials are mostly used for fabrication of electrochemical biosensors because of their unique properties. Carbon Nanotubes (CNTs) possess always high electrical and thermal conductivity, outstanding mechanical strength, large surface area, individual nanostructures in cylindrical form varying length to diameter ratios which is remarkable. CNTs are also linked with excellent conductivity, sensitivity of large value, great biocompatibility, and chemical stability with exceptional value [5-7]. The development of graphene (2D) nanomaterial showed best performance in the electrochemical sensor. Graphene is a single layer (atom thickness) having sp^2-hybridized carbon atoms. The flat hexagonal 2D lattice structures are formed by covalent bonds [8]. Although the graphene's electrochemical properties have proved that accelerated transfer of electrons between enzymes and electrodes can be achieved because of their unique electronic structure [9-11]. This essentially spreads from the pi-bonds which are delocalized in the basal plane above and below. These electrons which are delocalized generate within the plane electrical conductivities of high value for graphene [12]. CNTs- Graphene nanocomposites have a broad variety of applications for the electrochemical detection of toxic chemicals [13-15]. Diverse toxic chemicals are used in environmental samples, food, pharmaceutical, pathological, agriculture, chemical, and biological samples. This chapter focuses on the graphene-carbon nanotubes (GR-CNTs) nanocomposites used for electrochemical sensing of toxic chemicals.

2.   Carbon nanotubes (CNTs)

2.1   Carbon nanotubes (CNTs) structure

Carbon nanomaterials are capable of forming the different allotropes like fullerene. CNTs are most famous cylindrical fullerene molecules consisting of a hexagonal arrangement of sp^2-hybridized carbon atoms. The main classifications of CNTs as depicted in Figure 1. Single Walled Carbon Nanotubes can be formed by rolling up of single graphene layer in cylindrical form. MWCNTs are formed by arranging of two concentric graphene cylinders with inter distance of 0.34 nm. The properties of CNTs may vary based on their structure.

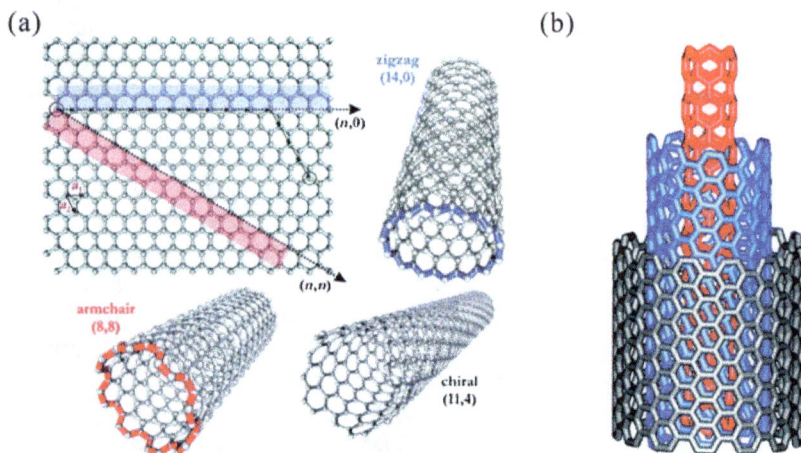

Figure 1 Structure of (a) single walled carbon nanotubes(SWCNTs) (b) multi -walled carbon nanotubes(MWCNTs). Reproduced permission from Ref. [16].

The zigzag, armchair or chiral structures are formed from SWCNTs (Figure 1) [16]. SWCNTs possess metallic or semiconducting form. The property of structures can be determined by the way of arrangement of atoms and thickness of nanotube [17]. The electrical properties of SWCNTs are defined by n, m integers i.e. roll-up vectors of the cylinder [18, 19]. If n-m=3q means, SWCNT poses metallic structure and n-m $\neq$ 3q means SWCNT possess semiconducting structure. The nanotubes of zigzag form have value of m is zero and if n=m, the cylindrical nanotubes are armchair and chiral structure. CNTs are having more mechanical strength, excellent chemical stability, and high electrical conductivity.

2.2   Synthesis of carbon nanotubes (CNTs)

Arc-discharge, laser-ablation, and chemical vapor deposition (CVD) are different manufacturing CNT methods. Arc-discharge is the simple effective method for producing

CNTs [20,21]. By this technique, synthesis of MWCNTs can be performed. The CVD synthesis is a popular and commonly used technique for producing higher yield and purity CNTs. By involving decomposition of gaseous carbon sources, the CVD method was employed to produce MWCNTs. There are two CVD methods. 1. Plasma-enhanced CVD (PECVD) 2. Thermal CVD [22]. The PECVD technique have been employed in new technologies like microwave PECVD, hot filament PECVD, DC glow discharge PECVD, inductively coupled plasma PECVD and radio frequency PECVD. The different manufacturing methods of CNT are shown in figure 2.

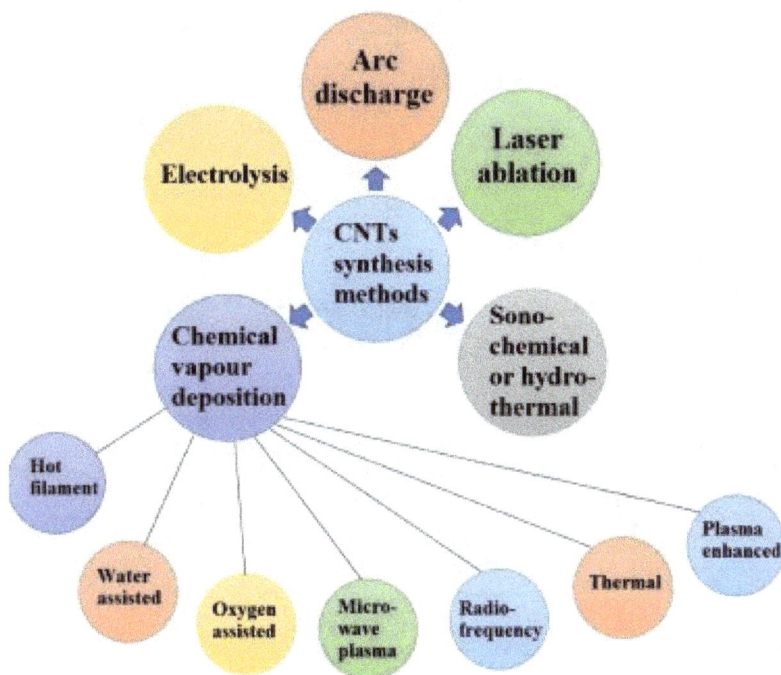

Figure 2 Different preparation methods of CNTs. Reproduced permission from Ref. [20].

2.3 Functionalization of CNTs (Covalent and non-Covalent F-CNTs)

Due to hydrophobic surfaces, pristine structure CNTs is insoluble in aqueous solutions, polymer resins and most of the solvents. The functional groups are either physically or chemically attached to the sidewalls of CNTs without reduction of properties is called as functionalization. Due to good biocompatibility and low toxicity, the functionalized CNTs are easily dispersible in liquids. Either covalent or non-covalent functionalization of CNTs can be made by forming bonds with walls of the nanotube. The functionalized CNTs can be effectively cross and penetrate the barriers of individual biological cells [24]. This mechanism of CNTs has been demonstrated as the bio-sensor [25]. Figure 3 gives several methods of CNTs functionalization.

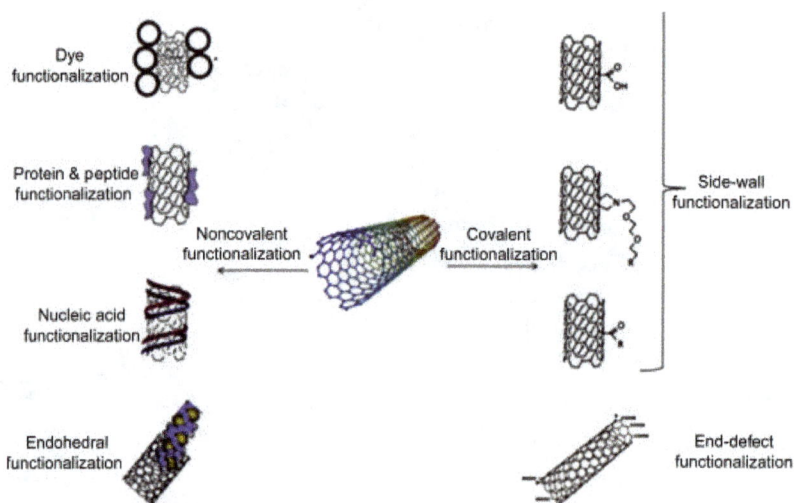

Figure 3 Illustration of functionalization of carbon nanotubes Reproduced permission from Ref. [23].

2.4 Role of carbon based nanomaterials in sensing of toxic chemicals

Carbon Nanotubes (CNTs) are more current creative material broadly emerged for sensing applications because of these attractive properties such as smaller size , large active surface area, volume ratio. The electrochemical technique has been extensively employed in the food analysis field, biomedical field and environmental science because it is saving time, low cost , extremely sensitive, and quite easy to work with; furthermore, it has fine reliability and quick response time [26–28]. The electrode material which is modified for sensing appilications is a crucial factor influencing and analysing the electrochemical performance of sensor. Amongst different nano carbon materials, graphene and CNTs are the most preferably electrode materials which are suitable for modified sensor applications. CNTs have drawn great attraction in the fields such as drug delivery, electrochemical sensor, gas adsorbent, catalyst supports for fuel cells, biological applications [29–31]. The CNTS based electrochemical sensors has been applied and successfully monitored the residue of toxic chemicals over the other materials. The recognition element is associated with a physical transducer (i.e. electrode) that transfers the analytical signal into electronic circuit to detect a target analyte called electrochemical sensor. It has been confirmed that CNTs materials can have large electrode surface area which enhances rapid electron transfer movement between electrode and analytes or catalysts. Due to fast electron transfer property, the efficiency of electrochemical reactions can be improved.

In environmental analysis, the applications of carbon nanomaterials have grown one of the most important areas (Figure 4). CNTs exhibited many advantages for the detection of environment pollutants over other metallic nanoparticles. This chapter has reviewed novel improvements in electrochemicals sensors based CNTs and graphene for detecting different environmental pollutants from various processes. Sensitive and selective determination of analytes have been extensively proved by an ever-increasing variety of systems in the area of sensing and biosensing due to its prominent physical and electrical properties. Additionally, due to some simple operation, inexpensive and facile fabrication process, portable or miniaturized devices, these carbon-based electrochemical sensors are becoming familiar for particular applications. Subsequently, electrochemical biosensors based on CNTs and graphene detecting the various biological pollutants in a promising way have gained significant interest in recent years due to carbon-based nanomaterials with a large surface area, excellent biocompatibility, and physiological balance can produce a suitable environment and long-term stability for biomolecule immobilization.

3.   Electrochemical biosensors involving carbon nanotubes (CNTs)

The electrochemical cell consisting of two and three electrodes cell, which transfers the biological event into a signal such as current or voltage is called an electrochemical biosensor (Figure 5). Electrochemical sensors are devices comprising a biological recognition element and physiochemical detector that are used to identify analyte. These devices have an extensive range of applications ranging from medical to environmental, agricultural and food industry. CNT-based electrochemical sensors execute an essential role because of their benefits such as sensitivity of high value, quick response time, simple to operate and convenient transportability. Biocatalytic sensors and bio affinity sensors are classified by the methods of recognition process in CNT electrochemical biosensors. The CNT bio catalytic sensors are employing the biological recognition element such as enzyme which provides the information about electroactive species . CNTs have essential benefits towards the fabrication of electrochemical sensors providing extra specific area with high sensitivity [32]. The cylindrical structural form and chemically stable of CNTs are allowing the fabrication of ultrasensitive electrochemical sensors.

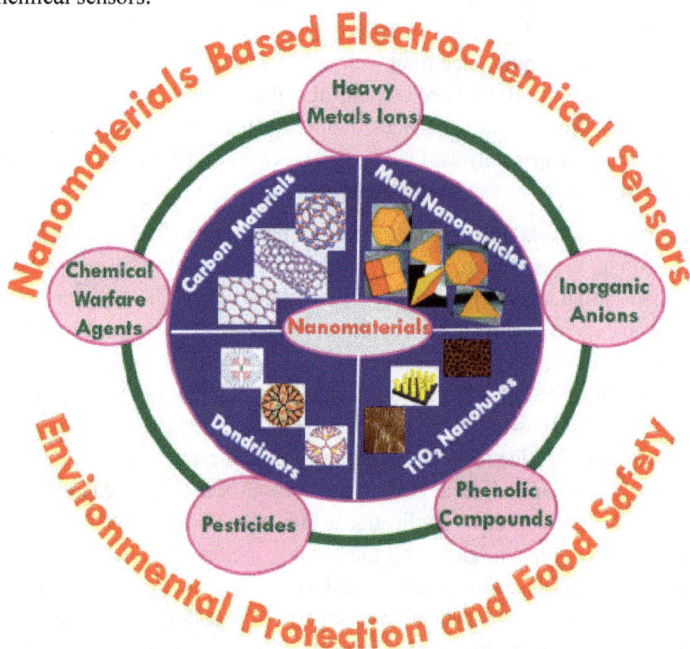

Figure 4 Electrochemical Biosensor for toxic chemicals. Reproduced with permission from Ref. [32].

4. Graphene

A one-atom thickness layer of hexagonally arranged, SP^2-bonded carbon atoms occurring within carbon material structure is characterized as graphene [33,34]. The graphene film thickness is in the order of 100 μm lateral size is perceived as carbon planes attached with Van der Waals forces. Semi-metal and stable properties are exhibited by graphene under ambient conditions. Among carbon nanomaterials, graphene possesses large active surface area [35, 36] providing accessible for direct interaction of wider area biomolecules [38]. The structural defects of graphene can be studied by low cost methods like oxidation, reduction etc. [39,40]. Newly, numerous excellent reviews have been focused on graphene and its derivatives based bio sensors [41]. The functionalized graphene based electrochemical sensors have been fabricated effectively for the detection of toxic chemicals sensitively. The electrochemical behavior of modified graphene and CNTs electrode has been studied with the help of cyclic voltammetry (CV) and square wave voltammetry (SWV) methods.

4.1 Graphene structure and its derivatives

The synthesis of graphene and its derivatives can be controlled depends on their properties and its applications. Figure 6 shows important kinds of graphene materials useful for designing biosensors. For making sensor fabrication, need of graphene oxidization under strong acid creates a large number of oxygen-contain functional groups (carboxyl, epoxide, and hydroxyl) on the graphene surface [42]. The graphene oxide (GO) is hydrophilic so it is capable of dissolving as one sheet into water or polar organic solvents. The structure and properties of the graphene oxide (GO) is regained again by either some compounds (like hydrazine) or heating in an atmosphere.

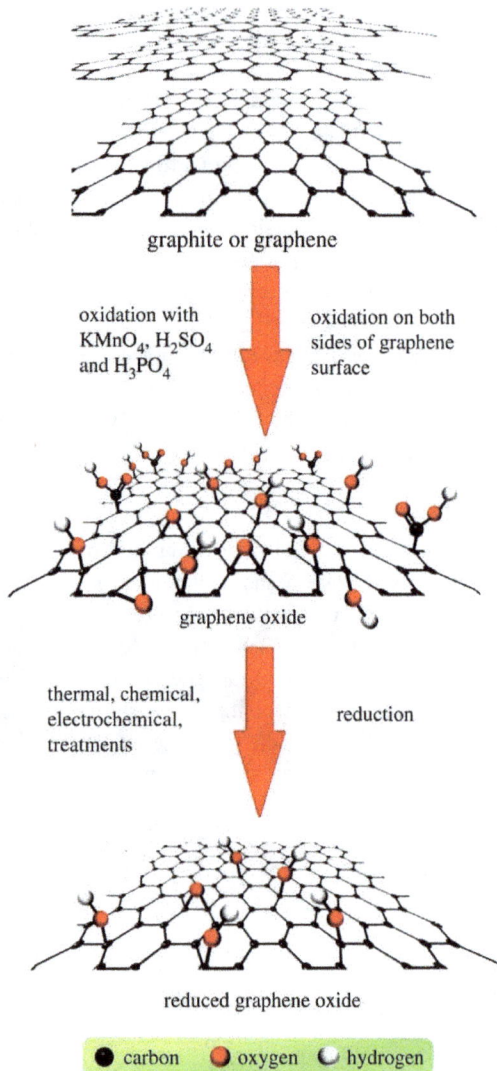

Figure 5 Graphene family structures (a) the pristine graphene (b) GO (c) reduced graphene oxide and (d) graphene quantum dot . Reproduced with permission from Ref. [42].

4.2   Role of graphene in electrochemical biosensors

Graphene and its derivatives (GO, rGO) are essentially important in the fabrication of electrochemical sensors. Graphene is the most popular two-dimensional materials which show excellent properties of graphene like semiconductor of zero-band gap, Dirac point of linear-like, relativistic-high charge velocity, large surface area and no toxicity. Graphene greatly possess proximity induced ability, ultra-high charge mobility, transparency, tremendous tensile strength and large thermal conductivity, sensing etc. Graphene, including accelerated electron transportation, outstanding mechanical properties, and bio compatibility properties are important to fabricate the electrochemical sensors. Due to specific surface area, graphene gives extra probability active sites for interactions of charge & bio molecular leading the improvement of selectivity and sensitivity. Based on high electro catalytic activity of graphene it makes an excellent electrode material for glucose-based biosensors. The functionalized area of graphene directly detects the biomolecules by its components of oxides, carboxyl, epoxides and hydroxyl groups on edge sites. Also, the functionalized graphene provides the binding of heteroatoms, proteins, antigens, nanoparticles, quantum dots, DNA, enzymes, antibodies, and other specific particles [43-45]. Graphene shows electrochemistry of enzyme, its electro catalytic activities toward small biomolecule such as NADH, dopamine, hydrogen peroxide, etc. These sensors have a wide range of detecting analyte including antioxidants, sugars, lipids, neuron transmitters, purine, biomarkers of cancer; toxins in food or drinking water are of great importance for human health.

5.   Toxic chemicals

Environmental pollution is growing with massive industrialization, rapid urbanization, and growing lifestyle of people. The GR-CNTs nanocomposites modified electrochemical sensors are having extremely high surface area, good biocompatibility and heterogeneous electron transfers kinetics which are extensively used in the area of environmental protection. This chapter concentrates on recent improvements of GR-CNT composite for detection of high toxic metal ions like Pb (II), Cd (II), Hg (II), Cr (VI), nitrite, nitrate, phenolic and pesticides compounds.

6.   Electrochemical sensors for toxic metal ions detection

Toxic heavy metal ions cause damage to human health and their contents destroy soil or water environment causing degradation of ecological and environment. To detect those chemicals, an essential simple, fast and sensitive method is needed. The electrochemical sensors are made up of graphene based materials such as rGO, GO have high sensitivity, selectivity, rapid detection, and high stability for measuring and removing of toxic metal ions. The sensing of Cu^{2+}, Pb^{2+}, Hg^{2+}, Cd^{2+}, Ag^{2+} ions was detected by graphene-based nanomaterials. Among heavy metals, Cd^{2+} and Pb^{2+} provides great risks to human health such as reproductive toxicity, human immune system damage, disorders of respiratory System and negative adverse causes on metabolism. Sensitivity, limit of detection (LOD), repeatability and reproducibility features have been evaluated the quality and analytical performance of sensors. Figure 6 shows the different hybrid nanomaterials in the sensing applications.

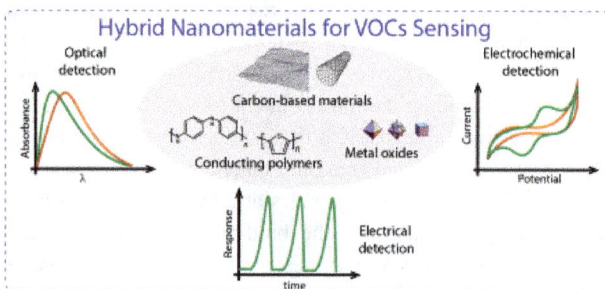

Figure 6 Schematic illustration of various hybrid nanomaterials for sensing applications. Reproduced with permission from [46].

7.   Carbon nanotubes-graphene nanocomposites (CNTs-Gr)

Due to resemblance properties (structural, physical) of Graphene (GR), Carbon nanotubes (CNTs) and their Sp^2 hybridization is giving synergistic effects in sensing applications [47,48]. The synergistic mechanism of CNTs-graphene composite are shown in Figure 7.

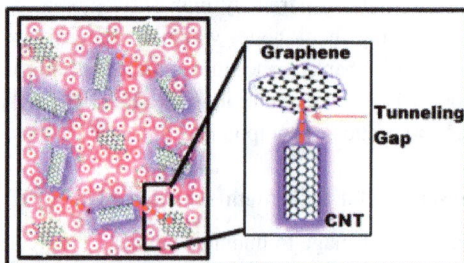

Figure 7 Schematic representation of Synergistic effect of graphene –CNT. Reproduced with permission from [49].

Graphene-Based Electrochemical Sensors for Toxic Chemicals   Materials Research Forum LLC
Materials Research Foundations **82** (2020) 211-242   https://doi.org/10.21741/9781644900956-8

7.1 Synthesis methods of CNTs-GR nanocomposite

The unique properties of graphene–carbon nanotubes (GR–CNTs) hybrid nanocomposite materials have been attracted in many applications such as electrochemical sensors, biosensors, energy storage materials, polymer composites, field effect transistors etc. [50-55]. Due to this combination, the outstanding mechanical, electrical and thermal properties of GR-CNT nanocomposite are demonstrated because of its synergistic effect between GR and CNT composite. There are many synthesis routes available for preparing 3D hybrids by using 1D (GR) and 2D (CNTs) nanostructures [56]. Due to three dimensional hierarchical arrangements of GR-CNT, the large active surface area, high charge density of graphene and 3D network CNT scan are increased than other carbon structures [57]. Different types of CNT-GR hybrid structures include nitrogen-doped aligned CNT-GR sandwiches [58], single-walled CNT-GR hybrids [59], MWCNT/rGO nanostructures [58] etc. The enhanced electrochemical performance has demonstrated than individual CNT-based or graphene based ones because of synergistic effect of 3D GR-CNTs network. Several researchers [60-62] proved that GO-CNT, GR-CNT hybrid nanomaterials exhibit large surface area, high electrical and catalytic properties than pristine CNTs, GO and graphene. The most popular synthesis methods of GR-CNT nanocomposite are sonication method [61, 63], CVD method [64] and electrostatic spray technique [65]. GR-CNT hybrid nanocomposites are showing good stability because of strong π-π stacking interaction of 3D GR and CNT [66].

Fan et al. [67] synthesized GR-CNT hybrid composite by growing the CNT pillar between two GOs using CVD method. Young Oh and coworkers [68] investigated the graphene–MWCNT/PDMS hybrid composites were prepared by solution method and represented in the figure 8.

7.2 CNTs-GR nanocomposite characterization

The structural, morphological, electrical, chemical and photocatalytic characterization of GR-CNT nanocomposite hybrid materials were studied by different characterization techniques such as X- ray diffraction (XRD), X-ray photoelectron spectroscopy (XPS), Field Emission - Scanning Electron Microscopy (FE-SEM), Transmission Electron Microscopy (TEM), Atomic Force Microscopy (AFM), cyclic voltammetry (CV), UV-Visible spectroscopy, Raman, FT-IR spectroscopes. Woo et al. [69] prepared and characterized GR-MWCNT nanocomposite by in situ reduction of graphene oxides. XRD pattern of graphite, MWCNT, Graphene oxide, Graphene and GR-MWCNT composite are represented in Figure 9. It found that the GR-MWNCT composite is increased the internal layer spacing comparable with graphene or CNTs based ones.

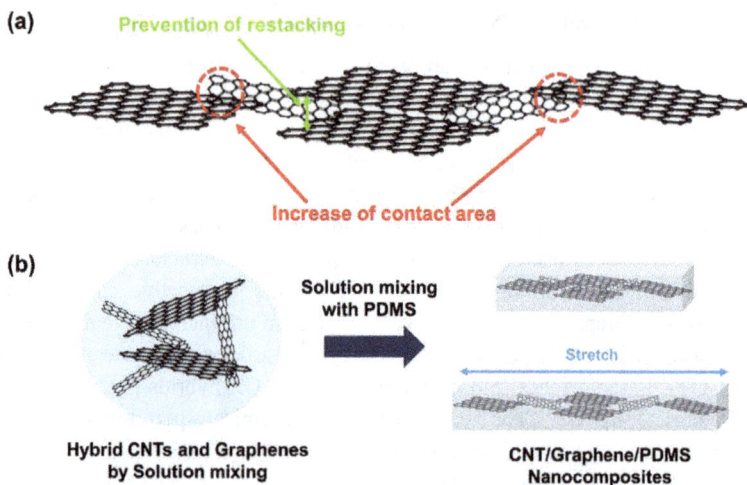

Figure 8 a) Preparation of GR-MWCNT by two step solution method. b) GR/CNT/polydimethylsiloxane nanocomposite. Reproduced with permission from [68].

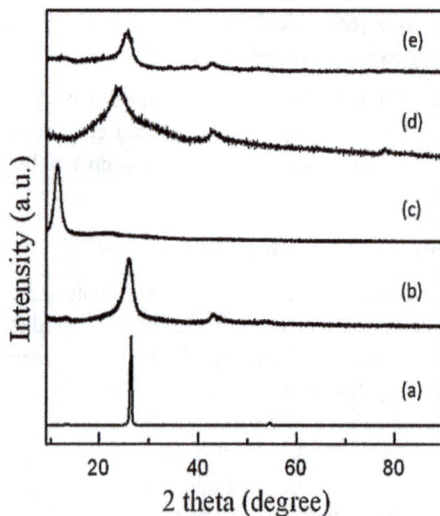

Figure 9 Different XRD patterns (a) graphite, (b) MWCNT, (c) graphene oxide, (d) graphene, and (e) graphene–MWCNT composite. Reproduced with permission from [69].

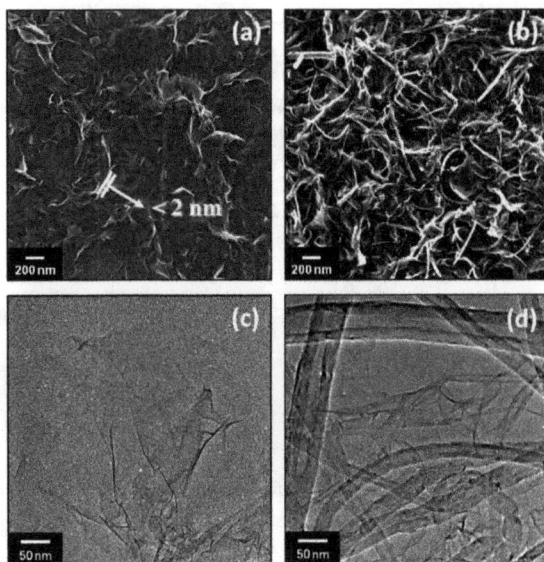

Figure 10 SEM images of (a) graphene and (b) graphene–MWCNT composite. HR-TEM images of (c) graphene and (d) graphene–MWCNT composite. Reproduced with permission from [69].

Figure 10 indicates the FE-SEM and HR -TEM images of GR-MWCNT composite. It tells that different chemical interaction exists of MWCNT and graphene sheets establishing the strong coupling between 3D GR-MWCNT composite found for various applications. Due to strong coupling between GR and MWCNT composite materials providing extra surface area, excellent reproducibility, high electron transfer rate they are used for fabricating electrochemical sensors.

7.3   GR-CNTs nanocomposite for electrochemical sensing of toxic chemicals

GR–CNTs hybrid nanocomposite material widely used as a sensor in the detection of environmental hazardous materials, food additives, pharmaceutical samples and biological substances are shown in Figure11.

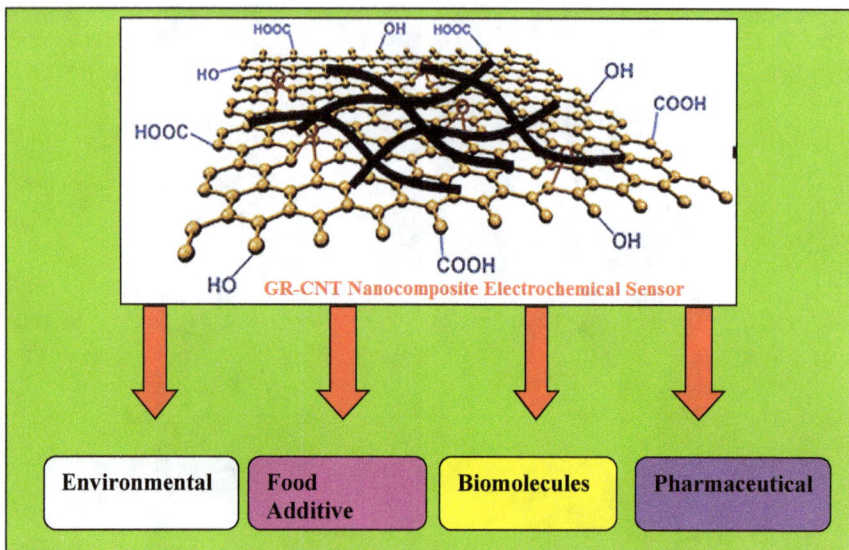

*Figure 11 Schematic illustration of graphene–carbon nanotube hybrid nanocomposite
sensing applications in various hazardous chemicals.*

7.3.1 Detection of metal ions

The heavy metal ions are released into water through industrial waste, chemical industry,
agriculture activities and other wastes which may lead to severe health issue by
generating free radicals. By considering several metal ions, the arsenic (As), cadmium
(Cd), lead (Pb), copper (Cu) and mercury (Hg) ions are highly toxic. The electrochemical
biosensors have been fabricated for detecting the heavy metal ions based on its fast
response, easy operation, low cost and portability. For detecting the metal ions, there is a
requirement of interfacing microorganisms, enzymes, microspheres, CNT's, metal
oxides, gold and silver nanoparticles. GR-CNT composite coupled with functional
nanomaterials showed the best selective and sensitive for detection of toxic chemicals.
The GR/MWCNT hybrid composite modified electrode prepared and its performance
was tested [70]. It was observed that the synergistic effect operating between
GR/MWCNT hybrid composite material enhance the electron transfer rate between
electrode interface, fast response and lower detection limit of heavy metals ions Pb^{2+} and
Cd^{2+} ions simultaneously. 3D GR/MWCNTs hybrid nanocomposites prepared by green
method and its simultaneous determination of heavy metal ions were discussed [70].

GR/MWCNTs modified GCE electrode is enhancing the sensitivity and detection of metal ions like Pb^{2+} and Cd^{2+} ions (for example Pb^{2+} ions are detected by lowest concentration range of 0.5 μg L^{-1} for Cd^{2+} and 0.5 μg L^{-1} with comparing of other poisonous electrodes such as HMDE, TMFE). The modified electrode of GR/MWCNT hybrid composite may be used to determine other heavy metal ions such as Zn^{2+}, Cd^{2+}, Pb^{2+}, Cu^{2+} ions simultaneously.

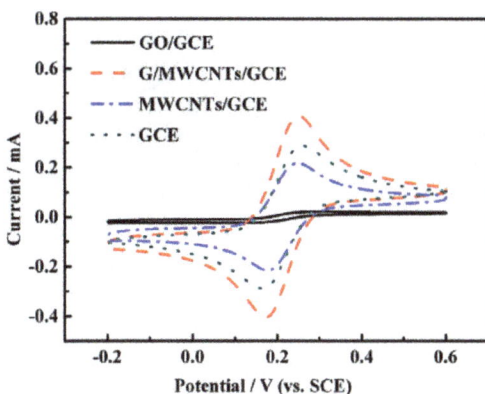

Figure 12 Cyclommetric Voltammetry curves (CVs) of modified and unmodified GR/MWCNT composite. Reproduced with permission from [70].

Marcein Musielak et al. [71] developed the GO/CNT membranes with varying pH values of 5,37,40,42,48,50 and 98 mg g−1 at different composition of CNTs for tracing heavy metal ions Cu(II), Cd(II), Pb(II), Co(II), Ni(II) and Zn(II). The GR/CNT membranes can be regenerated for many times compared to other membranes. The metal ions can be removed simultaneously depending upon q_{max} value and their affinity of the membranes. The order is likely to be Pb(II) > Cu(II)> Cd(II) > Zn(II) > Ni(II) > Co(II). The GO/CNTs membranes are equally important for adsorption of metal ions from wastewater.

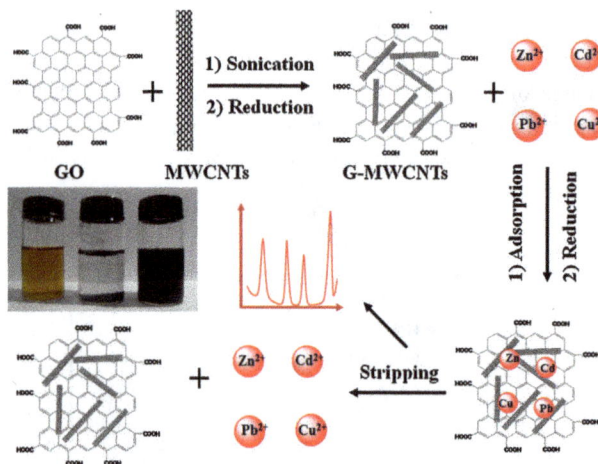

Figure 13 Fabrication of GO–MWCNTs hybrid nanocomposite in the electrochemical detection of different metal ions. Reproduced with permission from [70].

AL. Gahouari et al. [72] discovered and studied the GO-MWCNTs-L-Cysteine hybrid composites using drop casting method and the electrochemical detection of lead (Pb^{2+}) ions. Due to modification of ErGO–MWCNTs–L-cys/GCE electrode, the electrochemical conductivity is increased.

By comparing CV curves of all modified electrodes, as prepared ErGO MWNTs–L-cys/GCE modified electrode synergized because of large redox currents and lower voltage. Due to this reduction of modified ErGO–MWNTs–L-cys composite has large surface area and good electrical conductivity comparing with individual ones. Because of this property, the sensitivity and selectivity of modified ErGO–MWNTs–L-cys composite can be enhanced the detection of Pb^{2+}ions over other GR-CNTs composites.

Wang et al. [73] prepared the GR-CNT-TiO_2 and GR-TiO_2 nanocomposite using solvothermal method for removing the dye and Cr (VI) ions. The GR-MWCNT- TiO_2 composites shows enhanced performance of photocatalytic activity as compared to GR-TiO_2 composite. The amount of photocatalytic activity depends upon the percentage of MWCNTs added with graphene. The photocatalytic activity performance of Cr(VI) ions and dye was depressed when percentage of MWCNT content added exceeding beyond the optimum value known as "shielding effect" [74,75].

Zhan et al. [76] developed 3-D graphene/polydopamine modified MWCNT hybrid aerogel for adsorption of heavy metal ions (Cu(II) and Pb(II)). The adsorption mechanism of Pb(II) and Cu(II) ions as shown in figure 16.

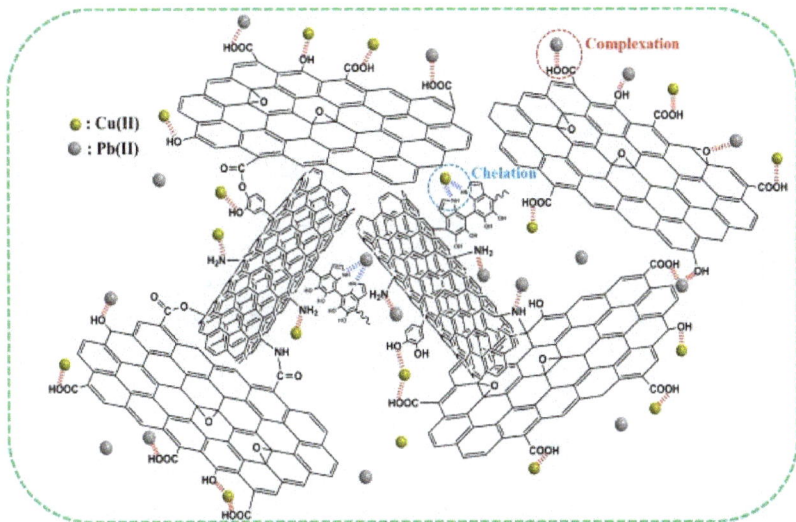

Figure 14 Schematic representation of GO/MWCNT-PDA hybrid aerogel for removal of metal ions Pb(II), Cu(II). Reproduced with permission from [76]

The adsorption ability toxic metal ions depend upon several factors such as initial range concentration of metal ions from waste water, adsorption time, pH value and etc. Figure 15 representing the adsorption capacity of GO/MWCNT hybrid aerogel to detect Cu(II) and Pb(II) metal ions with different pH values.

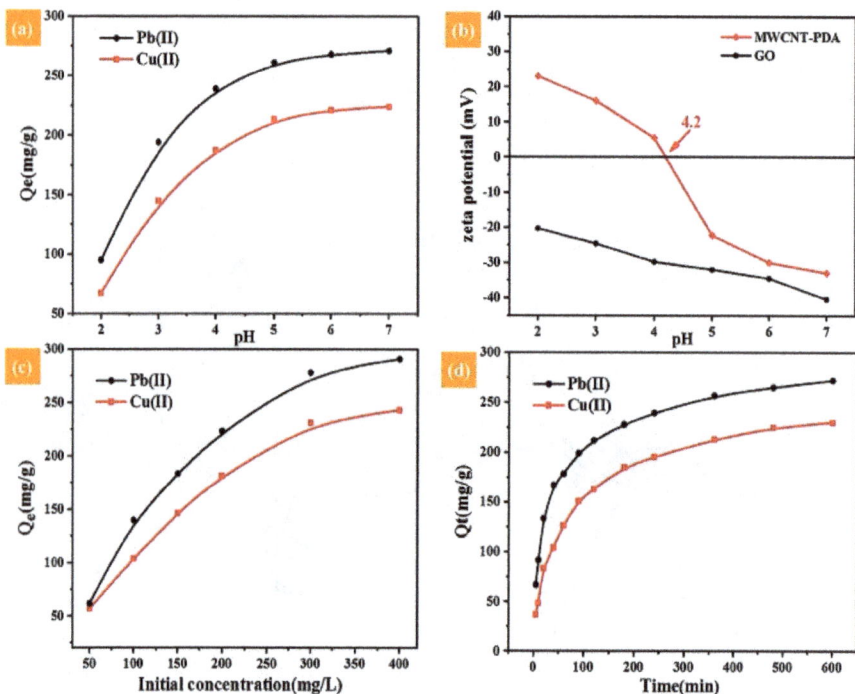

Figure 15 (a) The adsorption capacities of Cu(II) and Pb(II); (b) Zeta-potentials curves of MWCNT-PDA and GO under different pH values; (c) Initial metal ion concentration adsorption effect on Cu(II) and Pb(II); (d) Adsorption of Cu(II) and Pb(II) with varying contact time. Reproduced with permission from [76].

7.3.2 Detection of other toxic chemicals

In this section, the role of GR-CNTs nanocomposite for detection of other toxic chemicals such as phenolic compounds (nitrobenzene, nitrophenoletc), pesticides, hydrogen peroxide, antibiotics and pathogens etc. 3D GR/CNT hybrids are prepared by two step CVD and the hybrid electrode is modified with Nafion and horseradish peroxidase (HRP) used to detect hydrogen peroxide with improved sensitivity and detection limit. These studies are reported by Dong et al. [77]. They found that the three dimensional all-carbon hybrid electrode is able to detect the dopamine directly. But, modified HRP /Nafionthree dimensional all-carbon hybrid is used to detect the hydrogen peroxide indirectly. Wang et al. [78] investigated the electrocatalytic performance for methanol detection with carbon monoxide tolerance by fabricating PtRu catalysts decorated with 3D porous graphene-CNT composite. Yang and Li [79] developed CTAB (HexadecylTrimethyl Ammonium Bromide functionalized GO/MWCNT modified with glassy carbon electrode [CTAB-GO/MWCNT/GCE]for detecting the dopamine (DA), ascorbic acid (AA), uric acid (UA) and nitrite (NO_2^-) compounds simultaneously. They found that the modified CTAB-GO/MWCNT/GCE hybrid electrode showed best selectivity and sensitivity for detecting DA, AA and NO_2^- compounds simultaneously compared with CTAB-GO/GCE &MWCNT/GCE modified electrodes because of synergetic effect between the CTAB-GO and MWCNT. The SEM images of CTAB-GO/GCE and CTAB-GO/MWCNT/GCE are presented in figure 16. It tells that the graphene oxide sheets are having wrinkled texture and also MWCNTs are well distributed in the composite.

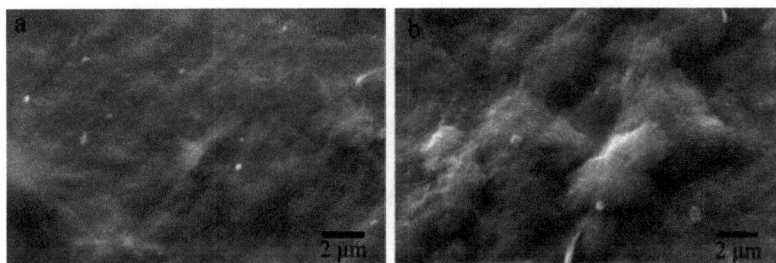

Figure 16: SEM image a) CTAB-GO/GCE b) CTAB-GO/MWCNT/GCE. Reproduced with permission from [79].

Bagheri et al. [80] developed the modified electrochemical sensor based on Cu-MWCNTs/GO hybrid for detection of nitrate and nitrite simultaneously. They found that 25-52 nm sized Cu nanoparticles decorated with MWCNT/rGO gives perfect matrix for tracing the nitrate and nitrite. The electrochemical parameters of GR-CNTs composites for the removal of various toxic chemicals are listed in Table 1.

Table 1 Parameters of GR-CNTs hybrid nanocomposites for sensing the toxic chemicals.

Material	Toxic Chemical Detected	Sensitivity	Limit of Detection	References
CNT/rGO/Bi	Pb(II) and Cd (II) ions	897 nA/ppbcm^2, 264 nA/ppbcm^2	0.2ppb, 0.6ppb	[81]
GR/MWCNTs/Bi	Pb(II) ions Cd(II) ions	0.5-30 µg L^{-1} 0.5-30 µg L^{-1}	0.2 µg L^{-1} 0.1 µg L^{-1}	[70]
rGO/MWCNTs/Pt	H$_2$O$_2$ Nitrite	1.99 µA pM^{-1} cm^{-2} 0.165 µApM^{-1} cm^{-2}	10pM–0.19nM 1 mM -12mM	[82]
GR-MWCNTs/Fe NPs	Nitrite	0.70 µA pM^{-1} cm^{-2}	75.6 nM	[83]
rGO/MWCNTs	Catechol Hydroquinone p-cresol Nitrite	- - - -	1.8 µM 2.6 µM 1.6 µM 25 µM	[84]
rGO/MWCNT	Nitrite	-	0.07 µM	[85]
GR/CNT/ZnO	Glucose	5.362 µAmM^{-1} cm^{-2}	4.5 µM	[86]
GR/CNT	Dopamine H$_2$O$_2$	470.7 mA M^{-1} cm^{-2} 137.9 mA M^{-1} cm^{-2}	20 nM 1 µM	[64]
CTAB-GO/MWCNTs	Dopamine Ascorbic acid Uric acid Nitrite	- - - -	1 µM 1.5 µM 1 µM 1.5 µM	[79]
Cu/MWCNT/rGO	Nitrite ions Nittrate ions	- -	30 nM 20 nM	[80]
PtNCs/MWCNTS/GR NPs	Dopamine Ascorbic acid	- -	0.5 µM 10.0 µM	[87]
PtNPs/rGO/MWCNT	Hydrazine	219.7 µA mM^{-1}	4.5 × 10^{-8}mol L^{-1}	[88]

Huang et al. [87] synthesized PtNCs (Platinum nanochains), MWCNTs, Graphene nanoparticles composite by using ultrasonic-assisted blending process as shown in Figure 17.

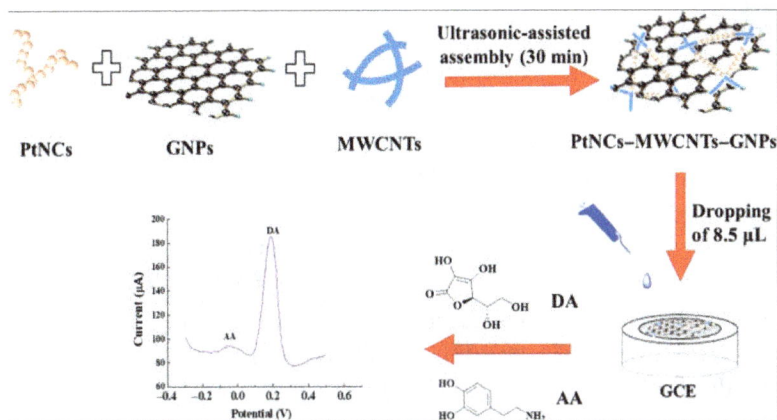

Figure 17 Preparation and electrochemical behavior of PtNCs/MWCNTs/GR NPs nanocomposite. Reproduced with permission from [87].

Modified glassy carbon electrode of PtNCs/MWCNTs/GR NPs nanocomposite [Pt NCs/MWCNTs/GR NPs/GCE] were prepared and detected both dopamine (DA) and Ascorbic acid (AA) by Cyclic Voltammetry (CV) and differential pulse voltammetry (DPV). It was found that Pt NCs/MWCNTs/GR NPs/GCE composite showed excellent reproducibility, good anti-interference ability and repeatability.

8.   Summary

The significant advances have been made in developing GR-CNTs based nanocomposite material possessing unique mechanical, thermal and electrical properties over other hybrids and also the synergistic effect operating between GR and CNTs providing extra active surface area, fast electron transfer and low detection limits for suitable electrochemical sensor applications. The development of GR-CNTS nanocomposite increased the detection of different toxic chemicals such as heavy metal ions, phenolic compounds, pesticides, biological substances. This chapter summarized the preparation and characterization of 3D GR-CNTs nanocomposite architectures for electrochemical sensor applications and also simultaneous detection of heavy metal ions, nitrite, nitrate, dopamine, ascorbic acid, hydrogen peroxide etc. are discussed. This chapter concludes that the existence of strong interaction of GR-CNTs composite matrix showed greatly improved selectivity, sensitivity and stability for sensing different toxic chemicals and also each combination GR-CNTs matrix composites possessing its own advantages and disadvantages.

References

[1] A. Touhami, Biosensors and Nanobiosensors Design and Applications. In Nanomedicine; One Central Press (OCP): Cheshire, UK, 2014, 374–403.

[2] A. P.Turner, Biosensors: Sense and sensibility.,Chem. Soc. Rev. 42 (2013) 3184–3196. https://doi.org/10.1039/c3cs35528d

[3] S. Vigneshvar, C.C. Sudhakumari, B.Senthilkumaran, H.Prakash,Recent Advances in Biosensor Technology for Potential Applications—An Overview. Front. Bioeng.Biotechnol., 4 (2016) 11. https://doi.org/10.3389/fbioe.2016.00011

[4] E.C.Alocilja, S. M. Radke,Market analysis of biosensors for food safety. Biosens.Bioelectron.,18 (2003) 841–846. https://doi.org/10.1016/S0956-5663(03)00009-5

[5] F.Liu, Y. Piao, K. S. Choi, T.S.Seo, Fabrication of free-standing graphene composite films as electrochemical biosensors. Carbon, 50 (2012) 123–133. https://doi.org/10.1016/j.carbon.2011.07.061

[6] Q.Sheng, M. Wang, J.Zheng, A novel hydrogen peroxide biosensor based on enzymatically induced deposition of polyaniline on the functionalized graphene-carbon nanotube hybrid materials. Sensor Actuat. B Chem., 160 (2011) 1070–1077. https://doi.org/10.1016/j.snb.2011.09.028

[7] Y. Yun, Z.Dong, V.N.Shanov, M.J. Schulz, Electrochemical impedance measurement of prostate cancer cells using carbon nanotube array electrodes in a microfluidic channel. Nanotechnology, 18 (2007) 465505. https://doi.org/10.1088/0957-4484/18/46/465505

[8] A.Bianco,H-M. Cheng, T.Enoki, Y.Gogotsi, R.H. Hurt, N. Koratkar, T. Kyotani, M. Monthioux, C.R. Park, J. M. D. Tascon, All in the graphene family—A recommended nomenclature for two-dimensional carbon materials. Carbon, 65 (2013) 1–6. https://doi.org/10.1016/j.carbon.2013.08.038

[9] L.Tang, Y. Wang, Y.Li, H. Feng, J.Lu, J. Li, Preparation, structure, and electrochemical properties of reduced graphene sheet films. Adv. Funct. Mater., 19 (2009) 2782–2789. https://doi.org/10.1002/adfm.200900377

[10] N.G.Shang, P. Papakonstantinou, M. McMullan, M. Chu, A. Stamboulis, A.Potenza, S.S. Dhesi, H. Marchetto, Catalyst-free efficient growth, orientation and biosensingproperties of multilayer graphene nanoflake films with sharp edge planes. Adv. Funct. Mater., 18 (2008) 3506–3514. https://doi.org/10.1002/adfm.200800951

[11] R.L.McCreery, Advanced carbon electrode materials for molecular electrochemistry, Chem. Rev., 108 (2008) 2646–2687. https://doi.org/10.1021/cr068076m

[12] W.Yang, K.R. Ratinac, S.P. Ringer, P. Thordarson, J.J Gooding, F. Braet, Carbon nanomaterials in biosensors: Should you use nanotubes or graphene? Angew. Chem. Int. Edit., 49(2010) 2114–2138. https://doi.org/10.1002/anie.200903463

[13] W. D. Zhang, B. Xu, L.-C. Jiang, Functional hybrid materials based on carbon nanotubes and metal oxides, J Mater Chem, 20 (2010) 6383-6391. https://doi.org/10.1039/b926341a

[14] W. Wu, M. Jia, Z. Zhang, X. Chen, Q. Zhang, W. Zhang, P. Li, L. Chen, Sensitive, selective and simultaneous electrochemical detection of multiple heavy metals in environment and food using a lowcost Fe_3O_4 nanoparticles/fluorinated multi-walled carbon nanotubes sensor, Ecotox Environ Safe, 175 (2019) 243-250. https://doi.org/10.1016/j.ecoenv.2019.03.037

[15] M.B.Wayu, J.E. King, J.A. Johnson, C.C. Chusuei, A zinc oxide carbon nanotube based sensor for in situ monitoring of hydrogen peroxide in swimming pools, Electroanalysis, 27 (2015) 2552-2558. https://doi.org/10.1002/elan.201500187

[16] C.Gao,Z.Guo,J-H. Liu, X-J. Huang, The new age of carbon nanotubes: An updated review of functionalized carbon nanotubes in electrochemical sensors, Nanoscale, 4 (2012) 1948-1963. https://doi.org/10.1039/c2nr11757f

[17] J.W.G.Wildoer, L.C. Venema, A.G. Rinzler, R.E. Smalley, C. Dekker, Electronically structure of atomically resolved carbon nanotubes, Nature. 39 (1998) 159-61.

[18] R.Saito, M. Fujita, G. Dresselhaus, M.S. Dresselhaus, Electronic structure of chiral graphene tubules, Appl. Phys. Lett. 60 (1992) 2204–2206. https://doi.org/10.1063/1.107080

[19] I. V. Zaporotskova, N.P.Boroznina, Y.N.Parkhomenko, L.V.Kozhitov, Carbon nanotubes: Sensor properties. A review, Modern Electronic Materials, 2 (2016) 95-105. https://doi.org/10.1016/j.moem.2017.02.002

[20] J. Prasek, J. Drbohlavova, J. Chomoucka, J. Hubalek, O. Jasek, V. Adam, R. Kizek, Methods for carbon nanotubes synthesis—review, J.Mater.Che. 21 (2011) 15872-15884. https://doi.org/10.1039/c1jm12254a

[21] K. Anazawa, K. Shimotani, C. Manabe, H. Watanabe, M. Shimizu, High purity carbon nano tube synthesis method, Applied Physics Letters, 81(2008) 739-741. https://doi.org/10.1063/1.1491302

[22] S. Marchesan, K. Kostarelos, A. Bianco, M. Prato, The winding road for carbon nano tubes in nano medicine, Mater.Today.18 (2014) 12–19. https://doi.org/10.1016/j.mattod.2014.07.009

[23] B. Singh, S. Lohan, P.S. Sandhu, A. Jain, S.K. Mehta, Chapter-15Functionalized carbon nanotubes and their promising applications in therapeutics and diagnostics, Nanobiomaterials in Medical Imaging, 8 (2016) 455-478. https://doi.org/10.1016/B978-0-323-41736-5.00015-7

[24] S. Niyogi, M.A. Hamon, H. Hu, B. Zhao, P. Bhowmik, R. Sen,. M.E. Itkis, R.C. Haddon, Chemistry of single walled carbon nanotubes, Acc. Chem. Res. 35(2002) 1105- 1113. https://doi.org/10.1021/ar010155r

[25] J. Wang, Carbon-nanotube based electrochemical biosensors: a review. Electroanal, 17 (2005),7 -14. https://doi.org/10.1002/elan.200403113

[26] D.W. Kimmel, G. LeBlanc, M.E. Meschievitz, D.E. Cliffel, Electrochemical sensors and biosensors, Anal. Chem. 84 (2012) 685–707. https://doi.org/10.1021/ac202878q

[27] H. Wang, Y. Liu, S. Yao, G. Hu, Fabrication of super pure single walled carbon nanotube electrochemical sensor and its application for picomole detection of olaquindox, Anal. Chim. Acta 1049 (2019) 82–90. https://doi.org/10.1016/j.aca.2018.10.024

[28] H. Wang, S. Yao, Y. Liu, S. Wei, J. Su, G. Hu, Molecularly imprinted electrochemical sensor based on Au nanoparticles in carboxylated multi-walled carbon nanotubes for sensitive determination of olaquindox in food and feedstuffs, Biosens. Bioelectron. 87 (2017) 417–421. https://doi.org/10.1016/j.bios.2016.08.092

[29] N. Karousis, I. Suarez-Martinez, C.P. Ewels, N. Tagmatarchis, Structure, properties, functionalization, and applications of carbon nanohorns, Chem. Rev. 116 (2016) 4850–4883. https://doi.org/10.1021/acs.chemrev.5b00611

[30] S. Zhu, G. Xu, Single-walled carbon nanohorns and their applications, Nanoscale 2 (2010) 2538–2549. https://doi.org/10.1039/c0nr00387e

[31] C. Jiang, Y. Yao, Y. Cai, J. Ping, All-solid-state potentiometric sensor using single-walled carbon nanohorns as transducer, Sens. Actuators B 283 (2019) 284–28926. https://doi.org/10.1016/j.snb.2018.12.040

[32] M. Govindhan, B-R. Adhikari, A. Chen, Nanomaterials-based electrochemical detection of chemical contaminants, RSC Adv. 4 (2014) 63741-63760. https://doi.org/10.1039/C4RA10399H

[33] X. Wang, Q. Li, J. Xie, Z. Jin, J. Wang, Y. Li, K. Jiang, S. Fan, Fabrication of ultralong and electrically uniform Single-Walled Carbon Nanotubes on clean substrates, Nano Letters. 9(2009), 3137–3141. https://doi.org/10.1021/nl901260b

[34] A. Bianco, H-M. Cheng, T. Enoki, Y. Gogotsi, R.H. Hurt, N. Koratkar, T. Kyotani, M. Monthioux, C.R. Park, J.M.D. Tascon, All in the graphene family—A recommended nomenclature for two-dimensional carbon materials. Carbon, 65(2013) 1–6. https://doi.org/10.1016/j.carbon.2013.08.038

[35] D.S. Meryl, P. Sungjin, Z. Yanwu, A. Jinho, A. Rodney, S.R. Graphene-Based Ultracapacitors. Nano Lett. 8 (2008) 3498–3502. https://doi.org/10.1021/nl802558y

[36] J. Kim, M. Ishihara, Y. Koga, K. Tsugawa, M. Hasegawa, S. Iijima, Low-temperature synthesis of large-area graphene-based transparent conductive films using surface wave plasma chemical vapor deposition. Appl. Phys. Lett. 98 (2011) 091502. https://doi.org/10.1063/1.3561747

[37] H.J. Williams, O. Richarde, Preparation of Graphitic Oxide. J. Am. Chem. Soc. 80 (1958) 1339. https://doi.org/10.1021/ja01539a017

[38] C.K. Chua, M. Pumera, Chemical reduction of graphene oxide: A synthetic chemistry viewpoint. Chem. Soc. Rev. 43 (2014) 291–312. https://doi.org/10.1039/C3CS60303B

[39] C.I.L. Justino, A.R. Gomes, A.C. Freitas, A.C. Duarte, T.A.P. Rocha-Santos, Graphene based sensors and biosensors. TrAC Trends Anal. Chem. 91 (2017) 53–66. https://doi.org/10.1016/j.trac.2017.04.003

[40] M. Carbone, L. Gorton, R. Antiochia, An Overview of the Latest Graphene-Based Sensors for Glucose Detection: The Effects of Graphene Defects. Electroanalysis 27 (2015) 16–31. https://doi.org/10.1002/elan.201400409

[41] D. Bitounis, H. Ali-Boucetta, B.H. Hong, D.H. Min, K. Kostarelos, Prospects and challenges of graphene in biomedical applications. Adv. Mater., 25 (2013) 2258–2268. https://doi.org/10.1002/adma.201203700

[42] A. N. Banerjee, Graphene and its derivatives as biomedical materials: future prospects
and challenges, Interface focus 8 (2018) 20170056. https://doi.org/10.1098/rsfs.2017.0056

[43] D. Li, W. Zhang, X. Yu, Z. Wang, Z. Su, G. Wei, When biomolecules meet graphene: From molecular level interactions to material design and applications. Nanoscale 8 (2016) 19491–19509. https://doi.org/10.1039/C6NR07249F

[44] A. Ambrosi, C.K Chua, A. Bonanni, M. Pumera, Electrochemistry of graphene and related materials. Chem. Rev., 114 (2014) 7150–7188. https://doi.org/10.1021/cr500023c

[45] D. Chen, H. Feng, J. Li, Graphene Oxide: Preparation, Functionalization, and Electrochemical Applications. Chem. Rev., 112(2012) 6027–6053. https://doi.org/10.1021/cr300115g

[46] R.S. Andre, R.C. Sanfelice, A. Pavinatto, L.H.C. Mattoso, D.S. Correa, Hybrid nanomaterials designed for volatile organic compounds sensors: A review, Materials & Design, 156 (2018) 154-166. https://doi.org/10.1016/j.matdes.2018.06.041

[47] P. Sharma, V. Bhalla, V. Dravid, G. Shekhawat, J-Wu, E. Senthilprasad, C. Raman Suri, Enhancing electrochemical detection on graphene oxide-CNT nanostructured electrodes using magneto-nanobioprobes, Scientific reports, 2 (2012) 877-883. https://doi.org/10.1038/srep00877

[48] I.A. Kinlcoh, J.Suhr, J. Lou, R. J. Young, P.M. Ajayan, Composites with carbon nanotubes and graphene: An outlook, Science, 362 (2018) 547-553. https://doi.org/10.1126/science.aat7439

[49] Q. Cheng, J. Tang. N. Shinya, L-C. Qin, Polyaniline modified graphene and carbon nanotube composite electrode for asymmetric supercapacitors of high energy density, J. Of. Power sources, 241(2013) 423-428. https://doi.org/10.1016/j.jpowsour.2013.04.105

[50] F. Liu, N. Hu, H. Ning, S. Atobe, C. Yan, Y. Liu, L. Wu, X. Liu, S. Fu, C. Xu, Y. Li, J. Zhang, Y. Wang, W. Li, Investigation of their interfacial mechanical properties of hybrid graphene-carbon nanotube/polymer nanocomposites, Carbon, 115 (2017) 694–700. https://doi.org/10.1016/j.carbon.2017.01.039

[51] H.A.-S. Mohammed, Electrical and mechanical properties of graphene/carbon nanotube hybrid nanocomposite,Synth. Met. 209 (2015) 41–46. https://doi.org/10.1016/j.synthmet.2015.06.023

[52] B. Li, S. Dong, X. Wu, C. Wang, X. Wang, J. Fang, Anisotropic thermal property of magneticallyoriented carbon nanotube/graphene polymer composites, Compos. Sci. Technol. 147 (2017) 52–61. https://doi.org/10.1016/j.compscitech.2017.05.006

[53] S.S. Jyothirmayee Aravind and S. Ramaprabhu, Sens. Actuators, B, 2011, 155, 679–686. https://doi.org/10.1016/j.snb.2011.01.029

[54] M.D. Stoller, S. Park, Y. Zhu, J. A, R.S. Ruoff, Graphene based ultracapacitors, Nano Lett., 8 (2008) 3498. https://doi.org/10.1021/nl802558y

[55] I. Meric, M.Y. Han, A.F. Young, B. Ozyilmaz, P. Kim and K.L. Shepard, Current saturation in zero-bandgap, top-gated graphene field-effect transistors, Nat. Nanotechnol., 3 (2008) 654. https://doi.org/10.1038/nnano.2008.268

[56] V. Eswaraiah, V. Sankaranarayanan, S. Ramaprabhu, Inorganic nanotubes reinforced polyvinylidene fluoride composites as low-cost electromagnetic interference shielding materials, Nanoscale Res. Lett., 6 (2011) 137. https://doi.org/10.1186/1556-276X-6-137

[57] S. Sasikaladevi, J. Aravind, V. Eswaraiah, S. Ramprabhu, Facile synthesis of one dimensional graphene wrapped carbon nanotube composites by chemical vapour deposition, J. Mat. Chem. 21 (2011) 15179. https://doi.org/10.1039/c1jm12731d

[58] C. Tang, Q. Zhang, M.Q. Zhao, J.Q. Huang, X.B. Cheng, G.L. Tian, H.J. Peng, F. Wei, Nitrogen-doped aligned carbon nanotube/graphene sandwiches: facile catalytic growth on bifunctional natural catalysts and their applications as scaffolds for high-rate lithium-sulfur batteries, Adv. Mater., 26(2014) 6100–6105. https://doi.org/10.1002/adma.201401243

[59] M.Q. Zhao, X.F. Liu, Q. Zhang, G.L. Tian, J.Q. Huang, W.C. Zhu, F. Wei, Graphene/single-walled carbon nanotube hybrids: one-step catalytic growth and applications for high-rate Li-S batteries, ACS Nano, 6 (2012) 10759–10769. https://doi.org/10.1021/nn304037d

[60] J. Xie, J. Yang, X.Y. Zhou, Y.L. Zou, J.J. Tang, S.C. Wang, F. Chen, Preparation of three-dimensional hybrid nanostructure-encapsulated sulfur cathode for high-rate lithium sulfur batteries, J. Power Sources, 253(2014) 55–63. https://doi.org/10.1016/j.jpowsour.2013.12.074

[61] V. Mani, B. Devadas, S.-M. Chen, Direct electrochemistry of glucose oxidase at electrochemically reduced graphene oxide-multiwalled carbon nanotubes hybrid material modified electrode for glucose biosensor, Biosens.Bioelectron. 41(2013) 309. https://doi.org/10.1016/j.bios.2012.08.045

[62] B. Devadas, V. Mani, S.-M. Chen, A Glucose/O_2 biofuel Cell Based on Graphene and Multiwalled Carbon Nanotube Composite Modified Electrode, Int. J. Electrochem. Sci.,7 (2012) 8064-8075.

[63] M.-Y. Yen, M-C. Hsiao, S.-H. Liao, P.-I. Liu, H.-M. Tsai, C.-C. M. Ma, N.-W. Pu, M.-D. Ger, Preparation of graphene/multi-walled carbon nanotube hybrid and its use

as photoanodes of dye-sensitized solar cells, Carbon, 49 (2011) 3597-3606. https://doi.org/10.1016/j.carbon.2011.04.062

[64] X. Dong, B. Li, A. Wei, X. Cao, M.B.C.-Park, H. Zhang, L.-J. Li, W. Huang, P. Chen, One-step growth of graphene–carbon nanotube hybrid materials by chemical vapor deposition, Carbon, 49 (2011) 2944-2949. https://doi.org/10.1016/j.carbon.2011.03.009

[65] P. Han, Y. Yu, Z. Liu, W. Xu, L. Zhang, H. Xu, S. Dong, G. Cui, Graphene oxide nanosheets/multi-walled carbon nanotubes hybrid as an excellent electrocatalytic material towards VO^{2+}/VO_2^+ redox couples for vanadium redox flow batteries, Energy Environ. Sci., 4 (2011) 4710-4717. https://doi.org/10.1039/c1ee01776d

[66] C. Zhang, L. Ren, X. Wang, T. Liu, Graphene Oxide-Assisted Dispersion of Pristine Multiwalled Carbon Nanotubes in Aqueous Media, J. Phys. Chem. C, 114 (2010) 11435-11440. https://doi.org/10.1021/jp103745g

[67] Z. Fan, J. Yan, L. Zhi, Q. Zhang, T. Wei, J. Feng, M. Zhang, W. Qian F. Wei, A three-dimensional carbon nanotube/graphene sandwich and its application as electrode in supercapacitor, Adv. Mater., 22 (2010) 3723-8. https://doi.org/10.1002/adma.201001029

[68] J. Young Oh, G.H. Jun, S. Jin, H.J. Ryu, S.H. Hong, Enhanced Electrical Networks of Stretchable Conductors with Small Fraction of Carbon Nanotube/Graphene Hybrid Fillers, ACS Appl. Mater. Interfaces, 8 (2016) 3319-3325. https://doi.org/10.1021/acsami.5b11205

[69] S. Woo, Y-R. Kim, T.D. Chung, Y. Piao, H. Kim, Synthesis of a graphene–carbon nanotube composite and its electrochemical sensing of hydrogen peroxide, Electrochimica Acta, 59 (2012) 509-514. https://doi.org/10.1016/j.electacta.2011.11.012

[70] H. Huang, T. Chen, X. Liu and H. Ma, Ultrasensitive and simultaneous detection of heavy metal ions based on three-dimensional graphene-carbon nanotubes hybrid electrode materials.

Analytica Chimica Acta, 852 (2014) 45-54. https://doi.org/10.1016/j.aca.2014.09.010

[71] M. Munishelak, A. Gargor, B. Zawisza, E. Talik, R. Sitko, Graphene Oxide/Carbon Nanotube Membranes for Highly Efficient Removal of Metal Ions from Water, ACS Appl. Mater.Interfaces, 11 (2019) 28582-28590. https://doi.org/10.1021/acsami.9b11214

[72] T. AL. Gahouari, G. Bodkhe, P. Sayyad, N. Ingle, S.M. Mahadik, S.M. Shirsat, M. Deshmukh, N. Musahwar and M. Shirsat, Electrochemical Sensor: L-Cysteine

Induced Selectivity Enhancement of Electrochemically Reduced Graphene Oxide–Multiwalled Carbon Nanotubes Hybrid for Detection of Lead (Pb^{2+}) Ions, Frontiers in materials, 7 (2020) 48. https://doi.org/10.3389/fmats.2020.00068

[73] C. Wang, M. Cao, P. Wang, Y. Ao, J. Hou, J. Qian, Preparation of graphene–carbon nanotube–TiO_2composites with enhanced photocatalytic activity for the removal of dye and Cr (VI), Applied catalyis A: general, 473 (2014) 83-89. https://doi.org/10.1016/j.apcata.2013.12.028

[74] J.G. Yu, T.T. Ma, S.W. Liu, Enhanced photocatalytic activity of mesoporous TiO_2 aggregates by embedding carbon nanotubes as electron-transfer channel, Phys. Chem. Chem. Phys. 13 (2011) 3491–3501. https://doi.org/10.1039/C0CP01139H

[75] Q. Li, B.D. Guo, J.G. Yu, J.R. Ran, B.H. Zhang, H.J. Yan, Highly efficient visible-light-driven photocatalytic hydrogen production of CdS-cluster-decorated graphene nanosheets, J. Am. Chem. Soc. 133(2011) 10878–10884. https://doi.org/10.1021/ja2025454

[76] W. Zhan, L. Gao, X. Fu, S.H. Siyal, G. Sui and X. Yang, Green synthesis of amino-functionalized carbon nanotube-graphene hybrid aerogels for high performance heavy metal ions removal, Appl. Sur. Sci., 467-468 (2019) 1122-1133. https://doi.org/10.1016/j.apsusc.2018.10.248

[77] X. Dong., Y. Ma, G. Zhu, Y.Huang, J. Wang, M B Chan-Park, L. Wang, W. Huang and P. Chen, Synthesis of graphene–carbon nanotube hybrid foam and its use as a novel three-dimensional electrode for electrochemical sensing, J. Mate. Che., 22 (2012) 17044. https://doi.org/10.1039/c2jm33286h

[78] Y-S. Wang, S-Y. Yang, S-M. Li, H-W. Tien, S-T. Hsiao, W-H Liao, C-H Liu, K-H. Chang, C-C. Ma, C-C. Hu, Three-dimensionally porous graphene–carbon nanotube composite-supported PtRu catalysts with an ultrahigh electrocatalytic activity for methanol oxidation, Electrochimica Acta (2013), 87, 261-269. https://doi.org/10.1016/j.electacta.2012.09.013

[79] Y.J. Yang, W. Lei, CTAB functionalized graphene oxide/multiwalled carbon nanotube composite modified electrode for the simultaneous determination of ascorbic acid, dopamine, uric acid and nitrite, Biosensors and Bioelectronics, 56 (2014) 300-306. https://doi.org/10.1016/j.bios.2014.01.037

[80] H. Bagheri, A. Hajian, M. Rezaei, A. Shirzadmehr, Composite of Cu metal nanoparticles-multiwall carbon nanotubes-reduced graphene oxide as a novel and high performance platform of the electrochemical sensor for simultaneous

determination of nitrite and nitrate, Journal of hazardous materials,324 (2017) 762-772. https://doi.org/10.1016/j.jhazmat.2016.11.055

[81] X. Xuan, J.Y. Park, Miniaturized flexible sensor with reduced graphene oxide/carbon nano tube modified bismuth working electrode for heavy metal detection, Sensors and Actuators B: Chemical, 255 (2018) 1220-1227. https://doi.org/10.1016/j.snb.2017.08.046

[82] V. Mani, B. Dinesh, S.M. Chen, R. Saraswathi. Direct electrochemistry of myoglobin at reduced graphene oxide-multiwalled carbon nanotubes-platinum nanoparticles nanocompositeand bio-sensing towards hydrogen peroxide and nitrite. Biosensors and Bioelectronics, 53 (2014) 420-27. https://doi.org/10.1016/j.bios.2013.09.075

[83] V. Mani, T.Y. Wu, and S.M. Chen. Iron nanoparticles decorated graphene-multiwalled carbon nanotubes nanocomposite-modified glassy carbon electrode for the sensitivedetermination of nitrite, Journal of Solid State Electrochemistry, 18 (2014) 1015–23. https://doi.org/10.1007/s10008-013-2349-z

[84] F. Hu, S. Chen, C. Wang, R. Yuan, D. Yuan, C. Wang, Study on the application of reduced graphene oxide and multiwall carbon nanotubes hybrid materials for simultaneous determination of catechol, hydroquinone, p-cresol and nitrite, Analytica Chimica Acta, 724 (2012) 40-46. https://doi.org/10.1016/j.aca.2012.02.037

[85] K. Deng, J. Zhou, H. Huang, Y. Ling, C. Li, Electrochemical Determination of Nitrite Using a Reduced Graphene Oxide–Multiwalled Carbon Nanotube-Modified Glassy Carbon Electrode, Analytical letters, 49 (2016) 2917-2930. https://doi.org/10.1080/00032719.2016.1163364

[86] K-Y Hwa, B. Subramani, Synthesis of zincoxide nanoparticles on graphene–carbon nanotube hybrid for glucose biosensor applications, Biosensors and Bio electronics, 62 (2014) 127-133. https://doi.org/10.1016/j.bios.2014.06.023

[87] Z-N, Huang, J. Zou, J. Teng, Q. Liu, M.M Yuan, F-P Jiao, X-Y. Jiang, J-G, Yu, A novel electrochemical sensor based on self-assembled platinum nanochains - Multi-walled carbon nanotubes-graphene nanoparticles composite for simultaneous determination of dopamine and ascorbic acid, Ecotoxicology and Environmental Safety, 172 (2019) 167–175. https://doi.org/10.1016/j.ecoenv.2019.01.091

[88] H. Yu, S-S. Wang, K-L Song, R. Li, A sensitive amperometric sensor for hydrazine based on Pt nanoparticles-reduced graphene oxide–multi-walled carbon nanotubes composite, International Journal of Environmental Analytical Chemistry, 99 (2019) 854-867. https://doi.org/10.1080/03067319.2019.1616707

Materials Research Foundations 82 (2020) 243-275   https://doi.org/10.21741/9781644900956-9

Chapter 9

Graphene-Carbon Nitride Based Electrochemical Sensors for Toxic Chemicals

S. Stanly John Xavier [1,*], T.S.T. Balamurugan [2], S. Ramalingam [3], R. Ramachandran [4,*],
N.S.K. Gowthaman [5*]

[1] Department of Chemistry, St. Xavier's College, Palayamkottai, India

[2] Department of Chemistry, Tsinghua University, Beijing, China

[3] Department of Chemistry, School of Advanced Sciences, Kalasalingam Academy of Research and Education, Krishnankoil-626126, Tamil Nadu, India

[4] Department of Chemistry, The Madura College, Vidya Nagar, Madurai – 625 011, Tamil Nadu, India

[5] Materials Synthesis and Characterization Laboratory, Institute of Advanced Technology, Universiti Putra Malaysia, 43400 UPM Serdang, Selangor, Malaysia

*-stanly.chem@gmail.com, ultraramji@gmail.com, nalla.perumal@upm.edu.my

Abstract

Developing cost effective, rapid and sensitive detection methods for the sensing of toxic chemicals is significant due to their potential application in chemistry like, clinical, industrial and environmental studies. Recently, Graphitic carbon nitrides (g-C_3N_4) become a new family of next generation in material chemistry courtesy of its peculiar physiochemical nature. The graphene-based two-dimensional layered structures with efficient intercalation, fine tunable surface, electronic and semiconductor properties of g-C_3N_4 provide enormous applications in a wide range of recent research. This unique nature of g-C_3N_4 has been explored in different fields such as sensors, bio-imaging, catalysis and energy storage devices. More specifically, g-C_3N_4 are extensively used in the detection of toxic chemicals owing to the alluring properties including high surface area, optoelectronic properties, physiochemical features, good water solubility, biocompatibility, non-toxicity etc. This chapter mainly summarizes the latest progress related on various synthetic methods and characterization techniques addressing the nature of g-C_3N_4 and its hybrids in detail. Furthermore, it deals with current applications of g-C_3N_4 in the electrochemical sensing of different toxic chemical contaminant in the

Graphene-Based Electrochemical Sensors for Toxic Chemicals   Materials Research Forum LLC
Materials Research Foundations **82** (2020) 243-275   https://doi.org/10.21741/9781644900956-9

environment. Finally, future prospects highlight the critical issues that provide innovative future development in this exciting research fields.

Keywords

Graphitic Carbon Nitride, 2D Layered Structures, Electrocatalysts, Electro Chemical Sensors, Toxic Chemicals

Contents

1. Introduction

Development of electrochemical sensing platforms over past decade is directly associated with the exploration of two-dimensional (2D) nano materials and their extensive utility in the design and fabrication of electrochemical sensing devices. In order to understand the tremendous amount of research focused on the development of nano-electrocatalysts to build sensing platforms, it is indispensable to know the history of 2D nanomaterials and their success in various electrocatalytic applications [1]. Today, 2D nanomaterials are the "gold standard" in material science; and actively pushing its boundaries to new horizons in various fields of research including electronics, energy conversion and storage, chemical catalysis, electrocatalysis, photo catalysis, drug delivery, nano medicine, biosensing, environmental and food analysis, etc. [1-3]. All these begun with a successful isolation of single layer of graphene nanosheets by Geim *et al.,* at Manchester University in 2004. Graphene is a single atomic layer of sp^2 hybridized carbons in a hexagonal arrangement resulting in a stable honeycomb 2D network structure. Graphene is the fundamental building blocks of 0-D fullerenes to mono and 3-D allotropes of carbons such nanotubes, and graphite, respectively [4]. Graphene possess some extraordinary physicochemical attributes such rapid electron transportability, large active surface area, high mechanical strength, and superior thermal stability over metal counterparts [5, 6]. This unique physiochemical attribute of graphene and its widespread revolution in the research community is a result of its single atomic thick 2D nanostructures. Thus, the hunt for other 2D nanostructures with excellent physiochemical parameters begun and lead to the discovery of graphene oxides (GO) and reduced graphene oxide (rGO) [5, 7].

The success story of graphene and its derivatives triggered the investigation towards the invention of graphene like 2D nanomaterials with elements other than carbon taking the atomic space in the honeycomb 2D network and their application in various fields [5-8]. This lead to the invention of boron nitrides nano sheets (BN), graphitic carbon nitrides (g-C_3N_4), transition metal oxide nano sheets and transition metal dichalcogenides (TMDS). In generic, these heteroatomic 2D nano sheets are in one (BN and, g-C_3N_4) to few atomic layers (metal oxides, and TMDS) in thickness. Further, all these heteroatomic 2D materials are able to reproduce the excellent physiochemical qualities of graphene and are employed as its substitute in diverse applications indexing, batteries, supercapacitors, electronics, catalysis, drug delivery and chemical and biosensing [9, 10].

Among the group of heteroatomic graphene-like 2D nanostructures g-C_3N_4 nanosheets (gCNNS) stood unique with high thermal stability, chemical resistance under acidic and alkaline environments, and unique electronic structure of a tunable semiconductor. A detailed study on gCNNS was traced back almost over two centuries into 1834 with Liebig, and Berzelius who first observed the formation of tri-s-triazines linear polymers connected through secondary nitrogen which they referred to as "melon" during their days [11] and people recognized the outstanding thermal stability of g-C_3N_4 allotropes over other carbon nitrides, this eventually kick starts further research in gCNNS [12]. The exceptional thermal and chemical stability and catalytic proficiency of gCNNS are attributed from its unique structure made of extremely condensed tri-s-triazine units. The 2D tri-s-triazines framework with donor nitrogen atoms offers the possibilities of localized electron clouds in gCNNS and the ternary nitrogen atom with no electron induces holes into the 2D electronic network and creates abundant active sites on the surface. The inclusion of donor and ternary (deficient) nitrogen into the 2D framework has led to formation of an exclusive electronic structure with a medium band gap of (2.7 eV) an indirect semiconductor which are easily tunable via complexion with other materials [12].

These excellent physiochemical ascribes of gCNNS soon made them as a prime material for various applications including nano drug delivery, chemical catalysis, fuel cells, energy conversion and storage devices, electrochemical biosensors, chemical sensors for environmental analysis, etc. Despite the widespread applications of gCNNS in hydrogen production, fuel cells, and photolytic dye degradations; it has made a significant impact in the development of electrochemical sensing devices utilizing gCNNS and its composites as effective electrocatalysts. They have been employed as sole or in composition with other nanomaterials as a electrocatalyst for the electrochemical sensing of biomolecules, drugs, food preservatives, aquatic organic pollutants, and toxic metals [13,14].

In the past decade, development of gCNNS based electrochemical sensing devices taking center stage courtesy of its exclusive advantages over graphene analogues such as, ease of large scale preparation, unique electronic surface properties, and simple tunable physiochemical behaviors. Further, the 2D framework with electron donor and receptor sites has made the electronic surface of gCNNS as an excellent site of interaction for other nano 2D frameworks (Gr, RGO, GO, TMDS, and TMO), metal nanoparticles (NPs) of various dimensions, biopolymers, and polycyclic organic species. The hybridizations of gCNNS with other nanostructures of its kind, or with other metal NPs assist gCNNS to overcome its inherent shortcomings via lowering the energies of highest occupied molecular orbitals (HOMO). The effect of hybridization on lowering the energies of HOMO is highly associated with coupling partners as other 2D networks predominantly relies on π-π stacking, and on the other side metal and metal oxide NPs achieving the same through strong electronic interactions. These strong π-π stacking and electronic interactions during hybridization tune the surface electronic structure and elevate the overall electrochemical performance of gCNNS and make them much feasible materials for sensing of biomolecules, drugs, and toxic pollutants and were well documented in the literature.

The triumph of gCNNS and its hybrids in photocatalytic dye degradation and optical sensing of organic pollutants and heavy metals were well discovered and documented over the years. However, the utilization of gCNNS for electrochemical sensing of toxic agents emerged only in the past decade yet crafted a sizable achievement in the short span of time. Periodic analysis of toxic substrates present in industrial and corporation wastes are essential, in order to control and regulate the emission of toxic substrates into the biosphere. gCNNS based electrochemical sensing platforms are emerging as an eventual alternative for the future rapid sensing of toxic pollutants from environmental samples. This revolution has been made possible by the excellency of gCNNS electrocatalyst coupled with the functional advantages of electrochemical sensing devices which includes portability, minimal to no pre-sampling, ease of use, low cost, rapid analysis, high sensitivity, accountable selectivity and reliability, high chemical, functional and storage stability. In this chapter, a comprehensive discussion about the synthesis, surface morphology, physiochemical characterizations, electronic behaviors, and electrochemical sensing performance of gCNNS and its hybrid electrocatalyst were discussed and given in subsequent sections.

2.   Preparation of g-C$_3$N$_4$ electrocatalysts

g-C$_3$N$_4$ is an attractive graphite-like structure with 2D-planar layers consisting of tris-s-triazine units stacked with one another by weak Van der Waals forces. There are numerous top down and bottom up protocols reported for the preparation of gCNNS. In the top down approach, bulk gCNNS solid materials are synthesized in large scale process, which are later delaminated into free standing single layers through various methods such as electrochemical oxidation, sonication, acidic oxidation, thermal oxidation and hydrothermal cutting. On the other hand, in the bottom up approach, assembly of smaller atoms or molecules leads to the formation of 2D layered structure through microwave assisted method, hydrothermal or solvothermal method and low temperature solid-phase method. Each synthetic routes and precursors have their own set of credits and limits on determining the shape, thickness, elemental composition of g-C$_3$N$_4$ and its hybrids.

2.1   Preparation of graphitic carbon nitride nanosheets (gCNNS) electrocatalyst

In electrochemical applications, different strategies have been incorporated in the preparation to increase the electrocatalytic performances of gCNNS. The metal free bulk polymeric gCNNS can be prepared by direct condensation of nitrogen rich precursors/organic compounds using urea, thiourea, melamine, dicyandiamide cyanamide and amino acids. The high degree of condensation process makes them attain chemical inertness and thermal stability. This type of condensation method is carried out in N_2, Ar, or in the air atmosphere. The stacked 2D gCNNS possessed several advantages including chemical, thermal stability, low cost and non-toxicity. However, the bulk materials obtained by this method have been restricted in electrochemical applications due to its low surface area (< 20 m^2 g^{-1}), poor conductivity, large particle size, poor water solubility. Hence, activation is adopted to improve the structural features and internal physical and chemical properties. Hatamine *et al.,* synthesized gCNNS with an ultra-thickness of 6 Å through liquid exfoliation by sonication [15]. The surface area of gCNNS modified electrodes is 20 times larger than the bare glassy carbon electrode (GCE). The exfoliated ultrathin gCNNS have high specific surface area and also provide more active sites in the structure. The increased surface area of gCNNS modified electrodes accelerating the migration rate of electrons results in greater sensitivity toward electrochemical sensing of Pb^{2+}. The activation of gCNNS by HCl mediated protonation further improves the substrate dispersion and also enables the ion mobility. These modifications enhanced the electrochemical sensing performances of gCNNS. Liu *et al.,* synthesized bulk gCNNS through thermal polycondensation method using urea as a precursor. Then, the bulk gCNNS was subjected to liquid exfoliation by ultrasonication and followed by activation with acidified HCl. The exfoliation and protonation treated gCNNS exhibited superior sensitivity compared to the untreated gCNNS and bare GCE [16]. Sadhukhan *et al.,* developed microwave assisted bottom up fabrication of g-CNNS from formamide [17]. Initially, microwave irradiation produces C_3N_4 nanodots (NDs) which are further utilized as precursor for the formation of g-CNNS through careful evaporation induced self-assembly and condensation of C_3N_4-NDs on a solid substrate. Wang *et al.,* have developed nano porous gCNNS using a soft template method using melamine and Triton X-100 with con.H_2SO_4 heated in an oil bath [18]. The obtained material is further heated to 500 °C in the muffle furnace. Thus, the introduced nanoporous morphology with remarkable surface area with excellent conductivity enables to tune the electrochemical properties of gCNNS. The gCNNS modified GCE was employed for the simultaneous detection of heavy metal ions with the use of square wave anodic stripping voltammetry (SWASV). Several porous morphologies of g-C_3N_4 were synthesized through a self-polymerization process using various surfactants such as P123, Brij 58, F127 or ionic liquid as soft templates.

2D ordered mesoporous C_3N_4 have been prepared using hard template SBA-15 mesoporous silica and melamine precursor at different pyrolysis temperatures [19]. The template assisted synthesis is supportive in developing controlled size nanostructures with appropriate surface area and desirable properties. The appropriate templates used in the fabrication affected the polymerization properties of the precursors, which resulted in different structures and surface morphology. The fabrication of C_3N_4 with different templates enables the development of different materials with required porous structure and specific morphology for the applications. Zhao and his co-workers prepared g-C_3N_4 products by thermal condensation of melamine at 723, 773, 823 and 873 K temperature for 4 h [20]. The bulk product obtained at 723 K exhibited irregular edge and flake-like stacking structure. The graphite-like structure with rough surface was resulted at 773, 823 and 873 K. Although the g-C_3N_4 exhibits bad electrochemical activity, the obtained g-C_3N_4 products at higher temperature directly used as electrochemical sensing material and exhibited increasing oxidation response towards tetrabromobisphenol-A (TBBPA) without any modification. Utilizing different temperatures in the synthesis of g-C_3N_4 predominantly affects the surface area and thickness of the material developed.

2.2 Preparation of graphitic carbon nitride nanosheets hybrid electrocatalysts

gCNNS has attracted much attention in photocatalysis and fluorescence applications due to its optical and photocatalytic properties. gCNNS material is also used in the field of electro catalysis and electrochemiluminescence. However, gCNNS exhibited poor conductivity and high rate of electron hole recombination, which restricted its electron transportation and electro catalytic activity. Hence, the electro-catalytic performance of gCNNS is further improved b compositing with metal/metal-oxide/carbon nanomaterials. These kinds of gCNNS hybrid materials have improved catalytic ability and strong interactions with the analyte.

Sheet-like g-C_3N_4 was synthesized from melamine by heating at 520 °C for 4 h [21]. The negatively charged chitosan is introduced in between the negatively charged g-C_3N_4. The chitosan is used to improve the film forming, binding ability, separating g-C_3N_4 nanosheets, mechanical and electrochemical properties. Graphene oxide doped g-CNNS was fabricated by mixing of 1 mL of 0.5 mg/mL exfoliated g-CNNS with 1 mL of 0.5 mg/mL of graphene oxide and then sonicated for 12 h. The CNNS is a fully covered with the high surface area of GO [22]. Synergistic effect of layer-by-layer structures and π-π or electrostatic interactions facilitated the conductivity, electrocatalytic acidity and selective electro oxidation process of the hybrid material.

One step simple pyrolysis technique was employed for the preparation of carbon supported gCNNS- ketjenblack carbon (KBC) composite by utilizing melamine and KBC at 550 °C for 4 h in a N_2 atmosphere. The prepared metal free carbon supported gCNNS oxidized hydrazine at lower over potential [23]. An ultrafast electrochemical sensing platform developed based on gCNNS-electrochemically deposited poly(3,4-

ethylenedioxythiophene) (PEDOT) composite which was fabricated through in-situ electropolymerization [24]. Initially gCNNS was synthesized from melamine using a pipe furnace at 550 °C. Then, gCNNS suspension was added with EDOT and electrochemically deposited by oxidation and polymerization on GCE. The PEDOT conducting matrix provides π-conjugation, high conductivity and fast electron properties to compensate for poor conductivity of gCNNS. The synergistic effect of gCNNS-EPEDOT on GCE exhibited superior electrocatalytic performance for the detection of acetaminophen.

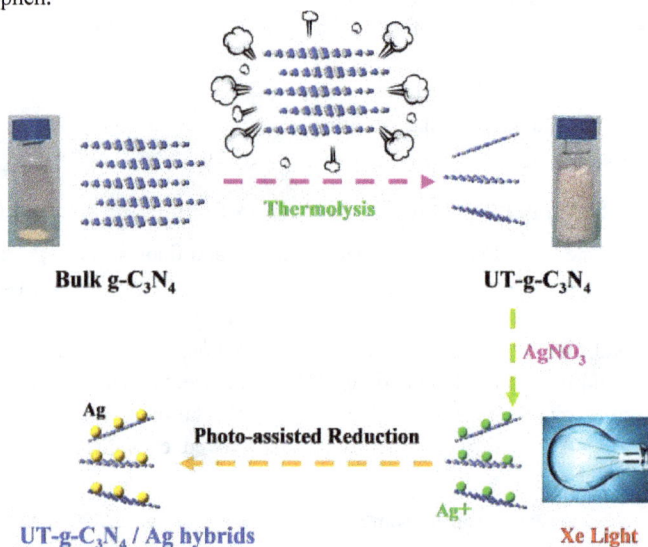

Figure 1. Schematic fabrication of ultrathin (UT) g-C₃N₄ by thermolysis (thickness of 2 nm) and UT-g-C₃N₄/Ag hybrid via photo-assisted reduction method. Reproduced with permission from [25].

Ultrathin gCNNS-Ag hybrid was prepared via thermal exfoliation and photo-assisted reduction technique. The fabricated hybrid displayed faster electron transfer rate comparatively with ultrathin and bulk gCNNS (Figure 1) [25]. The direct solid grinding and subsequent thermal polymerization method employed for the preparation of gCNNS-CNT nanocomposite and further Pd NPs modified gCNNS-CNT was attained by self-assembly method [26]. There are several carbon-based materials such as CNTs, Vulcan carbon, carbon black, mesoporous carbon, graphene sheets, N-doped graphene, RGO, N-rich RGO were utilized as conductive carbon support to advance the electron accumulation on the surface of the electrocatalysts.

PtNPs were anchored g-C_3N_4 through a simple one pot deposition reduction method and the prepared Pt-g-C_3N_4 modified carbon paper was used as an electrode material for the sensing of phenol [27]. ZnO nanosheets were grown on the exfoliated gCNNS using microwave assisted hydrothermal synthesis [28]. The noble metal free ZnO-gCNNS coated on the fluorine doped tin oxide (FTO) substrate was used for the electrochemical determination of H_2O_2. gCNNS was synthesized through a one-step electrochemical method utilizing voltammetric cells containing melamine and NaOH. The mesoporous polymeric gCNNS was stabilized on GCE by CV with scanning potential between -1.5 and 1.5 for 10 cycles at scan rate of 50 mV s^{-1}. Then, PANI and CdO-NPs were electrochemically deposited on the mesoporous gCNNS/GCE [29]. The aromatic type structure and π-conjugate system in polymeric g-C_3N_4 can easily couple with the other polymeric materials to form polymeric nanocomposites. Afshari *et al.*, developed PANI/g-C_3N_4/Ag nanocomposite electrocatalyst for the detection of hydrazine. In the first step, g-C_3N_4 was synthesized via stepwise polycondensation reaction of melamine and cyanuric acid in the presence of N,N-diisopropylethylamine. Then, the obtained material was sonicated for 10 h and 90 W powers to separate g-C_3N_4 layers without any involvement of solvents. PANI/g-C_3N_4 film was prepared by electropolymerization on the surface of FTO using cyclic voltammetry. Finally, AgNPs was electrodeposited on the surface of FTO/PANI/g-C_3N_4 through a potentiometric method in the solution of 0.2 M $AgNO_3$, and 0.1 M H_2SO_4 [30]. Similarly, V_2O_5/g-C_3N_4/PVA nanocomposite for electrochemical sensor applications was also portrayed [31].

Ultrathin g-C_3N_4 was collected from the bulk g-C_3N_4 through washing, ultrasonication, redispersion and centrifugation process (Figure 2) [32]. Sulphur doped g-CNSS with triclinic layered structure were prepared by thermal condensation of trithiocyanuric acid in a horizontal furnace at 550 °C for 5 h in static air. Doping of hetero atoms (C, N and S) provided more active sites in the nanosheets, which could enhance the electrical conductivity, fast electron transfer and electrocatalytic activity. Chemical doping is one of the effective methods to improve the surface and properties of the materials [33]. Graphene sheets are exhibiting outstanding electron collectors and transporters and have been used to improve the performance of the many electrochemical systems. The stacked 2-D structure of g-C_3N_4 is analogues to graphitic structure. Hence it is assumed that making composite of graphene and g-C_3N_4 will have excellent chemical, mechanical and electronic properties. Graphene supported Co-g-C_3N_4 novel metal-macrocyclic electrocatalyst was developed through slightly different pyrolysis techniques [34]. Firstly, the precursors were stirred and lyophilized and then subjected to pyrolysis in Ar atmosphere. The Co-doping and graphene support created many potential Co-N active sites, electron-hole puddles at the interface of graphene/g-C_3N_4, oxygen availability and facilitate charge transfer process [34]. Several spinel type magnetic nanoparticles were deposited on the gCNNS to form heterojunction structures that provide immobilization carriers for the electrochemical sensor [35-37]. Yuan *et al.*, synthesized ternary 3-D Au/g-C_3N_4/graphene nanocomposite for the electrochemical detection of antibiotics in food materials [38]. GO was synthesized through a modified Hummers method and 10 mg of GO was added with the 5 mL of melamine solution (2 mg mL^{-1}). After drying in the combustion boat the material was transferred to the pipe furnace and heated at 550 °C for 4 h. Finally, AuNPs synthesized through the simple reduction method were added with the solution of g-C_3N_4/graphene and subjected to the ultrasonication bath treatment. The synergy and interaction between graphene, g-C_3N_4 and AuNPs favour charge transfer and better electrocatalytic performance. Zhang *et al.*, developed a novel 3-D RuO_2/g-C_3N_4/rGO aerogel composite for supercapacitor applications [39]. g-C_3N_4 was prepared by heating melamine at 550 °C for 4 h with a heating rate of 10 °C min^{-1} in a muffle furnace. For obtaining protonated g-C_3N_4, the bulk product was milled into powder and treated with 37% HCl solution. RuO_2 NPs was obtained by the addition of 3 ml 30% H_2O_2 with 40 mL of $RuCl_3$ solution (0.1308 g, 0.63 mmol). Then the mixture was added in the oven at 95 °C for 5 h. Finally, 20 mL colloidal suspension of RuO_2 NPs (0.0838 g) was added into the mixture of GO (0.0838 g) and protonated g-C_3N_4 (0.0186 g) in 10 mL of deionized water. Then the mixture was transferred to the 50 mL Teflon lined stainless steel autoclave and treated at 150 °C for 20 h. The obtained cylindrical final black product was freeze dried.

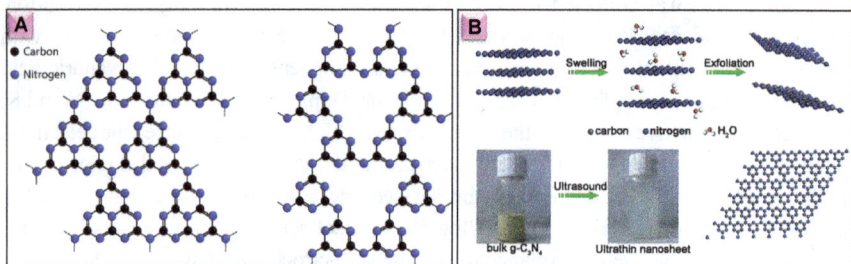

Figure 2. (A) Structure of g-C_3N_4 nanosheets; and (B) Ultrasonic exfoliation of g-C_3N_4 nanosheets from bulk g-C_3N_4. Reproduced with permission from [32].

3.   Characterization of electrocatalyst

Characterization techniques are essential to validate the formation of target nano composite at desired shape, dimension, composition and electronic states. Each aforementioned parameter plays a huge role in determining the catalytic activity, stability, electrical and electronic properties of a nanomaterial. An inclusive discussion about the various characterization tools employed to study the successful formation of g-C_3N_4 and its hybrids and what are the key parameters and findings of each characterization tools are given in brief at subsequent subsections.

3.1   Spectrophotometric and spectrofluorimetric characterizations

The optical property of g-CNNS was studied by UV-vis and fluorescence spectroscopy (Figure 3). g-CNNS typically shows a strong absorption band in the region of 250 nm to 320 nm. The UV-vis peak around 250 nm is corresponding to the characteristic of π-π* electronic transition in the s-triazine compounds (Figure 3A) [40]. It's also further confirming the presence of s-triazine rings in the g-CNNS. In certain hybrids, it exhibited strong absorption around 317 nm and broad absorption peak around 267 nm. The absorption was strengthened on the modification of g-CNNS with CNT. The strengthened absorption peak may be correlated to the interaction of g-CNNS with CNTs via π-π interaction [41]. It also exhibits the absorption pattern of a semiconductor with an obvious band gap at 460 nm. Hence, g-CNNS have shown slightly yellow color appearance. The electronic structure of g-CNNS is also characterized by UV-vis diffuse reflectance spectroscopy (UV-DRS), the absorption band observed around 340 nm indexed to the band gap energy of about 2.72 eV and also assigned to [42].

The g-CNNS have displayed attractive fluorescence properties. Generally, g-CNNS exhibits green and blue fluorescent colors. Fluorescence spectra of g-CNNS are usually broad and predominantly dependent on the excitation wavelengths. The g-CNNS exhibited strong fluorescence emission peaks at 430 nm when excited at 320 nm (Figure 3B). The fluorescence characteristics of g-C_3N_4 is correlated to the π-π* electronic transitions in π conjugated polymeric units [40]. The fluorescence spectrum of g-CNNS is strongly dependent on the degree of condensation during the formation and packing of layers in the g-CNNS. The fluorescence spectrum of uncondensed melom shows fluorescence maximum around 360 nm. However, there is a shift in the case of melon condensed at 550 °C for 4 h. The fluorescence is again blue shifted for the condensation at 600 °C owing to the perfect packing by extended tampering at high temperature [43]. The modification on the g-CNNS drastically quenches the fluorescence intensity owing to the shift in highest occupied molecular orbital (HOMO) of g-CNNS to lower energies. Doping can influence the fluorescence properties of g-CNNS and the O and S doped carbon nitrides shows red shift with excitation wavelength from 420 nm to 520 nm and the maximum fluorescence intensity observed with the excitation of 430 nm [44]. In certain cases, the fluorescent emission of g-C_3N_4 nanosheets exhibits excitation–independent behavior, which suggests that fluorescence properties of g-CNNS are dependent on their surface properties rather than the property of morphology [45]. The effect of pH toward the fluorescence stability of g-CNNS were studied with pH range from 2 to 12. The fluorescence intensity of g-CNNS was almost stable for a wide pH range from 2 to 9 and only minor change was observed for pH from 10 to 14 [46]. Several g-CNNS strongly emitted blue fluorescence with the quantum yield of 40 % relative to the fluorescence of quinine sulfate [47]. Comparatively, doped C_3N_4 have higher quantum yield than bulk and blank C_3N_4 [48]. The strong fluorescence characteristics enable the fluorescence sensor probe systems for various molecular species.

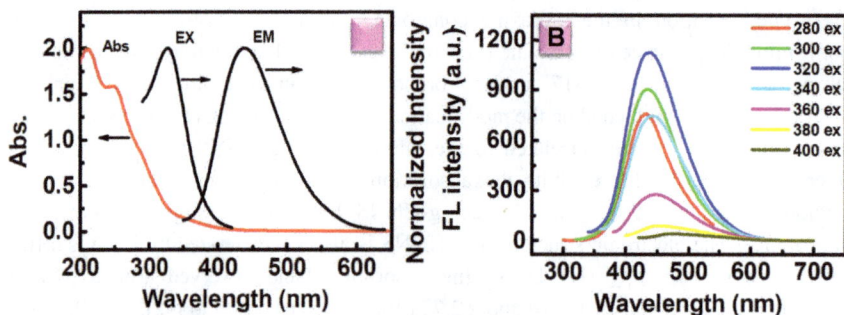

Figure 3. (A) UV and (B) Fluorescence spectra of g-C_3N_3 nanosheets. Reproduced with permission from [40].

3.2 FT-IR spectral characterization

FT-IR spectroscopy is an effective tool to understand changes in chemical structures including change in bonding, formation, and/or rearrangement of functional groups during the preparation of nanomaterials. The FT-IR spectra of bulk g-C_3N_4 displays a wide absorption band at the wavelength of 3500-3000 cm^{-1} which is characteristic to the N-H bonds found in the uncondensed terminal (primary and, secondary) amines profound in bulk g-C_3N_4 [49-51]. While the peak intensity drops with elevation in temperature signifies the elimination of terminal amines. Further, a handful of peaks can be observed in the range of 1700-1000 cm^{-1} which are advocated by the C-N stretching ternary amines present in bulk g-C_3N_4 and the intensity of these peaks increase with an increasing thermal treatment temperature. A distinct sharp peak at around the wavelength of 810 cm^{-1} is characteristic to the vibration mode of triazine skeleton of g-CNNS. A study of increase in the temperature up to 800 K during thermal treatment showcased a study increase in the peak intensities at 1700-1000 cm^{-1} and 810 cm^{-1} indicates the smooth exfoliation of g-CNNS. Further increase in temperature over 875 K resulting in a drop in the peak intensities at aforementioned wavelengths implies thermal deformation of g-CNNS [52].

In addition, FT-IR spectroscopy can be employed to study the interaction of other elemental doping, 2D materials (graphene, GO, RGO, TMDS), CNTs, biopolymers, metal oxides, and metal NPs with g-C$_3$N$_4$ nanosheets. The inclusion of B atoms into the atomic framework results in disappearance C-N linkages as the peak intensities at 1700-1000 cm^{-1} and 810 cm^{-1} diminished with increasing B doping [53], whereas, the incorporation of S into the framework results in minimizing N-H stretching intensities at 3500-3000 cm^{-1} indicates terminal occupation of S atoms in g-CNNS [54]. Aleksandrzak *et al.*, has studied the interaction between 2D carbon materials (GO, rGO, GONP, rGONP) and g-CNSS [55]. They explored that the interaction of such carbon materials has a significant effect on the chemical structure of g-CNSS. In specific, GO and GONP on hybridization with gCNNS results in a clear drop in the absorption peak at 3500-3000 cm^{-1} which is corresponding to N-H stretching of terminal amines. This drop in peak intensity can be justified by the hydrogen bonding between the N-H hydrogens and oxygen functionalities of GO and GONP. A similar effect of minimum change in peak intensities are observed in presence of rGO, and rGONP justify the effective hydrogen bonding in the case of its presiders. The hybridizing of GO and GONP with gCNNS resulted in a larger amount of change in chemical nature compared to the effects of the same between hybridization of gCNNS with rGO and rGO NP [55]. Ong *et al.*, have studied the effect of metal NPs loading on gCNNS via FT-IR. They have found that N-H stretching dropped rapidly with increasing metal loading while the C-N and triazine N peaks displayed minimum distortion (Figure 4A) [56]. Xavier *et al.*, have studied the consequence of metal oxide loading on gCNNS chemical structures through decorating MnO$_2$ nanotubes over gCNNS displayed a sizable drop in N-H stretching intensity with notable redshift and deformation of C-N peaks [57].

3.3   Raman spectroscopy

The Raman spectroscopy is a valuable tool to measure the ratio of sp^2 carbon and degree of carbon materials. The Raman spectra of GO shows two prominent peaks at 1344 and 1595 cm^{-1} denoted as D band and G band, respectively [58]. A ratio of peak intensities of D and G bands will define the degree of distortion and ratio of catalytic graphitic sites on the surface of carbon materials [59] and the value of I_D/I_G ratio will be altered with hybridization with other nano materials [60]. Hence, Raman spectra of gNNCS-carbon hybrids will be pivotal to understand their degree of distortion. N-doped graphene with I_D/I_G value of 0.94; the I_D/I_G ratio showed a sharp increase on coupling with gCNNS and calculated to be 0.99. The elevation of I_D/I_G signifies the increase in graphitic carbon domains NDC on coupling with gCNNS [61]. Raman spectrum of pristine $g-C_3N_4$ displayed three sizable peaks at 708, 1233, and 1561 cm^{-1}. The peak at 708 cm^{-1} have been ascribed to the breathing modes of the s-triazine rings; while, the other two peaks are attributed to the defects in graphite structure, and stretching modes of C=N, respectively [62-63]. Raman spectra of Au/g-C$_3$N$_4$ resulted in disappearance of the peaks at 1233 and 708 cm^{-1} with a new peak emerging 1563 cm^{-1} characteristic to C=C stretching. A complex of CNTs/g-C$_3$N$_4$ has shown a definite D and G bands along with a sizable peak accountable to breathing mode s-triazine at 710 cm^{-1} [64] (Figure 4B). In brief, the Raman analysis of gCNNS and its derivatives states that degree of distortion (i.e., catalytic activity) of gCNNS can be fine-tuned the carbon structures and nonmetallic doping resulted notable change in surface activity; while, doping of metals showcased a drastic change in the chemical nature and activity of the gCNNS-composites.

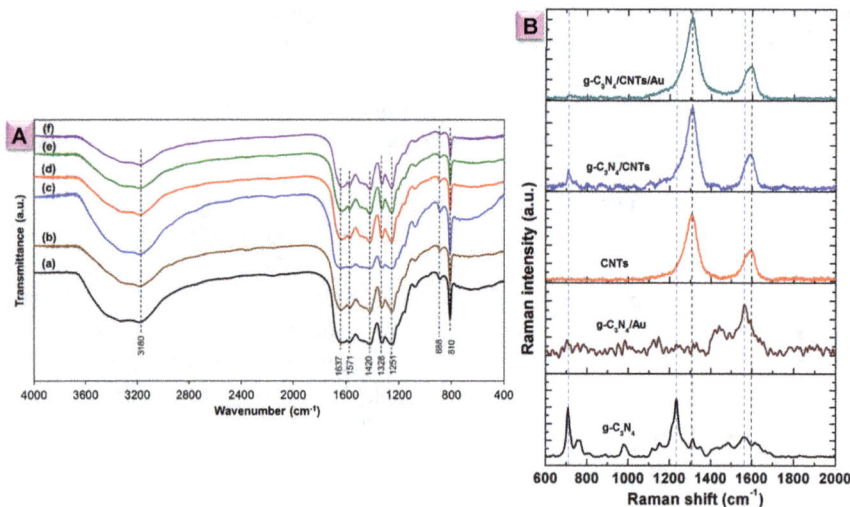

Figure 4. (A) FT-IR spectra of g-C$_3$N$_4$ nanosheets on increasing loading of Pt NPs. Reproduced with permission from [56]. (B) Raman spectra of g-C$_3$N$_4$, g-C$_3$N$_4$-CNTs, and g-C$_3$N$_4$-CNTs-Au. Reproduced with permission from [64].

3.4   Morphological characterization

Scanning electron microscope (SEM) and transmission electron microscopes (TEM) are the appropriate tools to analyze the surface morphology and shape of the nanomaterials. The surface morphology of gCNNS and its hybrids are analyzed through SEM and TEM techniques. While SEM is limited to observe the surface, shape and arrangement of layers; TEM give a much clear details about number of layers, thickness of layers, interlayer distance, and interaction modes of the layers with other nanomaterials such metals, metal oxides, GO and CNTs [65]. Sadhukhan *et al.*, have used SEM and TEM tools to observe the effect of concentration on the bottom up synthesis of gCNNS from carbon nitride dots (CNDs). His approach of self-assembled condensation of CNDs have produced thin sheets of gCNNS as envisioned through SEM which is duly verified via TEM analysis of the same [66]. Zhao *et al.*, have employed SEM and TEM techniques to monitor the effect of temperature on the bottom up synthesis of gCNNS under four different operating temperatures [52] (Figure 5). An increase in operating temperature to 773 K results in a drop in the number of layers arranged together and much regular sheet like structure confirmed via SEM and TEM analysis. Further hike in temperature to 823 K resulted in fine sheets (SEM) and few layers of much transparent sheets (TEM). A raise in temperature to 873 K resulted in the formation of fakes broken out of sheets like structures as observed under SEM and TEM which displayed the image of ruptured g-CNNS. This study clearly portrays the role of SEM and TEM analysis on controlled synthesis of gCNNS [52]. Ramalingam *et al.,* sourced HR-TEM to observe the interaction between porous g-C_3N_4 nanosheets with MWCNTs which showcased a uniform overlap and intercalation of g-C_3N_4 and MWCNTs to yield a g-C_3N_4 decorated MWCNTs bounded strong with each other [67]. Aleksandrzak *et al.*, used SEM and TEM to study the interaction of g-C_3N_4 with GO, GONPs, rGO, and rGO NPs. In the TEM images of the four complexes GONPs- g-C_3N_4 displayed maximum intercalation of the two layered materials as other combinations showing some accumulations of the layers [55]. Cheng *et al.,* have studied the interaction of AuNPs with g-C_3N_4 under TEM analysis. The TEM of pristine g-C_3N_4 displayed a clear single layer of g-C_3N_4 nanosheets and AuNPS/g-C_3N_4 portrayed the uniform decoration of AuNPs over the surface of g-CNNS [14]. These observations under SEM and TEM signify the necessity of these characterizations on controlled synthesis of g-CNNS and its hybrids.

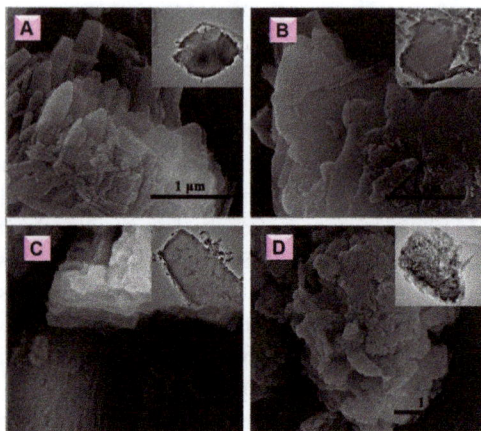

Figure 5. The SEM and TEM images (inset) of the g-C₃N₄ products synthesized at (A) 723 K, (B) 773 K, (C) 823 K, (D) 873 K. Reproduced with permission from [52].

3.5 EDX and XPS analysis

Energy dispersive X-ray (EDX) spectroscopy is a primitive tool to analyze the elemental composition of the nano materials. On the other hand, X-ray photoelectron spectroscopy (XPS) is an effective and accurate tool to measure the elemental composition along with their electronic states. The EDX spectra have been used extensively to study the elemental ratio and distribution of elements on the surface of g-C₃N₄ nanosheets and its hybrids. Pumera and co-workers have used EDX to study the elemental distribution of g-CNNS prepared from various resources. Their studies reveal that the gCNNS prepared from triazinetrihydrazine, dicyandiamide, and melamine yield g-C₃N₄ nanosheets only consist of carbon and nitrogen with no other elemental doping. However, g-CNNS prepared from trithiocyanuric acid, cyanuric acid, cyanuric chloride precursors are doped with sulfur, oxygen and chlorine respectively [68]. Mitra *et al.*, and Sharma *et al.*, employed EDX, and XPS to study the elemental distribution and electronic state of the elements present in Cu_2ONPs/g-C_3N_4 nano hybrid [69,70]. Further, the XPS survey of the same exhibited characteristic peaks of carbon, nitrogen, oxygen and copper at the binding energies of Cu, C, N and O with binding energies at 530.6, 286, 397.6 and 931.4 eV, respectively [69,70]. Further, analysis of decoupled XPS spectra confirms the electronic stats of copper (I), and existence of C-C, C-N, C=C, and C=N-C linkages in the framework [70]. Dou *et al.*, studied the elemental propositions of their prepared Fe, S doped g-C₃N₄ nanosheets. The elemental mapping of pristine g-C₃N₄ nanosheets displayed the presence of carbon and nitrogen atoms in the structure. Inclusion of $FeCl_2$

in the synthesis to induce Fe doping resulted Fe-g-C_3N_4 nanosheets as the presence of Fe is imaged in EDX mapping and spectra [71-73]. In detail, the C1s spectra of carbon at 287.90 eV is symbolize the presence of C-C, C=C linkages on the gCNNS framework. The peak corresponding to N is decoupled into three peaks with the binding energies of 398.40 eV (N1), 399.70 eV (N2), 400.80 eV (N3), which are representative of sp^2-N connected with carbon (C-N=C), and sp^3-N connected to carbon (-C-N-C_2), and hydrogen atoms (-NH_2) respectively [32]. Further, 2p spectra of Fe displayed its characteristic peaks at 710.7 and 723.8 eV corresponding to orbital split of Fe $2p^{3/2}$ and $2p^{1/2}$, justified the Fe (II) oxidation state of iron. Finally, decoupled 2p spectra of S atom displayed a peak at binding energy of 163.1 eV epitomize the C-S linkage in the g-C_3N_4 frameworks as S atoms substitute N atoms in the aromatic rings [72]. This study emphasizes the function of EDX and XPS to understand the chemical structure of the nanocomposite which determines the physiochemical attributes of the gCNNS and its hybrids [68,72,73,74].

3.6 XRD analysis

X-ray diffraction (XRD) is an instrumental technique to study the crystalline nature and number of layers/crystalline planes exists in the prepared nanomaterials and interactions between the constituents at a hybrid nanomaterial. Fan *et al.,* have utilized XRD to study the thermal exfoliation of gCNNS from bulk g-C_3N_4 at various temperatures [75]. The XRD pattern of bulk and g-CNNS have shown two typical diffraction peaks at 27.83° and 12.94°. The sharp peak with maximum intensity observed at 27.83° can be indexed to the (002) lattice plane, resulted by the periodic interlayer stacking along the c-axis of g-C_3N_4. The weak diffraction peak at 12.94° can be assigned to the (100) lattice plane of g-C_3N_4 with a planar spacing of 0.680 nm symbolize the in-plane packing of heptazine units in g-CNNS [75]. An overlapping the diffraction patterns of bulk g-C_3N_4 and gCNNS does not shown any significant change in angle of diffraction planes while the diffraction intensities varied with g-CNNS prepared at different temperatures. In particular, the intensity of the (002) peak has been dropped during liquid stripping; which demonstrates the formation of few layers of g-CNNS. A new diffraction peak has been observed at 6.22° on analyzing the g-CNNS prepared at 60 °C can be annexed to the (001) lattice plane with the d-spacing of 1.421 nm [76]. The intensity of the (001) peak increases with increasing temperature from 60 to 100 °C; to yield sharp peaks corresponding to (001) and (100) lattice planes of g-C_3N_4 nanosheets prepared at (100 °C) [77]. Ho *et al.,* have studied XRD patterns of g-CNNS prepared form various amount of precursors loading [78]. A typical diffraction spectra displayed two diffraction peaks at 13.0° and 27.20° corresponding to the crystal planes (100) and (002), respectively [79]. The diffraction angle of (002) peak was found to be increased from 27.20° to 27.42° with decreasing loading of thiourea precursor under minimized loading and drop in interplanar distance [78,80]. Xu *et al.,* have studied the interaction of g-CNNS with rGO under different ratio ranging from 1 to 5 (g-C$_3$N$_4$ /rGO) (Figure 6) [81]. The XRD of rGO shows a broad diffraction peak at 25.0° with an interlayer distance of 0.36 nm corresponding to 002 plane. The g-C$_3$N$_4$ displayed distinct diffraction peaks at about 13.2° and 27.4°, which can be indexed to the (100) and (002) planes of hexagonal g-C$_3$N$_4$ (JCPDS 87-1526) [82]. The XRD of g-C$_3$N$_4$ /rGO nanocomposite with various ratio of g-C$_3$N$_4$, exhibited a diffraction peak between 25.0 and 27.4° signifies an intermediate inter-layer distance between pure g-C$_3$N$_4$ and rGO, respectively [81]. These XRD patterns relates he hybridizations of other nanostructure on determine interlayer distance of g-CNNS and catalytic proficiency [79,81].

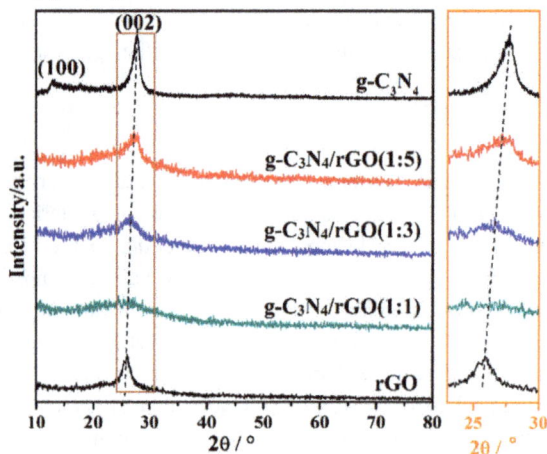

Figure 6. X-ray diffraction spectra of g-C$_3$N$_4$ /rGO. Reproduced from [81].

4.   Electrochemical behavior and sensing performance

A detailed analysis on the change in surface, morphology, elemental and electronic nature of g-C$_3$N$_4$ and its hybrids via diverse spectroscopic tools unveils g-CNNS can blend with other nanomaterials to improve its shortcomings on conductivity while preserving their excellent catalytic activity, stability, and mechanical strength. Electron portability and electrocatalytic activity, selectivity towards target analyte, signal reliability and stability of the sensing electrode are the key parameters in the design of electrocatalytic sensors for any applications. In this section, we give a short look on the electron portability and electrocatalytic sensing performance of g-C$_3$N$_4$ and its hybrids.

4.1 Electrochemical Impedance analysis

Electrochemical impedance spectroscopy (EIS) is an instrumental tool to study interfacial electronic properties of an electrocatalytic material. It has been widely employed to study the electron transfer resistance at the interface of electrode and electrolyte [83]. The electron portability and electrochemical impedance of the g-CNNS and its hybrids are studied through EIS analysis. Amiri *et al.,* have studied the EIS of g-C$_3$N$_4$-chitosan complex over bare SPCE has shown that the g-C$_3$N$_4$-chitosan/SPCE electrode displayed minimized electron transfer resistance over unmodified SPCE dissipate the excellent blending of the g-C$_3$N$_4$-chitosan hybrid rapid electron transfer on the surface of the electrode [84]. Chen *et al.,* have studied the EIS of g-C$_3$N$_4$ and S doped g-CNNS. The charge transfer resistance of the two catalysts are in the order of g-C$_3$N$_4$ (600 Ω) > S-g-C$_3$N$_4$ (150 Ω). The drop in charge transfer resistance (i.e., rapid electron transfer) can be accounted by the substitution of nitrogen atoms by electron rich S in the g-C$_3$N$_4$ framework [54, 72]. Zhang *et al.,* have studied the interactions of g-CNNS with CNTs and their electron transportability through EIS. The recorded impedance spectrum of GCE, g-C$_3$N$_4$, CNTs, COOH-CNTs, and g-C$_3$N$_4$-CNTs resulted the Nyquist plots with the diameters (i.e., charge transfer resistance) of 950, 10^4, 160, 135 and 125 Ω respectively. The study suggested that the self-assembly of g-C$_3$N$_4$-CNTs nanocomposites resulted in maximum overlapping and eventually reduced the interfacial electron transfer resistance of the resulting composite over other tested electrode materials (Figure 7) [85]. Tonda *et al,* have studied the effect of metal doping in altering the electron transfer resistance of g-CNNS. An EIS spectrum of bulk g-C$_3$N$_4$, g-CNNS, and Fe-g-CNNS delivered the charge transfer resistance of approximately 180, 175 and 110 Ω respectively. The drop in the charge transfer resistance on Fe doping can be justified by the loading of Fe metal in the voids of g-C$_3$N$_4$ nanosheets [86]. These aforementioned studies have clarified that sluggish electron portability and semiconducting nature of the g-CNNS can be fine-tuned under hybridization with other nanostructures with excellent synergy to yield g-C$_3$N$_4$-hybrids of rapid electron portability while preserving its inherent catalytic proficiency.

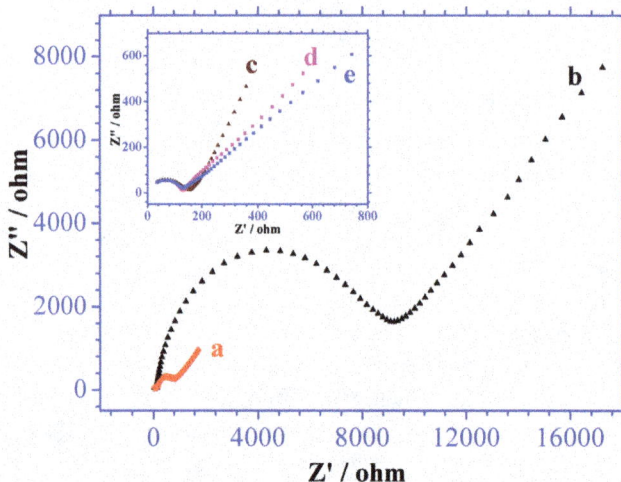

Figure 7. EIS spectra of (a) GCE, (b) CNNS/GCE, (c) CNT/GCE, (d) CNT-COOH and (e) CNNS-CNT/GCE. Reproduced with permission from [85].

4.2 Electrochemical pollutant sensors based on graphitic carbon nitride nanosheets

The easily tunable electrical properties and excellent synergy of g-CNNS with other nanomaterials in g-C_3N_4 -hybrids studied through various characterizations techniques earlier come handy in the development of g-C_3N_4 based toxic sensors. The g-CNNS plays vital role in the design of electrochemical sensors to assay toxic pollutants such as heavy metals, organic nitro compounds, bisphenol, etc. A comprehensive study of g-C_3N_4 based electrochemical pollutant sensors are given as follows. Berman *et al.*, have employed this gCNNS prepared via self-condensation of CNDs for the successful detection of toxic Hg^{2+} ions via anodic voltammetric stripping technique. His gCNNS/GCE sensor has delivered a broad working range with a calculated detection limit of 91 pM [66]. The gCNNS/CNT nano hybrid prepared by Zhang *et al.*, has served as a prolific electrocatalyst for the DPV based electrochemical detection of wide spread industrial polyphenolic pollutants catechol and hydroquinone with a good sensitivity with a accountable detection limits of 90, and 130 nM respectively (Figure 8A) [85]. Lu *et al.*, have developed a differential pulse voltammetric (DPV) sensing platform for the rapid high sensitive detection of tetrabromobisphenol (TBBP) using gCNNS as electrocatalytic material. The gCNNS based TBBP sensor offered a notable sensing performance at a working range of 20 -1000 nM with a LOD of 5 nM (Figure 8B) [52]. Up next, an amperometric electrochemical sensor to assay industrial oxidizing and bleaching agent H_2O_2 from various samples was developed by Tian *et al.*, using gCNNS/ZnO/FTO as a sensing electrode. Tian's H_2O_2 sensor based on gCNNS/ZnO/FTO electrode showcased a broad dynamic range of 0.05 to 14.15 mM, and calculated detection limit of 1.7 µM. Further, the sensor delivered a meaningful selectivity and stability towards H_2O_2 sensing [87]. Amiri *et al.*, has proposed an electrochemical sensor for the rapid, high sensitive and selective detection of Hg^{2+}ions using gCNNS/Chitosan nano hybrid under DPV techniques. His sensor delivered adequate selectivity towards Hg^{2+} ion in presence of Fe^{2+} and Cu^{2+} ions. The gCNNS/Chitosan based Hg^{2+} sensor has delivered a detection limit of 10 nM [85]. Dai *et al.*, have developed a chronoamperometric sensor for the high sensitive detection of hydrazine (HZ) a renowned industrial reducing agent and environmental pollutant using Co_3O_4/g-C_3N_4 electrocatalyst. The Co_3O_4/g-C_3N_4 based hydrazine sensor delivers high sensitivity, selectivity and stability over other electrochemical sensors reported for the same. Dai's HZ sensor owns a dynamic range of 5 to 1000 µM and calculated detection limit of 1 µM [88]. In 2019, Ramalingam *et al.* developed a DPV based electrochemical sensor for the rapid simultaneous detection of multiple toxic metal pollutants using gCNNS/MWCNT electrocatalyst. The gCNNS/MWCNT based electrochemical sensor was successful in the simultaneous detection of Cd(II), Hg(II), Pb(II) and Zn(II) metal ions with a distinct electrochemical signals at -0.78, $+0.35$, -0.5 and -1.16 V (*vs.* Ag/AgCl), correspondingly. The gCNNS/MWCNT based heavy metal sensor delivered high sensitivity and reliably towards the sensing of each metals from various real-world samples. The calculated detection limits for the four analytes are in the range of 8–60 ngL^{-1} [67]. The aforementioned reports are some key discoveries of gCNNS based electrochemical sensors to assay toxic pollutants. These studies have shown the individual catalytic excellency and synergic blending of gCNNS and its materials towards electrocatalytic sensing of toxic substance form ecological samples. A comprehensive data about the g-CNNs based electrochemical sensors developed for the high sensitive rapid detection of toxic chemicals are given in Table 1 [52,54,66,67,84-88].

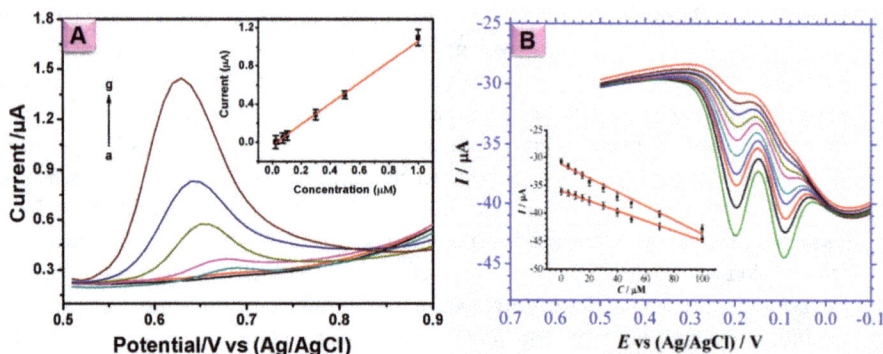

Figure 8. (A) g-C₃N₄ nanosheets based TBBP sensor. Reproduced permission from Ref. [52] and (B) g-CNNS/CNTs based electrochemical sensor was successful in the simultaneous detection of hydroquinone and catechol. Reproduced with permission from [85].

Table 1. g-CNNS based electrochemical sensors for toxic substances.

S.No	Catalyst	Analyte	Technique	Dynamic range	LOD	Reference
1	g-C₃N₄	TBBP	DPV	20 -1000 nM	5 nM	[52]
2.	S-g-C₃N₄	4-NP	DPV	0.05–90 µM	16 nM	[54]
3	g-C₃N₄	Hg^{2+}	ASV	0.1 nM- 1 µM	91 pM	[66]
4	g-C₃N₄/Chitosan	Hg^{2+}	DPV	0.8 – 10 µM	10 nM	[84]
5	g-C₃N₄/CNT	HQ	DPV	1 to 250 µM	130 nM	[85]
	g-C₃N₄/CNT	Catechol	DPV	1 to 200 µM	90 nM	
6	g-C₃N₄/ZnO/FTO	H_2O_2	Amperometry	0.05 -14.15 mM	1.7 µM	[87]
7	Co₃O₄/g-C₃N₄	HZ	Chronoamperometry	5 - 1000 µM	1 µM	[88]
8	g-C₃N₄/MWCNT	Hg^{2+}	DPV	4.8 - 93.0 µg/L	40 ng/L	[67]
		Pb^{2+}	DPV	0.35 -110 µg/L	8 ng/L	
		Cd^{2+}	DPV	4.25 -251 µg/L	30 ng/L	
		Zn^{2+}	DPV	4.2 -202 µg/L	60 ng/L	

5. Summary and future prospects

This chapter provides a brief information about the history of 2D nanomaterials based electrochemical sensors, evolution of g-CNNS in the design of electrochemical sensors, available protocols for the controlled synthesis of g-CNNS, role of characterization techniques in analyzing the chemical nature and catalytic efficacy of an electrocatalyst and its application in electrochemical sensing of toxic species from environmental samples. From the discussions, it has been evident that the g-CNNS own unique electronic structure enriched with active sites and indirect semiconductor behavior. The 2D framework consisting of donor acceptor nitrogen along with voids provides g-CNNS an ample possibility towards synergic blending with other nanostructures results a greater refinement in its chemical, electronic nature and improved catalytic proficiency. However, the studies about chemical functionalization of g-CNNS frameworks with other bio materials, impacts on chemical electronic nature and its applications are yet to be uncovered. Even though the development of toxic substance sensors based on g-CNNS are very promising, the research is in primitive states; and have to be improved in the design of futuristic strips based rapid electrochemical sensing chips. On observing the growth of electrochemical sensing platforms, developing a g-CNNS based environmental pollutant sensing devices are greatly possible in the near future with more research works invested on this prospect.

References

[1] P.K. Kannan, D.J. Late, H. Morgan, C.S. Rout, Recent developments in 2D layered inorganic nanomaterials for sensing, Nanoscale 7 (32) (2015) 13293-13312. https://doi.org/10.1039/C5NR03633J

[2] A.H. Khan, S. Ghosh, B. Pradhan, A. Dalui, L.K. Shrestha, S.A. Charya, K. Ariga, Two dimensional (2D) nanomaterials towards electrochemical nanoarchitectonics in energy-related applications, B. Chem. Soc Jpn. 90 (6) (2017) 627-648. https://doi.org/10.1246/bcsj.20170043

[3] W. Wen, Y. Song, X. Yan, C. Zhu, D. Du, S. Wang, A.M. Asiri, Y. Lin, Recent advances in emerging 2D nanomaterials for biosensing and bioimaging applications, Mater. Today Commun. 21 (2) (2018) 164-177. https://doi.org/10.1016/j.mattod.2017.09.001

[4] K.S. Novoselov, A.K. Geim, S.V. Morozov, D. Jiang, Y. Zhang, S.V. Dubonos, I.V. Grigorieva, A.A. Firsov, Electric field effect in atomically thin carbon films, Science 306 (5696) (2004) 666-669. https://doi.org/10.1126/science.1102896

[5] Y. Huang, J. Liang, Y. Chen, The application of graphene based materials for actuators, J. Mater. Chem. A. 22 (9) (2012) 3671-3679. https://doi.org/10.1039/c2jm15536b

[6] G. Jo, M. Choe, S. Lee, W. Park, Y.H. Kahng, T. Lee, The application of graphene as electrodes in electrical and optical devices, Nanotechnology 23 (11) (2012) 112001. https://doi.org/10.1088/0957-4484/23/11/112001

[7] Y. Chen, C. Tan, H. Zhang, L. Wang, Two-dimensional graphene analogues for biomedical applications, Chem. Soc. Rev. 44 (9) (2015) 2681-2701. https://doi.org/10.1039/C4CS00300D

[8] D. Chimene, D.L. Alge, A.K. Gaharwar, Two-dimensional nanomaterials for biomedical applications: Emerging trends and future prospects, Adv. Mater. 27 (45) (2015) 7261-7284. https://doi.org/10.1002/adma.201502422

[9] J.N. Coleman, M. Lotya, A. O'Neill, S.D. Bergin, P.J. King, U. Khan, K. Young, A Gaucher, S. De, R.J. Smith, I.V. Shvets, Two-dimensional nanosheets produced by liquid exfoliation of layered materials, Science 331 (2011) 568-571. https://doi.org/10.1126/science.1194975

[10] A. Gupta, T. Sakthivel, S. Seal, Recent development in 2D materials beyond graphene, Prog. Mater. Sci. 73 (2015) 44-126. https://doi.org/10.1016/j.pmatsci.2015.02.002

[11] J.V. Liebig, About some nitrogen compounds, Ann. Pharm. 10 (1834) 10. https://doi.org/10.1002/jlac.18340100103

[12] J. Zhu, P. Xiao, H. Li, S.A. Carabineiro, Graphitic carbon nitride: synthesis, properties, and applications in catalysis, ACS Appl. Mater. Interfaces 6 (2014) 16449-16465. https://doi.org/10.1021/am502925j

[13] M. Xiong, Q. Rong, H.M. Meng, X.B. Zhang, Two-dimensional graphitic carbon nitride nanosheets for biosensing applications, Biosens. Bioelectron. 89 (2017) 212-223. https://doi.org/10.1016/j.bios.2016.03.043

[14] N. Cheng, J. Tian, Q. Liu, C. Ge, A.H. Qusti, A.M. Asiri, A.O. Al-Youbi, X. Sun, Au-nanoparticle-loaded graphitic carbon nitride nanosheets: green photocatalytic synthesis and application toward the degradation of organic pollutants, Appl. Mater. Interfaces 5 (15) (2013) 6815-6819. https://doi.org/10.1021/am401802r

[15] A. Hatamie, P. Jalilian, E. Rezvani, A. Kakavand, A. Simchi, Fast and ultra-sensitive voltammetric detection of lead ions by two-dimensional graphitic carbon nitride (g-C_3N_4) nanolayers as glassy carbon electrode modifier, Measurement 134 (2019) 679-687. https://doi.org/10.1016/j.measurement.2018.10.082

[16] Y. Liu, G.L. Wen, X. Chen, R. Weerasooriya, Z.Y. Hong, L.C. Wang, Z.J. Huang, Y.C Wu, Construction of electrochemical sensing interface towards Cd (II) based on activated g-C_3N_4 nanosheets: considering the effects of exfoliation and protonation treatment, Anal. Bioanal. Chem. 412 (2020) 343-353. https://doi.org/10.1007/s00216-019-02240-z

[17] M. Sadhukhan, S. Barman, Bottom-up fabrication of two-dimensional carbon nitride and highly sensitive electrochemical sensors for mercuric ions, J. Mater. Chem. A. 1 (8) (2013) 2752-2756. https://doi.org/10.1039/c3ta01523h

[18] W. Wang, J. Zhao, Y. Sun, H. Zhang, Facile synthesis of g-C_3N_4 with various morphologies for application in electrochemical detection, RSC Adv. 9 (2019) 7737-7746. https://doi.org/10.1039/C8RA10166C

[19] Y. Zhang, X. Bo, A. Nsabimana, C. Luhana, G. Wang, H. Wang, M. Li, L. Guo, Fabrication of 2D ordered mesoporous carbon nitride and its use as electrochemical sensing platform for H_2O_2, nitrobenzene, and NADH detection, Biosens. Bioelectron. 53 (2014) 250-256. https://doi.org/10.1016/j.bios.2013.10.001

[20] Q. Zhao, W. Wu, X. Wei, S. Jiang, T. Zhou, Q. Li, Q. Lu, Graphitic carbon nitride as electrode sensing material for tetrabromobisphenol-A determination, Sensor Actuat. B-Chem. 248 (2017) 673-681. https://doi.org/10.1016/j.snb.2017.04.002

[21] M. Amiri, H. Salehniya, A. Habibi-Yangjeh, Graphitic carbon nitride/chitosan composite for adsorption and electrochemical determination of mercury in real samples, Ind. Eng. Chem. Res. 55 (29) (2016) 8114-8122. https://doi.org/10.1021/acs.iecr.6b01699

[22] H. Zhang, Q. Huang, Y. Huang, F. Li, W. Zhang, C. Wei, J. Chen, P. Dai, L. Huang, Z. Huang, L. Kang, Graphitic carbon nitride nanosheets doped graphene oxide for electrochemical simultaneous determination of ascorbic acid, dopamine and uric acid, Electrochim. Acta. 142 (2014) 125-131. https://doi.org/10.1016/j.electacta.2014.07.094

[23] K. Ramanujam, T. Thirupathi, Carbon supported g-C_3N_4 for electrochemical sensing of hydrazine, J. Electrochem. Energy 4 (1) (2018) 21-31. https://doi.org/10.1515/eetech-2018-0003

[24] Y. Xu, W. Lei, J. Su, J. Hu, X. Yu, T. Zhou, Y. Yang, D. Mandler, Q. Hao, A high-performance electrochemical sensor based on g-C_3N_4-E-PEDOT for the determination of acetaminophen, Electrochim. Acta. 259 (2018) 994-1003. https://doi.org/10.1016/j.electacta.2017.11.034

[25] J. Zou, D. Mao, A.T.S. Wee, J. Jiang, Micro/nano-structured ultrathin g-C$_3$N$_4$/Ag nanoparticle hybrids as efficient electrochemical biosensors for L-tyrosine, Appl. Surf. Sci. 467 (2019) 608-618. https://doi.org/10.1016/j.apsusc.2018.10.187

[26] Z. Xiang Zheng, M. Wang, X. Zhao Shi, C. Ming Wang, Palladium nanoparticles/graphitic carbon nitride nanosheets-carbon nanotubes as a catalytic Amplification Platform for the selective determination of 17α-ethinylestradiol in Feedstuffs, Sci. Rep. 9 (1) (2019) 1-9. https://doi.org/10.1038/s41598-019-50087-2

[27] B.B. Song, Y.F. Zhen, H.Y. Yin, X.C. Song, Electrochemical sensor based on platinum nanoparticles modified graphite-like carbon nitride for detection of phenol, J. Nanosci. Nanotechnol. 19 (7) (2019) 4020-4025. https://doi.org/10.1166/jnn.2019.16297

[28] H. Tian, H. Fan, J. Ma, L. Ma, G. Dong, Noble metal-free modified electrode of exfoliated graphitic carbon nitride/ZnO nanosheets for highly efficient hydrogen peroxide sensing, Electrochim. Acta. 247 (2017) 787-794. https://doi.org/10.1016/j.electacta.2017.07.083

[29] S. Bonyadi, K. Ghanbari, M. Ghiasi, All-electrochemical synthesis of a three-dimensional mesoporous polymeric g-C$_3$N$_4$/PANI/CdO nanocomposite and its application as a novel sensor for the simultaneous determination of epinephrine, paracetamol, mefenamic acid, and ciprofloxacin, New J. Chem. 44 (8) (2020) 3412-3424. https://doi.org/10.1039/C9NJ05954G

[30] M. Afshari, M. Dinari, M.M. Momeni, The graphitic carbon nitride/polyaniline/silver nanocomposites as a potential electrocatalyst for hydrazine detection, J. Electroanal. Chem. 833 (2019) 9-16. https://doi.org/10.1016/j.jelechem.2018.11.022

[31] A. Karthika, A. Suganthi, M. Rajarajan, An in-situ synthesis of novel V$_2$O$_5$/g-C$_3$N$_4$/PVA nanocomposite for enhanced electrocatalytic activity toward sensitive and selective sensing of folic acid in natural samples, Arab. J. Chem. 13 (2) (2020) 3639-3652. https://doi.org/10.1016/j.arabjc.2019.12.009

[32] X. Zhang, X. Xie, H. Wang, J. Zhang, B. Pan, Y. Xie, Enhanced photoresponsive ultrathin graphitic-phase C$_3$N$_4$ nanosheets for bioimaging, J. Am. Chem. Soc. 135 (1) (2013)18-21. https://doi.org/10.1021/ja308249k

[33] C. Rajkumar, P. Veerakumar, S.M. Chen, B. Thirumalraj, K.C. Lin, Ultrathin sulfur-doped graphitic carbon nitride nanosheets as metal-free catalyst for electrochemical

sensing and catalytic removal of 4-nitrophenol, ACS Sustain. Chem. Eng. 6 (12) (2018) 16021-16031. https://doi.org/10.1021/acssuschemeng.8b02041

[34] Q. Liu, J. Zhang, Graphene supported Co-g-C$_3$N$_4$ as a novel metal–macrocyclic electrocatalyst for the oxygen reduction reaction in fuel cells, Langmuir 29 (11) (2013) 3821-3828. https://doi.org/10.1021/la400003h

[35] Y. Mei, C.Y. Tang, Recent developments and future perspectives of reverse electrodialysis technology: A review. Desalination 425 (2018) 156-174. https://doi.org/10.1016/j.desal.2017.10.021

[36] S. Ju, T.Y. Cai, H.S. Lu, C.D. Gong, Pressure-induced crystal structure and spin-state transitions in magnetite (Fe$_3$O$_4$), J. Am. Chem. Soc. 134 (33) (2012) 13780-13786. https://doi.org/10.1021/ja305167h

[37] M. Hassannezhad, M. Hosseini, M.R. Ganjali, M. Arvand, A graphitic carbon nitride (g-C $_3$N$_4$/Fe$_3$O$_4$) nanocomposite: an efficient electrode material for the electrochemical determination of tramadol in human biological fluids, Anal. Methods. 11 (15) (2019) 2064-2071. https://doi.org/10.1039/C9AY00146H

[38] Y. Yuan, F. Zhang, H. Wang, L. Gao, Z. Wang, A Sensor Based on Au Nanoparticles/Carbon Nitride/Graphene Composites for the Detection of Chloramphenicol and Ciprofloxacin, ECS J. Solid State Sci. Technol. 7 (12) (2018) 201-208. https://doi.org/10.1149/2.0111812jss

[39] J. Zhang, J. Ding, C. Li, B. Li, D. Li, Z. Liu, Q. Cai, J. Zhang, Y. Liu, Fabrication of novel ternary three-dimensional RuO$_2$/graphitic-C$_3$N$_4$@reduced graphene oxide aerogel composites for supercapacitors, ACS Sustain. Chem. Eng. 5 (6) (2017) 4982-4991. https://doi.org/10.1021/acssuschemeng.7b00358

[40] Q. Lu, J. Deng, Y. Hou, H. Wang, H. Li, Y. Zhang, One-step electrochemical synthesis of ultrathin graphitic carbon nitride nanosheets and their application to the detection of uric acid, Chem. Comm. 51 (61) (2015) 12251-12253. https://doi.org/10.1039/C5CC04231C

[41] H. Zhang, Y. Huang, S. Hu, Q. Huang, C. Wei, W. Zhang, W. Yang, P. Dong, A. Hao, Self-assembly of graphitic carbon nitride nanosheets–carbon nanotube composite for electrochemical simultaneous determination of catechol and hydroquinone, Electrochim. Acta. 176 (2015) 28-35. https://doi.org/10.1016/j.electacta.2015.06.119

[42] C. Rajkumar, P. Veerakumar, S.M. Chen, B. Thirumalraj, K.C. Lin, Ultrathin sulfur-doped graphitic carbon nitride nanosheets as metal-free catalyst for electrochemical

sensing and catalytic removal of 4-nitrophenol, ACS Sustain. Chem. Eng. 6 (12) (2018) 16021-16031. https://doi.org/10.1021/acssuschemeng.8b02041

[43] A.J. Wang, H. Li, H. Huang, Z.S. Qian, J.J. Feng, Fluorescent graphene-like carbon nitrides: synthesis, properties and applications, J. Mater. Chem C 4(35) (2016) 8146-8160. https://doi.org/10.1039/C6TC02330D

[44] A. Thomas, A. Fischer, F. Goettmann, M. Antonietti, J.O. Müller, R. Schlögl, J.M. Carlsson, Graphitic carbon nitride materials: variation of structure and morphology and their use as metal-free catalysts. J. Mater. Chem. 18 (41), (2008) 4893-4908. https://doi.org/10.1039/b800274f

[45] Y.T. Wang, N Wang, M.L. Chen, T Yang, J.H. Wang, One step preparation of proton functionalized photoluminescent graphitic carbon nitride and its sensing applications, RSC Adv. 6 (101) (2016) 98893-98898.

[46] S.S.J. Xavier, G. Siva, M. Ranjani, S.D. Rani, N. Priyanga, R. Srinivasan, M. Pannipara, A.G. Al-Sehemi, Turn-on fluorescence sensing of hydrazine using MnO_2 nanotube- decorated g-C_3N_4 nanosheets, New. J. Chem. 43 (33) (2019) 13196-13204. https://doi.org/10.1039/C6RA22829A

[47] M. Rong, X. Song, T. Zhao, Q. Yao, Y. Wang, X. Chen, Synthesis of highly fluorescent P, O-g-C_3N_4 nanodots for the label-free detection of Cu^{2+} and acetylcholinesterase activity, J. Mater. Chem. C 3 (41) (2015) 10916-10924. https://doi.org/10.1039/C5TC02584B

[48] S. Sun, X. Gou, S. Tao, J. Cui, J. Li, Q. Yang, S Liang, Z. Yang, Mesoporous graphitic carbon nitride (g-C_3N_4) nanosheets synthesized from carbonated beverage-reformed commercial melamine for enhanced photocatalytic hydrogen evolution, Mater. Chem. Front. 3 (4) (2019) 597-605. https://doi.org/10.1039/C8QM00577J

[49] J. Fang, H. Fan, H. Zhu, L.B. Kong, L. Ma, Dyed graphitic carbon nitride with greatly extended visible-light-responsive range for hydrogen evolution, J. Catal. 339 (2016) 93-101. https://doi.org/10.1016/j.jcat.2016.03.021

[50] B. Fahimirad, A. Asghari, M. Rajabi, Magnetic graphitic carbon nitride nanoparticles covalently modified with an ethylenediamine for dispersive solid-phase extraction of lead (II) and cadmium (II) prior to their quantitation by FAAS, Microchim. Acta. 184 (8) (2017) 3027-3035. https://doi.org/10.1007/s00604-017-2273-5

[51] F. Dong, Y. Sun, L. Wu, M. Fu, Z. Wu, Facile transformation of low cost thiourea into nitrogen-rich graphitic carbon nitride nanocatalyst with high visible light

photocatalytic performance, Catal. Sci. Technol. 2 (7) (2012) 1332-1335. https://doi.org/10.1039/c2cy20049j

[52] Q. Zhao, W. Wu, X. Wei, S. Jiang, T. Zhou, Q. Li, Q. Lu, Graphitic carbon nitride as electrode sensing material for tetrabromobisphenol-A determination, Sens. Actuators B Chem. 248 (2017) 673-681. https://doi.org/10.1016/j.snb.2017.04.002

[53] S. Thaweesak, S. Wang, M. Lyu, M. Xiao, P. Peerakiatkhajohn, L. Wang, Boron-doped graphitic carbon nitride nanosheets for enhanced visible light photocatalytic water splitting, Dalton Trans. 46 (32) (2017) 10714-10720. https://doi.org/10.1039/C7DT00933J

[54] C. Rajkumar, P. Veerakumar, S.M. Chen, B. Thirumalraj, K.C. Lin, Ultrathin sulfur-doped graphitic carbon nitride nanosheets as metal-free catalyst for electrochemical sensing and catalytic removal of 4-nitrophenol, ACS Sustain. Chem. Eng. 6 (12) (2018) 16021-16031. https://doi.org/10.1021/acssuschemeng.8b02041

[55] M. Aleksandrzak, W. Kukulka, E. Mijowska, Graphitic carbon nitride/graphene oxide/reduced graphene oxide nanocomposites for photoluminescence and photocatalysis, Appl. Surf. Sci. 398 (2017) 56-62. https://doi.org/10.1016/j.apsusc.2016.12.023

[56] W.J. Ong, L.L. Tan, S.P. Chai, S.T. Yong, Heterojunction engineering of graphitic carbon nitride (g-C_3N_4) via Pt loading with improved daylight-induced photocatalytic reduction of carbon dioxide to methane, Dalton Trans. 44 (3) (2015) 1249-1257. https://doi.org/10.1039/C4DT02940B

[57] S.S.J. Xavier, G, Siva, M. Ranjani, S.D. Rani, N. Priyanga, R. Srinivasan, M. Pannipara,. A.G. Al-Sehemi, G.G. Kumar, Turn-on fluorescence sensing of hydrazine using MnO_2 nanotube-decorated g-C_3N_4 nanosheets, New. J. Chem. 43 (33) (2019) 13196-13204. https://doi.org/10.1039/C9NJ01370A

[58] G. Yasin, M. Arif, M. Shakeel, Y. Dun, Y. Zuo, W.Q. Khan, Y. Tang, A. Khan, M. Nadeem, Exploring the Nickel–Graphene Nanocomposite Coatings for Superior Corrosion Resistance: Manipulating the Effect of Deposition Current Density on its Morphology, Mechanical Properties, and Erosion-Corrosion Performance, Adv. Eng. Mater. 20 (7) (2018) 1701166. https://doi.org/10.1002/adem.201701166

[59] Z. Yang, X. Zheng, Z. Li, J. Zheng, A facile one-pot synthesis of carbon nitride dots–reduced graphene oxide nanocomposites for simultaneous enhanced detecting of dopamine and uric acid, Analyst 141 (15) (2016) 4757-4765. https://doi.org/10.1039/C6AN00640J

[60] G. Yasin, M. Arif, M. Shakeel, Y. Dun, Y. Zuo, W.Q. Khan, Y. Tang, A. Khan, M. Nadeem, Exploring the Nickel–Graphene Nanocomposite Coatings for Superior Corrosion Resistance: Manipulating the Effect of Deposition Current Density on its Morphology, Mechanical Properties, and Erosion-Corrosion Performance, Adv. Eng. Mater. 20 (7) (2018) 1701166. https://doi.org/10.1002/adem.201701166

[61] A.C. Ferrari, J. Robertson, Interpretation of Raman spectra of disordered and amorphous carbon, Phys. Rev. B. 61 (20) (2000) 14095. https://doi.org/10.1103/PhysRevB.61.14095

[62] P.V. Zinin, L.C. Ming, Sharma, S.K. Khabashesku, V.N. Liu, X. Hong, S. Endo, S.T. Acosta, Ultraviolet and near-infrared Raman spectroscopy of graphitic C_3N_4 phase, Che. Phys. Lett. 472 (1-3) (2009) 69-73. https://doi.org/10.1016/j.cplett.2009.02.068

[63] C.Wu, Z. Wang, L.Wang, P.T Williams, J. Huang, Sustainable processing of waste plastics to produce high yield hydrogen-rich synthesis gas and high quality carbon nanotubes, RSC Adv. 2 (10) (2012) 4045-4047. https://doi.org/10.1039/c2ra20261a

[64] R.C Pawar, S. Kang, S.H. Ahn, C.S. Lee, Gold nanoparticle modified graphitic carbon nitride/multi-walled carbon nanotube (g-C_3N_4/CNTs/Au) hybrid photocatalysts for effective water splitting and degradation, RSC Adv. 5 (31) (2015) 24281-24292. https://doi.org/10.1039/C4RA15560B

[65] B.J. Inkson, Scanning electron microscopy (SEM) and transmission electron microscopy (TEM) for materials characterization, Material Characterization using nondestructive evaluation (NDE) methods, Elsevier, (2016) 17-43. https://doi.org/10.1016/B978-0-08-100040-3.00002-X

[66] M. Sadhukhan, S. Barman, Bottom-up fabrication of two-dimensional carbon nitride and highly sensitive electrochemical sensors for mercuric ions, J. Mater. Chem. A 1 (8) (2013) 2752-2756. https://doi.org/10.1039/c3ta01523h

[67] M. Ramalingam, V.K. Ponnusamy, S.N Sangilimuthu, A nanocomposite consisting of porous graphitic carbon nitride nanosheets and oxidized multiwalled carbon nanotubes for simultaneous stripping voltammetric determination of cadmium (II), mercury (II), lead (II) and zinc (II), Microchim. Acta. 186 (2) (2019) 69. https://doi.org/10.1007/s00604-018-3178-7

[68] H.L. Lee, Z. Sofer, V. Mazánek, J. Luxa, C.K. Chua, M. Pumera, Graphitic carbon nitride: effects of various precursors on the structural, morphological and

electrochemical sensing properties, Appl. Mater. Today 8 (2017) 150-162.
https://doi.org/10.1016/j.apmt.2016.09.019

[69] A. Mitra, P. Howli, D. Sen, B. Das, K.K. Chattopadhyay, Cu_2O/g-C_3N_4
nanocomposites: an insight into the band structure tuning and catalytic efficiencies,
Nanoscale 8 (45) (2016) 19099-19109. https://doi.org/10.1039/C6NR06837E

[70] A.S. Sharma, V.S Sharma, H. Kaur, Graphitic carbon nitride decorated with Cu_2O
nanoparticles for the visible light activated synthesis of ynones, aminoindolizines and
pyrrolo [1, 2a] quinoline, ACS Appl. Nano Mater. 3 (2) (2020) 1191-1202.
https://doi.org/10.1021/acsanm.9b01928

[71] Q. Liu, Y Guo, Z. Chen, Z. Zhang, X. Fang, Constructing a novel ternary Fe
(III)/graphene/g-C_3N_4 composite photocatalyst with enhanced visible-light driven
photocatalytic activity via interfacial charge transfer effect, Appl catal. 183 (2016)
231-241. https://doi.org/10.1016/j.apcatb.2015.10.054

[72] L. Ge, C. Han, X. Xiao, L. Guo, Y. Li, Enhanced visible light photocatalytic
hydrogen evolution of sulfur-doped polymeric g-C_3N_4 photocatalysts,
Mater. Res. Bull. 48 (10) (2013) 3919-3925.
https://doi.org/10.1016/j.materresbull.2013.06.002

[73] J. Gao, Y. Wang, S. Zhou, W. Lin, Y. Kong, A facile one-step synthesis of Fe-doped
g-C_3N_4 nanosheets and their improved visible-light photocatalytic performance,
Chem Cat. Chem. 9 (9) (2017) 1708-1715. https://doi.org/10.1002/cctc.201700492

[74] H. Dou, S. Zheng, Y. Zhang, The effect of metallic Fe (II) and nonmetallic S
codoping on the photocatalytic performance of graphitic carbon nitride, RSC Adv. 8
(14) (2018) 7558-7568. https://doi.org/10.1039/C8RA00056E

[75] L.Ge, Synthesis and photocatalytic performance of novel metal-free g-C_3N_4
photocatalysts, Mater. Lett. 65 (17-18) (2011) 2652-2654.
https://doi.org/10.1016/j.matlet.2011.05.069

[76] X. Bai, L. Wang, R. Zong, Y. Zhu, Photocatalytic activity enhanced via g-C_3N_4
nanoplates to nanorods, J. Phys. Chem. 117 (19) (2013) 9952-9961.
https://doi.org/10.1021/jp402062d

[77] C. Fan, J. Miao, G. Xu, J. Liu, J. Lv, Y. Wu, Graphitic carbon nitride nanosheets
obtained by liquid stripping as efficient photocatalysts under visible light, RSC Adv.
7 (59) (2017) 37185-37193. https://doi.org/10.1039/C7RA05732F

[78] Z. Zhao, Y. Sun, Q. Luo, F. Dong, H. Li, W.K. Ho, Mass-controlled direct synthesis of graphene-like carbon nitride nanosheets with exceptional high visible light activity. Less is better, Sci. Rep. 5 (2015) 14643. https://doi.org/10.1038/srep14643

[79] W. Zhang, Q. Zhang, F. Dong, Z. Zhao, The multiple effects of precursors on the properties of polymeric carbon nitride, Int. J. Photoenergy (2013) 1-9. https://doi.org/10.1155/2013/685038

[80] M. Tahir, C. Cao, N. Mahmood, F.K. Butt, A. Mahmood, F. Idrees, S. Hussain, M Tanveer, Z. Ali, I. Aslam, Multifunctional g-C_3N_4 nanofibers: a template-free fabrication and enhanced optical, electrochemical, and photocatalyst properties, ACS Appl. Mater. Interfaces 6 (2) (2014) 1258-1265. https://doi.org/10.1021/am405076b

[81] J. Xu, D. Li, Y. Chen, L. Tan, B. Kou, F. Wan, W. Jiang, F. Li, Constructing sheet-on-sheet structured graphitic carbon nitride/reduced graphene oxide/layered MnO_2 ternary nanocomposite with outstanding catalytic properties on thermal decomposition of ammonium perchlorate, J. Nanomater. 7 (12) (2017) 450. https://doi.org/10.3390/nano7120450

[82] L. Tan, J. Xu, X. Zhang, Z. Hang, Y. Jia, S. Wang, Synthesis of g-C_3N_4/CeO_2 nanocomposites with improved catalytic activity on the thermal decomposition of ammonium perchlorate, Appl. Surf. Sci. 356 (2015) 447-453. https://doi.org/10.1016/j.apsusc.2015.08.078

[83] B.A. Mei, O. Munteshari, J. Lau, B. Dunn, L. Pilon, Physical interpretations of Nyquist plots for EDLC electrodes and devices, J. Phys. Chem. 122 (1) (2018) 194-206. https://doi.org/10.1021/acs.jpcc.7b10582

[84] M. Amiri, H. Salehniya, A. Habibi-Yangjeh, Graphitic carbon nitride/chitosan composite for adsorption and electrochemical determination of mercury in real samples, Ind. Eng. Chem. Res. 55 (29) (2016) 8114-8122. https://doi.org/10.1021/acs.iecr.6b01699

[85] H. Zhang, Y. Huang, S. Hu, Q. Huang, C. Wei, W. Zhang, W. Yang, P. Dong, A. Hao, Self-assembly of graphitic carbon nitride nanosheets–carbon nanotube composite for electrochemical simultaneous determination of catechol and hydroquinone, Electrochim. Acta 176 (2015) 28-35. https://doi.org/10.1016/j.electacta.2015.06.119

[86] S. Tonda, S. Kumar, S. Kandula, V. Shanker, Fe-doped and-mediated graphitic carbon nitride nanosheets for enhanced photocatalytic performance under natural

sunlight, J. Mater. Chem. A. 2(19) (2014) 6772-6780.
https://doi.org/10.1039/c3ta15358d

[87] H. Tian, H. Fan, J. Ma, L. Ma, G. Dong, Noble metal-free modified electrode of exfoliated graphitic carbon nitride/ZnO nanosheets for highly efficient hydrogen peroxide sensing, Electrochim. Acta, 247 (2017) 787-794.
https://doi.org/10.1016/j.electacta.2017.07.083

[88] G. Dai, J. Xie, C. Li, S. Liu, Flower-like Co_3O_4/graphitic carbon nitride nanocomposite based electrochemical sensor and its highly sensitive electrocatalysis of hydrazine, J. Alloys Compd. 727 (2017) 43-51.
https://doi.org/10.1016/j.jallcom.2017.08.100

Chapter 10

Graphene-Metal Organic Framework Composite Based Electrochemical Sensors for Toxic Chemicals

P. Arul [a], N.S.K. Gowthaman [b]*, S. Abraham John [c]*, Hong Ngee Lim [d], Sheng-Tung Huang [a] and Govindasamy Mani [e]

[a] Institute of Biochemical and Biomedical Engineering, Department of Chemical Engineering and Biotechnology, National Taipei University of Technology, Taipei, 106, Taiwan

[b] Materials Synthesis and Characterization Laboratory, Institute of Advanced Technology, University Putra Malaysia, 43400 UPM Serdang, Selangor, Malaysia

[c] Centre for Nanoscience and Nanotechnology, Department of Chemistry, The Gandhigram Rural Institute (Deemed to be University), Gandhigram-624 302, Dindigul, Tamilnadu, India

[d] Department of Chemistry, Faculty of Science, University Putra Malaysia, 43400 UPM Serdang, Selangor, Malaysia

[e] Department of Material Science and Engineering, National Taipei University of Technology, Taipei, 106, Taiwan

*nalla.perumal@upm.edu.my (N.S.K. Gowthaman), s.abrahamjohn@ruraluniv.ac.in (S.A. John)

Abstract

Metal organic frameworks (MOFs) are a class of porous materials designed by coordination chemistry between metal ions and secondary organic building units (linkers). They emerged as an extensive class of crystalline materials with higher porosity than other framework materials like zeolites, activated carbon and metal-complex hydrides, respectively. Besides, they have high thermal stability, well-organized structure, low density, large internal surface area, ease in synthesis and broad-spectrum properties which makes them suitable for diverse applications. On the other hand, bulky structure of MOFs having some limitations like poor solubility, lacking electronic conductance and surface to volume ratio is minimal. To fulfill the above shortcoming is to introduce properties of other active materials like carbon nanostructure, metal oxide, metal nanoparticles, graphene carbon nitrite so on. Among the different composite materials especially carbon-based nanocomposite like graphene oxide (GO) and its derivatives have gained much attention because GO exhibiting 2D amphiphilic contains huge hydroxyl, epoxy and carboxylic acid functional groups on its conjugated planes.

The co-existence of aromatic sp^2 feature and oxygen functionalities allow GO in wide bonding interactions. Due to the solubility, sheet with basal like structure of GO can easily functionalized with other active materials. The obtained composite materials could enhance the optical, electrical, thermal and mechanical properties and can then be utilized for electrocatalytic applications. This chapter deals with the introduction of MOF with different synthetic methods and their characterization. Then these composite materials are utilized for electrochemical determination of toxic components including heavy metals, toxic anions, pesticides, aromatic nitro compounds, phenolic compounds and toxic solvents. The described MOF with graphene-based composites are well-known electrocatalyst for determination of toxic compounds.

Keywords

Metal-Organic Frameworks, Graphene Oxide, Electrocatalyst, Electrochemical Sensors, Toxic Components, Aromatic Nitro Compounds

Contents

1.   Introduction

In recent years, researchers have made great progress in constructing new chemical design structures from zero-dimensional (0D) to 3D by assembling the different building units [1]. A well-defined framework assembly of building units to the architectures have been attained by diverse connecting points like weak π-π stacking and strong covalent bond interactions [1,2]. The metal-organic frameworks (MOFs) could be obtained via coordination between metal ions and secondary building units (SBUs) [3,4] (Figure.1A). The SBUs are different functionalities, *i.e.* bi-dentate to poly-dentate carboxylate functional groups have been utilized to develop a new type of organic frameworks [2-5]. The structural designs of MOFs are based on selection of metal ions and organic linkers to form 3D network frameworks (Figure.1B).

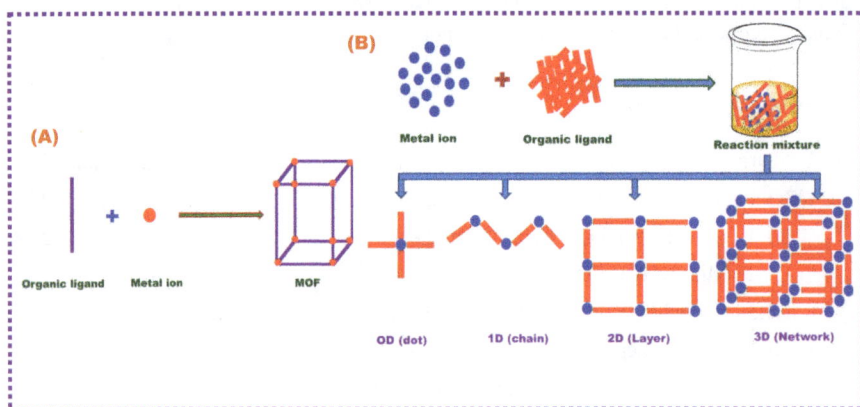

Figure 1. Scheme showing the formation of MOF (A) and MOFs with different dimensionalities (B).

In addition to that, organic linkers can also be important for connecting units in the assembly of MOF structures [6]. It is difficult to predict the MOF structure on the basis of the organic linkers and metal ions used [5,6]. MOFs are specific kinds of flexible porous crystalline materials with different dimensionalities and various topologies could be obtain on the basis of metal ion geometry and secondary organic building units [2-6]. The MOFs can be categorized based on the guest species and also depending upon the structural dimensionality. Figure 1B displays the different dimensional organic frameworks with respect to metal ion and ligands. In 1D frameworks, the coordination bonds are spread over the polymer in one direction and possible cavities are accommodated with small sized molecules [7-9]. In 2D frameworks, a single type layers are superimposed through either edge to edge or staggered type of stacking where weak interactions exist between the layers [7,8]. The modifications of ligand which constitute the layers can control the way of stacking and functionality [8]. In 3D MOFs, frameworks are highly porous and stable due to the coordination bonds spread in three directions. The 3D pillared layers and grids are found in most of the MOFs [2-8].

1.1 Synthesis of MOF

1.1.1 Solvothermal and hydrothermal synthesis of MOFs

In the synthesis of MOFs, the building units and synthetic routes must be carefully selected based on the requirement of framework structures. The MOFs were synthesized via hydrothermal and solvothermal, microwave, sonochemical and electrochemical methods (Figure 2). The solvothermal and hydrothermal methods are one of the important synthetic methods for highly crystalline organic framework structures [10-13]. Generally, organic solvents such as N,N-dimethylformamide (DMF), N,N-diethylformamide (DEF), N-methylpyrrolidine (NMP) and dimethyl sulfoxide (DMSO) were frequently used in the synthesis of solvothermal method [12,13]. The solvothermal synthesis of different organic and inorganic framework materials often takes several hours and requires heating based on the solvents (80-180°C) within a sealed vessel. The pressure inside the sealed vessel is very important, which may affect the yield significantly. These synthetic methods are similar with only difference being the utilization of solvent. Recently, much effort has been devoted to the hydrothermal synthesis of different metallic organic frameworks in green (water used as solvent) and facile ways. The varieties of synthetic reports are available in the literature for solvothermal and hydrothermal for metal organic frameworks [10-13].

1.1.2 Microwave synthesis of MOFs

Microwave synthesis method has been extensively used for quick and rapid synthesis of highly porous crystalline under protic conditions [14]. Besides fast crystallization and tuning the structures, potential advantages of this technique include phase selectivity, particle size distribution, and controlling the growth of the structures [15]. Commercial microwave equipment provides adjustable power outputs and has a fiber optic temperature and pressure controller. In microwave synthesis, a substance mixture in a suitable solvent and then transferred into a Teflon linked vessel, sealed and placed in the microwave unit, and heated for the appropriate time at an optimized temperature [16]. In this technique, where an applied electric field is coupled with the permanent dipole moment of the molecules in synthesis medium inducing molecular rotations, results in rapid heating of the liquid phase [15,16]. The MW synthetic technique has been used for the synthesis of diverse metal organic frameworks [14-16].

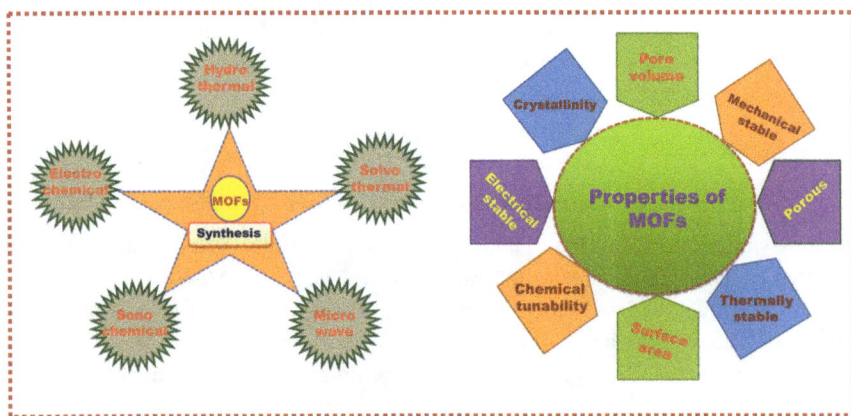

Figure 2. Scheme showing the different synthetic routes of MOFs and its properties.

1.1.3 Sonochemical synthesis of MOFs

A Sonochemical method is a one of the easiest and effective synthesis routes for crystalline MOF. An accelerated nucleation can also achieve a reduce in crystallization time and significantly smaller particles size than other conventional methods [17,18]. The mixture of substance solution was introduced into a reactor fitted to a sonicator bath with variable power output without external cooling. Initially, formation and collapse of bubbles was formed in the solution after sonication, termed acoustic cavitation, produces very high local temperatures and pressures, and results in producing fine crystallites [19]. These are some common example for the synthesis of highly crystalline MOFs sonochemically [17-19].

1.1.4 Electrochemical synthesis of MOFs

An electrochemical synthesis is classified as two main categories like anodic and cathodic electro deposition [20-22]. The anodic electrochemical synthesis metal ions continuously supplied through anodic dissolution as a metal source instead of metal salts, which react with the dissolved organic linkers and a conducting salt in the reaction medium [23]. The metal deposition on the cathode is avoided by employing protic solvents, but in the process H_2 is generated. The electrochemical route is also possible to run a continuous process to obtain a higher solids content compared to conventional batch reactions [24]. The first electrochemical synthesis of Cu-MOF using bulk copper plates are arranged as the anodes in an electrochemical cell with the 1,3,5-benzene tricarboxylic acid (H_3BTC) dissolved in MeOH as solvent and a copper cathode. During a period of 150 min at 12-19 V and 1.3 A, a greenish blue precipitate was formed. After activation, a dark blue colored powder (octahedral crystals in 0.5 to 5 μm) having surface area of 1,820 m^2/g was obtained. There are some common examples for synthesis of different MOFs by electrodeposition [20-24].

1.1.5 Mechanochemical synthesis of MOFs

Mechanical synthesis is a green technology and breakage of intra-molecular bonds followed by a chemical transformation taking place [25,26]. Highly porous nature of MOF synthesis by mechanochemical was reported first in 2006. Mechanochemical reactions can takes place at room temperature under solvent-free conditions, which is especially advantageous when organic solvents can be avoided [26]. Quantitative yields of small MOF particles can be obtained in short reaction times, normally in the range of 10-60 min. In many occasions, metal oxides were found to be preferred than metal salts as a starting material, which results in water as the only side product [26]. The addition of small amounts of solvents in liquid-assisted grinding can lead to acceleration of mechanochemical reactions due to an increase of mobility of the reactants on the molecular level [25-27]. Already many analytical reports are available for the synthesis of MOFs via mechanochemical method [25-27].

1.2   Composition of MOFs

In recent years, MOF has received arousing interest due to their high surface area, pore volume and excellent thermal conductivities [1-5]. The synthesized MOFs are potentially utilized in various fields such as catalysis, gas storage, adsorption, drug delivery, chemical and electrochemical sensors [1-5,7,9,12]. But limitation of MOF is its poor electrical conductivity due to the presence of non-conducting organic ligands and huge molecular weight of the framework structures [28]. To fulfill the above limitation, MOF was prepared as composites with different functional materials. The combination of MOFs with active functional materials has been offered recently to enhance the functional activity than MOF alone [7]. The result of the composite material provides fabrication protocols for high-performance composites with sophisticated architectures [7]. The hybrid composites are combination of two or more distinct constituent materials with properties noticeably different from those of the individual components [3,7]. In composite materials shows good structural flexibility, high porosity with ordered crystalline structures and various kinds of functional properties like unique optical, electrical, magnetic and catalytic [3-6]. Therefore, new physical and chemical properties and enhanced the electro catalytic performance than individual components [8]. Consequently, the remarkable features of composites resulting from the synergistic combination of both MOF and other active components make them suitable for a wide range of applications [3-8]. To date, MOF composites have been successfully made with active species, including metal nanoparticles (NPs), metal oxides (MOx), quantum dots (QDs), polyoxometalates (POMs), polymers, graphene, carbon based materials [3,5,6-8] Figure 3.

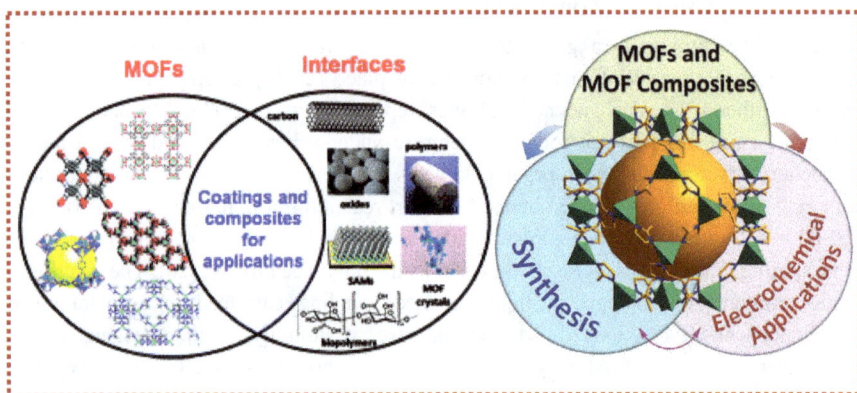

Figure 3. Scheme showing the composition of MOFs with diverse functional material.

1.2.1   MOF with carbon-based composites

MOFs with carbon-based composites are most important because it possesses various allotropes like graphite, fullerenes, nanotubes and diamond, micro textures with different degrees of graphitization, dimensionality from 0D to 3D [3,7,8]. As the most promising candidates for functional applications especially graphene and carbon nanotubes (CNTs), are gaining increasing attention owing to their outstanding properties of in-plane graphite with a high surface area [29]. Graphene is a monolayer of carbon atoms arranged in a 2D honeycomb lattice and can be seen as an individual atomic plane pulled out of bulk graphite [7]. Graphene oxide (GO) is a graphene derivative derived from the oxidative exfoliation of graphite, which is solution-dispersible and can act as the precursor of graphene after reduction [29,30]. On the other hand, CNTs are well-ordered, high aspect ratio allotropes of carbon. The two main variants, single-walled carbon nanotubes (SWCNTs, diameters of 0.4 to 2 nm) and multi-walled carbon nanotubes (MWCNTs, diameters of 2 to 100 nm), both possess a high tensile strength, ultra-light weight, and excellent chemical and thermal stability [29-31]. The exceptionally mechanical, electrical and thermal properties of graphene commend them as valuable nanostructured fillers in MOF composites [29,31]. The MOF composites, which combines graphene based materials with functional inorganic materials resulting in a combination of the individual properties, is even more predestined for applications concerning sustainable energy and environment [3,7,8,29,31]. To date, numerous MOF–nanocarbon composites have been successfully made with graphene oxide intensively explored for diverse applications.

1.2.2 Electrochemical sensor using MOFs

Electrochemical sensors is one of the important analytical detection tool and they can provide high sensitivity and selectivity with relatively inexpensive equipment compared to conventional techniques. The MOF with graphene oxide composites have been extensively used for electrochemical sensing applications (Figure 4). The redox and catalytically active sites were introduced through the use of active metal ions or ligands in MOFs, thereby detection of metal ions and nitro compounds was achieved [31,32]. The MOFs were rarely used for conductance-based sensing applications because of their poor electrical conductance [28]. At present, the research on pure MOFs in electrochemical sensors is still shortcoming. Therefore, researchers prepared composites with graphene oxide to be used to detect nitro compounds, toxic anions, hazards compounds and environmental pollutants [32]. The superior electrocatalytic activity of the MOF with graphene oxide composites make them feasible for various sensing applications. The presence of MOF with graphene based composites on electrode surface facilitates the sensitivity, selectivity, anti-interference activity and lowers the limit of detection (LOD). Further, the oxidation/reduction potentials of the analytes can favorably shift to the lower values during electrochemical sensing studies.

This chapter concludes the electrocatalytic activity of MOF-graphene based composites reported for electrochemical sensing of toxic chemicals, environmental pollutants, toxic metal ions, nitro aromatic compounds and toxic solvents. It also discusses various advantages of the MOF with graphene based composites for electrochemical sensing applications. The MOF with graphene oxide based composites have been applied for electrocatalytic application of various environmentally important analytes like toxic metal ions, nitroaromatic compounds, phenolic isomers. Information is given in this chapter for the synthesis, properties and applications of MOF with graphene based composites as electrocatalyst for sensing approaches.

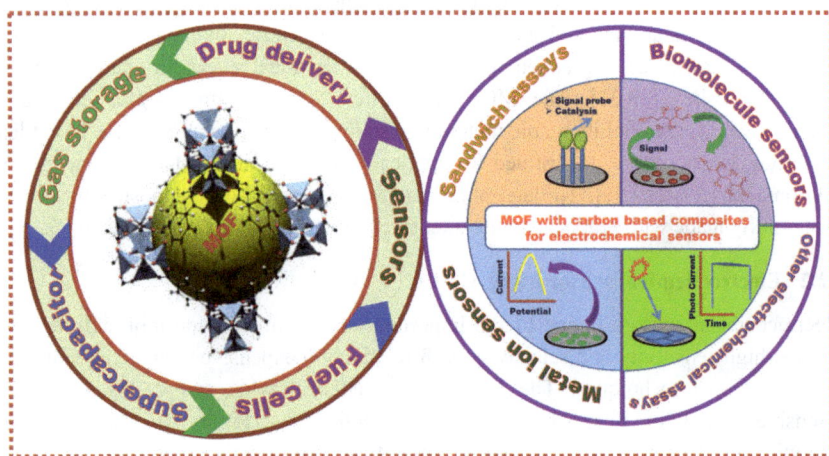

Figure 4. Scheme showing the MOF with composites for electrochemical application.

2. MOFs based Electrochemical sensors

The metal organic framework structures have been extensively used for electrochemical sensing applications. The redox and catalytically active sites were introduced through the use of active metal ions or ligands in MOFs, thereby electrochemical detection of toxic metal ions, nitro compounds and hazards compounds [33-50]. In recent years, MOFs were rarely used for conductivity-based sensing applications because most of the MOFs are less conductivity. However, the application of MOFs in non-enzymatic electrochemical sensors has aroused widespread concern in recent years. Cui *et al.* reported the synthesis of 4,4'-bipyridine based Cu-MOF [Cu(bpy)(H$_2$O)$_2$(BF$_4$)$_2$(bpy)] by deposition. The synthesized Cu-MOF was modified on carbon paste electrode (CPE) and then utilized for 2,4-Dichlorophenol, an exhibited a wide linear range concentration from 4 to 100 × 10^{-6} M with LOD 1.1 × 10^{-6} M [33]. Song *et al.* reported the bimetallic CoNi-MOF using mixed organic linkers 4-(1H-tetrazol-5-yl)benzoic acid (H$_2$TZB) and 2,4,6-tri(4-pyridyl)-1,3,5-triazine (TPT). The bimetallic-MOF was utilized for trace level determination of hazardous compounds deoxynivalenol (DON) and salbutamol (SAL) by

EIS. The sensor showed linear wide range from 1 to 500×10^{-12} mL^{-1} with LOD 0.05×10^{-12} mL^{-1} [34].

Li *et al.* demonstrated the sensing of heavy metal ions (Cd^{2+}) using the Mg-MOF from 3,3'-(pyridine-2,5-diyl)dibenzoicacid and $MnCl_2 \cdot 4H_2O$ by reflux. The obtained sensor showed wide-linear range from 0 to 60 µg L^{-1} by SWASV with LOD 0.12 ppb (S/N = 3) [35]. Sohail *et al.* displayed the water stable Zn-MOF and then utilized for oxidation of hydrazine. It showed wide linear range from 20 to 350×10^{-3} M (R^2 = 0.9922) for hydrazine by linear sweep LSV with LOD of 2×10^{-3} M (n = 3) was observed [36]. Wang *et al.* reported simultaneously detection of multiple heavy metal ion pollutants (Cd^{2+}, Pb^{2+} and Cu^{2+}) using ferrocenecarboxylic acid functionalized Zr-MOF (Fc-NH$_2$-UiO-66). The LODs of the ratiometric sensor were estimated to be 8.5×10^{-9} M for Cd^{2+}, 0.6×10^{-9} M for Pb^{2+} and 0.8×10^{-9} M for Cu^{2+}, respectively [37]. Zhang *et al.* reported the Ce-MOF modified with cationic surfactant of cetyltrimethylammonium bromide (CTAB). The CTAB/Ce-MOF was modified on GCE and then utilized for Bisphenol A by DPV and wide linear range from 0.005 to 50×10^{-6} M with LOD 2.0×10^{-9} M (S/N=3) [38].

Ji *et al*, reported the Cu based MOF and then utilized for Sunset yellow (SY) and tartrazine (TZ). It showed wide linear range concentration from 0.3 to 50 nM and 1.0 to 100 nM for SY and TZ with LOD 0.05 and 0.14 nM [39]. Wang *et al.* reported the burkholderia cepacia lipase (BCL) modified with Cu-MOF nanofibers for hazards pesticides. It showed good sensitivity for wide range from 0.1 to 38×10^{-6} M with LOD 0.067×10^{-6} M for methyl parathion. Finally it showed good selectivity and practical application was demonstrated in residues in vegetable samples [40]. Figure 5 shows the various electrchemical sensors using MOF modified eectrodes. Roushani *et al.* synthesized the Zn based MOF and then utilized for hazards Cd^{2+}. The obtained sensor based on interaction between Cd^{2+} and –N complexation by soft-soft interaction. The obtained sensor exhibited wide range concentration from 0.7 to 120 with LOD 0.2 µg L^{-1} [41].

Zhang *et al.* reported the core–shell nanostructured composite of Fe$^{(III)}$-MOF with mesoporous Fe$_3$O$_4$@C nanocapsules. The Fe-MOF@mFe$_3$O$_4$@mC showed an excellent electrochemical activity towards the Pb^{2+} and As^{3+}, good water stability and high specific surface area. The wide range concentration from 0.01 to 10×10^{-9} M with LOD 2.27 and 6.73×10^{-12} M toward detecting Pb^{2+} and As^{3+} was achieved [44]. Guo *et al.* reported the amino-functionalized Ni(II)-MOFs using 2-aminobenzenedicarboxylic acid as an organic linker by one-pot hydrothermal method. The prepared Ni-MOFs were found to be an effective material for selective detection of microgram levels of Pb^{2+} in an aqueous solution performed by SWASV [45]. Kung *et al.* demonstrated the porphyrin based Zr-MOF (MOF-525) thin films are grown on conducting glass substrates by solvothermal approach. The modified glass substrate was utilized for determination of nitrite by AMP and wide linear range concentration from 20 to 800×10^{-6} M with LOD 2.1×10^{-6} M, respectively [46]. Prathap *et al.* reported the highly porous ZIF-8 in an aqueous solution at room temperature and then utilized for electrochemical sensing of trinitrotoluene (TNT). The interaction between the electron-deficient aromatic core of TNT and the electron-rich ZIF-8 is considered to favor the formation of donor–acceptor electron-transfer mechanism, and the electron conductivity of ZIF-8 facilitates the effective reduction of TNT. The sensor performance is highly linear, with very low LOD 346×10^{-12} M [47].

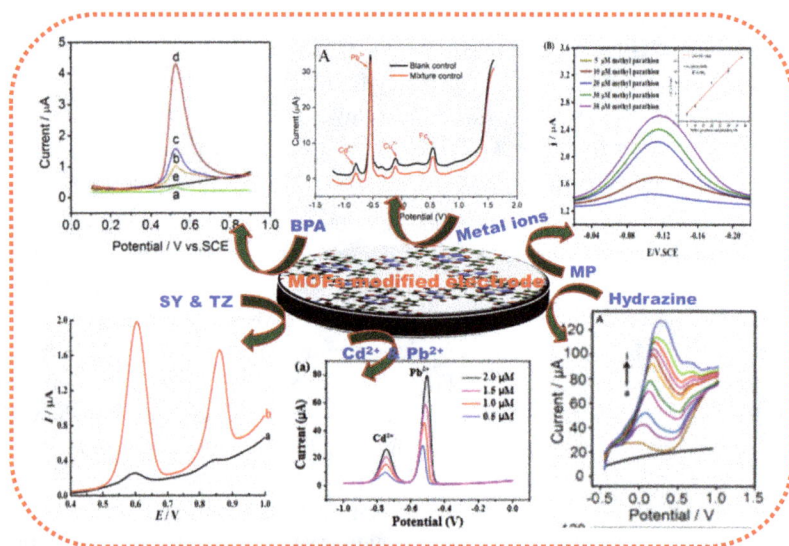

Figure 5. Electrocatalytic application of different toxic ions, environmental pollutants, heavy metal ions, nitro compounds using various MOFs modified electrodes. Reprinted with permission from [36-40,42] Copyright (2018), (2020), (2018), (2016), (2019), (2019) Elsevier & Royal Society of Chemistry.

Su *et al.* reported the Zr-MOF (MOF-525) with crystal sizes ranging from 100 to 700 nm were synthesized by adjusting the content of benzoic acid by solvothermal synthetic process. The synthesized crystals showed similar surface area of 2500 m^2/g with a unique pore size of 1.85 nm. The MOF thin films were applied for electrocatalytic nitrite oxidation and great detection limit of 0.72×10^{-6} M with high sensitivity of 40.6 μA/mM-cm^2 [48]. Bao *et al.* reported the ultrathin Ni-MOF nanobelts, $[Ni_20(C_5H_6O_4)_2 (H_2O)_8]$ 40 H_2O (Ni-MIL77), have been exploited for non-enzymatic urea sensor. Ni-

MOF ultrathin nanobelts exhibit a high sensitivity of 118.77 mA mM^{-1} cm^{-2}, wide linear range of 0.01 to 7.0 × 10^{-3} M with LOD 2.23 × 10^{-3} M (S/N = 3) [49]. Arul *et al.* synthesized the Ni-MOF capped with polyvinyl pyrrolidone (PVP) using 2-amino-1,4-benzene dicarboxylic acid (NH$_2$-1,4-BDC) by solvothermal method. The Ni-MOF-PVP/GCE exhibited a wide range concentration from 0.2 × 10^{-6} M to 1 × 10^{-3} M with the LOD of 97 × 10^{-9} M (S/N = 3) was achieved by AMP [50] Figure 6 shows the various electrchemical sensing of toxic metal, nitrocompounds using MOFs modified eectrodes. The electrocatalytic performance of the MOFs fabricated electrode was utilized towards the detection of toxic chemicals and environmental pollutants is summarized in Table 1 [33-50]. The use of one-component MOFs as electrode material in electrochemical sensors often results in wide linear ranges, a low sensitivity, and poor stability due to their less electron conductivity with poor mechanical properties. Hence, MOFs with graphene based composites to improve the sensitivity, stability and linear range detection.

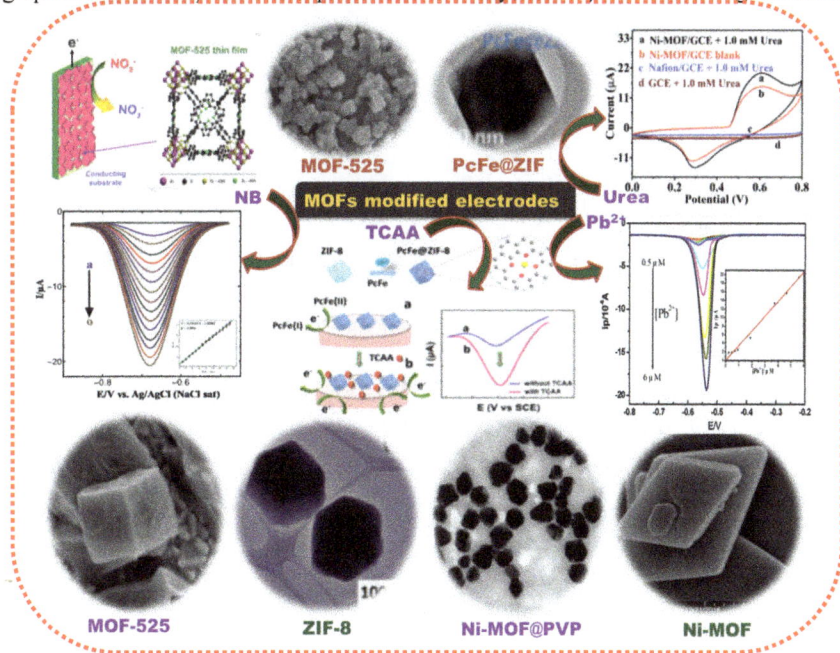

Figure 6. Electrocatalytic application of toxic anions, environmental pollutants, heavy metal ions and urea using MOFs modified electrodes. Reprinted with permission from [43,45,46,48-50] Copyright (2017), (2017), (2015), (2018), (2018), (2016), (2019) Elsevier & Royal Society of Chemistry and American Chemical Society.

Table 1. *Comparison of performance of the MOFs fabricated electrochemical sensors towards the detection of toxic ions and environmental pollutants.*

Sensing material	Technique	Analytes	Linear range (M)	Detection limit (M)	Ref
Cu-MOF	DPV	DCP	$4 - 100 \times 10^{-6}$	1.1×10^{-6}	33
CoNi-MOF	EIS	DON & SAL	$1 - 500 \times 10^{-2}\,\text{M L}^{-1}$	$0.05 \times 10^{-12}\,\text{M L}^{-1}$	34
Mg-MOF	SWASV	Cd^{2+}	$0 - 60 \times \mu\text{g L}^{-1}$	0.12 ppb	35
Zn-MOF	DPV	Hydrazine	$20 - 350 \times 10^{-3}$	2×10^{-3}	36
Zr-MOF	DPV	$Cd^{2+}, Pb^{2+}, Cu^{2+}$	$1 \times 10^{-9} - 2 \times 10^{-6}$	$8.5, 0.6, 0.8 \times 10^{-9}$	37
CTAB-Ce-MOF	DPV	BSA	$0.005 - 50 \times 10^{-6}$	2.0×10^{-9}	38
Cu-MOF	DPV	SY & TZ	$0.3 - 50, 1.0 - 100 \times 10^{-9}$	$0.05\ \&\ 0.14 \times 10^{-9}$	39
Cu-MOF	SWASV	MP	$0.1 - 38 \times 10^{-6}$	0.067×10^{-6}	40
Zn-MOF	DPV	Cd^{2+}	$0.7 - 120 \times \mu\text{g L}^{-1}$	$0.2 \times \mu\text{g L}^{-1}$	41
Yb-MOF	SWASV	$Cd^{2+}\ \&\ Pb^{2+}$	$0.1 - 1.0 \times 10^{-6}$	0.10×10^{-9}	42
Zn-MOF	DPV	TCAA	$0.02 - 1 \times 10^{-6}$	1.89×10^{-6}	43
Fe-MOF	EIS	$Pb^{2+}\ \&\ As^{3+}$,	$0.01 - 10 \times 10^{-9}$	$2.27, 6.73 \times 10^{-12}$	44
MOF-525	Amp	Nitrite	$20 - 800 \times 10^{-6}$	2.1×10^{-6}	46
ZIF-8	DPV	TNT	$0.4 - 460 \times 10^{-9}$	346×10^{-12}	47
Zr-MOF	CV	Nitrite	$10 - 800 \times 10^{-6}$	0.72×10^{-6}	48
Ni-MOF	DPV	Urea	$0.01 - 7 \times 10^{-3}$	2.23×10^{-3}	49
Ni-MOF-PVP	DPV	NB	$0.2 \times 10^{-6} - 1 \times 10^{-3}$	97×10^{-9}	50

2,4 dichlorophneol ***(DCP)***; *deoxynivalenol* ***(DON)***; *Salbutamol* ***(SAL)***; *Bisphenol A* ***(BSA)***; *Sunset yellow* ***(SY)*** *and Tartrazine* ***(TZ)***; *Methylparathion* ***(MP)***; *Trinitro-toluidine* ***(TNT)***; *Nitrobenzene* ***(NB)***, *Square-Wave Anodic Stripping Voltammetry* ***(SWASV)***; *Electrochemical impedance spectroscopy* ***(EIS)***, *Differential pulse voltammetry* ***(DPV)***.

3. MOFs with graphene-based composites

In recent years, an increasing interest has been focused on metal nanoparticles (MNPs), graphene and graphene-based nanocomposites [7,8]. Owing to high mechanical strength and chemical stability as well as electrical resistivity, graphene and its counterparts including graphene oxide (GO) and nitrogen doped graphene (N-GO) etc. It has been demonstrated as an excellent catalyst for various nano-architectures and catalyst specific reactions. Hybridization with different nanocrystals including noble metals, metal oxide and sulfides and metal coordination compounds has proven effective for enhanced physicochemical properties and chemical functions, ranging from gas adsorption, catalysis, photovoltaic device and sensors [1,3-5,7,8,29]. Li *et al.* reported the novel synthesis of Cu-MOFs with electrochemically reduced graphene oxide (ERGO) based composites. The Cu-MOFs/ERGO composites displayed significantly enhanced electrocatalytic activity towards BSA. It showed wide linear range concentration from $0.02 - 90 \times 10^{-6}$ M with LOD of 6.7×10^{-9} M (S/N=3) by DPV [51].

Saraf *et al.* reported the simple ultra-sonication of slow diffusion driven Cu-MOF with chemically rGO. The synergistic effects between Cu-MOF with rGO Nano-sheets exhibited high charge storage efficiency (685.33 F.g^{-1} at 1.6 A.g^{-1}) and high energy (137.066 W h kg^{-1}). Additionally, electrode modified with Cu-MOF/rGO hybrid performs exceptionally towards the electrochemical detection of nitrite in a wide linear range from $3 - 40000 \times 10^{-3}$ M ($R^2 = 0.99911$), with a notable LOD of 33×10^{-9} M and a high sensitivity of 43.736 mA mM^{-1} cm^{-2} [52]. Karimian *et al.* demonstrated the Zr-MOF with TiO_2 functionalized graphene oxide@UiO-66 (TGO@UiO-66) for simultaneous detection of paraoxon and chlorpyrifos. The square wave voltammogram (SWV) of TGO@UiO-66/GCE in presence of paraoxon and chlorpyrifos showed two characteristic peaks at 0.45 and 1.3 V. The designed sensor exhibited low detection limits of 0.2 and 1.0×10^{-9} M within the linear ranges of $1 - 100 \times 10^{-9}$ M and $5 - 300 \times 10^{-9}$ M for paraoxon and chlorpyrifos [53]. Bhardwaj *et al.* reported the three-phase composite material consisting of SiO_2-coated Cu-MOF, single layer graphene, and aniline was synthesized. In presence of ammonium persulfate as an oxidant, the aniline component of this mixture was polymerized to polyaniline to bridge Cu-MOF with graphene. The sensor showed wide linear range from 1 - 100 ppm with LOD of 1 ppm [54].

Lu *et al.* demonstrated the graphene aerogel (GA) with Zr-based MOF composites was developed for simultaneous detection of multiple heavy-metal ions in aqueous solutions. The sensor showed lower LODs for simultaneously detecting multiple metal ions were 9×10^{-9} M for Cd^{2+}, 1×10^{-9} M for Pb^{2+}, 8×10^{-9} M for Cu^{2+}, and 0.9×10^{-9} M for Hg^{2+}. Finally, the practical applications in detecting Cd^{2+}, Pb^{2+}, Cu^{2+}, and Hg^{2+} in river water and the leaching solutions of soil and vegetable also demonstrated [55] Figure 7. Li *et al.* reported the Cu-based MOF using 1,3,5-benzenetricarboxylic acid (BTC) with high-conductivity ball-mill-exfoliated graphene (Cu-BTC@GS) by a simple method. The composite materials showed better electrocatalytic activity towards the biomolecules (xanthine and hypoxanthine) and phenolic pollutants (BSA and p-chlorophenol) by double potential step chronocoulometry [56]. Li *et al.* reported the Cu centered MOF with graphene composites (Cu-MOF-GN) [Cu-MOF = $Cu_3(BTC)_2$ (BTC = 1, 3, 5-

benzenetricarboxylicacid)] and then modified on GCE (Cu-MOF-GN/GCE). The electrochemical sensor exhibited a wide linear concentration from 1.0×10^{-6} to 1.0×10^{-3} M with the detection limits of 5.9×10^{-7} M for HQ and 3.3×10^{-7} M for CT (S/N = 3) [57].

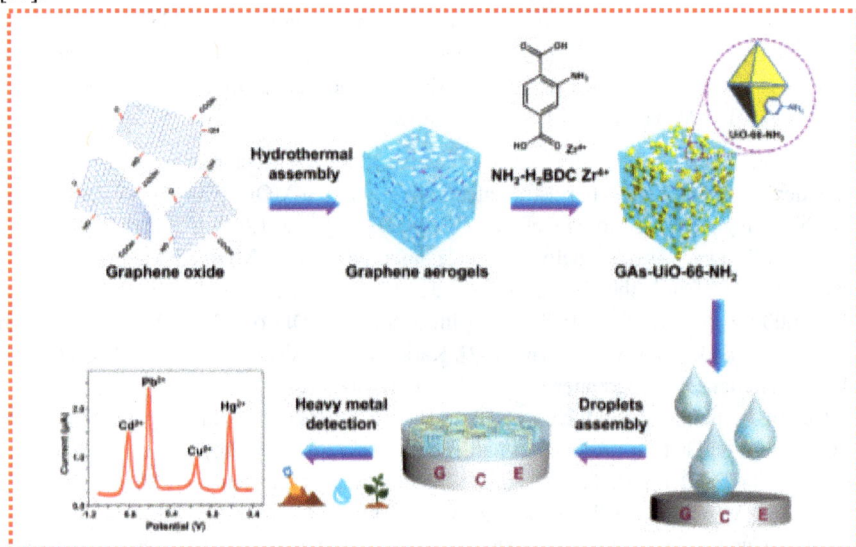

Figure 7. Scheme showing the synthesis and fabrication of GAs-UiO-66-NH₂ for heavy metal ions sensor [55]. Reprinted with permission from [55] Copyright (2019) ACS.

Baghayeri *et al*. demonstrated the novel nanocomposite of graphene oxide with zinc-based MOF (GO/Zn-MOF) was prepared by a simple solvothermal method. The electrochemical As(III) sensing capability of the nanocomposite was explored by casting the GO/Zn-MOF on GCE, followed by an electrochemically reduction of GO. The present sensor showed excellent electrochemical performance such as a wide linear range from 0.2 to 25 ppb (mg/L), LOD (S/N = 3) of 0.06 ppb and good reproducibility [58]. Peng *et al*. reported the porphyrinic Zr-MOF with reduced graphene oxide (PCN-224/rGO) nanocomposite for *p*-arsanilic acid (*p*-ASA). The PCN-224/rGO nanocomposite had strong affinity to p-ASA via Zr–O–As coordination and π–π stacking. The developed composite sensor had a wide detection range from 10 ng L^{-1} to 10 mg L^{-1}, with a LOD 5.47 ng L^{-1}. Finally, the sensor was successfully applied to monitor *p*-ASA in simulated natural water and swine manure lixivium [59]. Gan *et al*. reported the highly dispersed Au nanorods and GO-wrapped microporous ZIF-8 were successfully encapsulated inside the ZIF-8 (AuNRs@ZIF-8) by epitaxial growth or nucleus coalescence. The microporous ZIF-8 shell functions as a protective coating to effectively prevent AuNRs from dissolution, aggregation, and migration during the electrochemical testing of four pesticides as niclosamide, dichlorophen, carbendazim, and diuron [60] Figure 8.

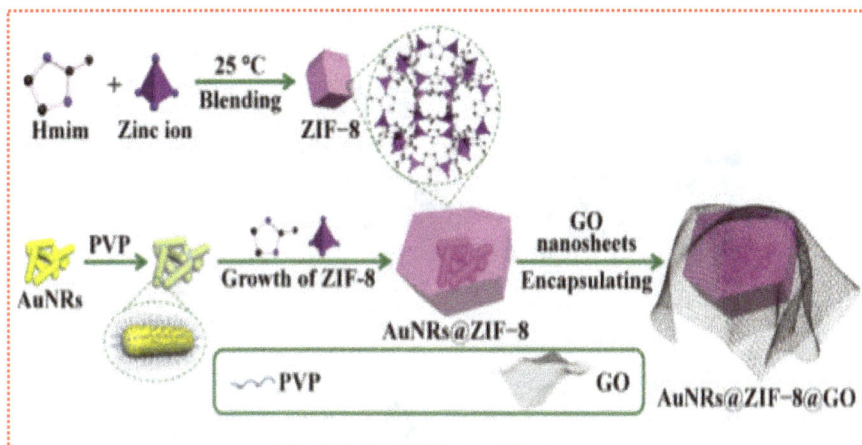

Figure 8. Scheme showing the synthesis of AuNRs@ZIF-8@GO [60]. Reprinted with permission from [60] Copyright (2019) Royal Society Chemistry.

Ding *et al.* reported the newly developed chemically functionalized 3D graphene oxide hydrogels (FGH) decorated with MOFs-derived Co_3O_4 nanostructures and then utilized for sensing of acetone. The sensor showed good sensitivity with linear range concentration from 4.01/1 ppm to 81.2/50 ppm with response time ~ 20 s [61]. Travlou *et al.* reported the hybrid material consisting of Cu-based MOF with aminated graphite oxide for ammonia sensor. A hybrid material with the smallest content of graphene phase exhibited the largest signal change upon exposure to ammonia [62].

Chen et al. demonstrated the preparation of Cu-based MOF-199 with GO by solvothermal and then used for electrochemical determination of CT and HQ. It showed wide linear range from 0.1 to 566 × 10^{-6} M for CT and 0.1 to 476 μM for HQ with the same LOD of 0.1 μM (S/N = 3) [63]. Wang et al. reported the incorporation of rGO into a MIL-101(Cr) for the modification of CPE and then utilized for electrochemical sensors. The resulted electrodes exhibited high sensitivity and reliability in the simultaneous identification and quantification of CC and HQ. It shows linear wide range concentration from 10 to 1400×10^{-6} M and 4 to 1000×10^{-6} M, and LODs were 4 and 0.66×10^{-6} M (S/N=3) for CC and HQ [64]. Wang et al. demonstrated the new type of Cu-BTC on electro-reduced GO (ERGO) by electrodeposition. The modified electrode was used for highly sensitive determination of 2,4,6-trinitrophenol (TNP) by DPV. It shows three reduction peaks, the first at a potential of -0.42 V (vs. SCE) and wide linear range concentration from 0.2 to 10×10^{-6} M TNP with LOD 0.1×10^{-6} M (S/N=3) [65]. The performance of the MOF with graphene based composite fabricated electrochemical sensors was compared and the results are given in Table 2 [51-53,55,57-59,63-65].

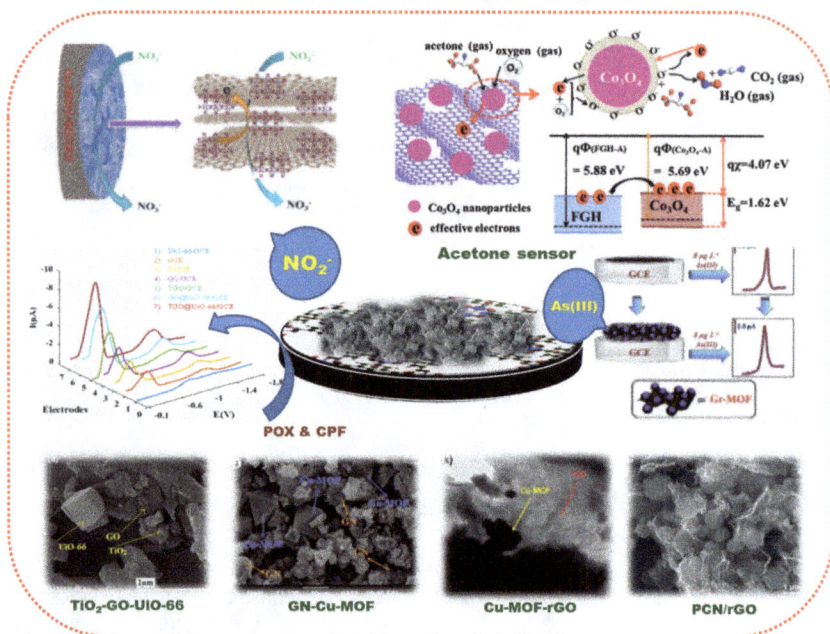

Figure m9. Electrocatalytic application of toxic anion, environmental pollutants, acetone using various MOFs-GO based composites. Reprinted with permission from [52,53,56,58,61] Copyright (2016), (2019), (2019), (2020), (2018), (2018) Elsevier & Royal Society of Chemistry and ACS.

Table 2. *Comparison of performance of the various MOFs-graphene based electrochemical sensors towards the detection of toxic chemicals, environmental pollutants and isomeric phenolic compounds.*

Sensing material	Technique	Analytes	Linear range (M)	Detection limit (M)	Ref
Cu-MOF/ERGO	DPV	BSA	$0.02 - 90 \times 10^{-6}$	6.7×10^{-9}	51
Cu-MOF/rGO	AMP	Nitrite	$3 - 40000 \times 10^{-6}$	33×10^{-9}	52
TGO@UiO-66	SWV	Paraoxon	$0 - 100 \times 10^{-9}$	0.2×10^{-9}	53
		Chlorpyritos	$5 - 300 \times 10^{-9}$	0.1×10^{-9}	
GA-UiO-66-NH$_2$	DPV	Cd^{2+}	$0.01 - 1.5 \times 10^{-6}$	9×10^{-9}	55
		Pb^{2+}	$0.001 - 2 \times 10^{-6}$	1×10^{-9}	
		Cu^{2+}	$0.01 - 1.6 \times 10^{-6}$	8×10^{-9}	
		Hg^{2+}	$0.001 - 2.2 \times 10^{-6}$	0.9×10^{-9}	
Cu-MOF-GN	DPV	HQ	$1.0 \times 10^{-6} - 1.0 \times 10^{-3}$	5.9×10^{-7}	57
		CT	$1.0 \times 10^{-6} - 1.0 \times 10^{-3}$	3.3×10^{-7}	
GO-Zn-MOF	DPV	As(III)	$0.2 - 25$ ppb	0.06 ppb	58
PCN-224/rGO	PEC	p-ASA	10 ng L^{-1} -10 mg L^{-1}	5.47 ng L^{-1}	59
Cu-MOF-199/GO	DPV	CT	$0.1 - 566 \times 10^{-6}$	0.1×10^{-6}	63
		HQ	$0.1 - 476 \times 10^{-6}$	0.1×10^{-6}	
MIL-101-rGO	DPV	CC	$10 - 1400 \times 10^{-6}$	4×10^{-6}	64
		HQ	$4 - 1000 \times 10^{-6}$	0.66×10^{-6}	
Cu-MOF/ERGO	DPV	TNP	$0.2 - 10 \times 10^{-6}$	0.1×10^{-6}	65

*Bisphenol A (**BSA**); Hydroquinone (**HQ**); Catechol (**CT**); Photoelectrochemical (**PEC**); p-arsanilic acid (**p-ASA**); Trinitrophenol (**TNP**), Square-Wave Voltammetry (**SWV**); Differential pulse voltammetry(**DPV**).*

Rani et al. reported the porous Zn-MOF@rGO, synthesized via solvothermal technique and then utilized for the amperometric determination of hydrazine. The Zn-MOF@rGO solution was used to modify on gold electrode (AuE) for the superior electrocatalytic oxidation of hydrazine in real water sample. TEM studies revealed prismatic shape of Zn-MOF with size ranging between 160 and 180 nm and were well-dispersed on the rGO sheets. The hydrazine sensor using Zn-MOF@rGO as electrocatalyst displayed a low LOD, high sensitivity and a fast response time (< 2s) [66]. Tung et al. reported the graphene hybrid nanocomposite with MOFs including copper, benzene-1,3,5-tricarboxylate (pG-Cu BTC), zirconium 1,4-dicarboxybenzene (pG-UiO 66) and 2-methylimidazole zinc salt (pG-ZIF 8), were investigated to enhance the sensing performance and capability of distinguishing different VOC biomarkers (e.g., methanol, ethanol, chloroform, acetone, acetonitrile and THF). It showed that the pG-Cu BTC sensor has the highest sensitivity and selectivity towards chloroform and methanol VOCs at 2.82 - 22.6 ppm level [67] Figure 10.

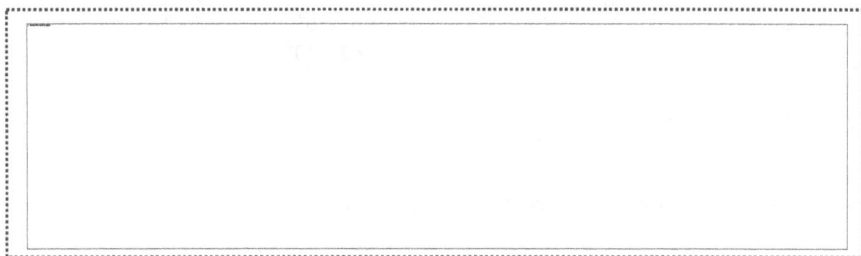

Figure 10. Schematic showing the concept of VOC biomarkers detection for human health monitoring by using graphene-based MOFs hybrid nanocomposites [67]. Copyrighted by Elsevier (2020).

Gu et al. reported the Cr-based MOF (MIL-101(Cr))/ rGO electrocatalyst was prepared through a one-pot hydrothermal method. The obtained results imply that, compared to the single components, it exhibits higher electrocatalytic activity for the reduction of metronidazole and stripping of Cd^{2+} and Pb^{2+}. Optimizing the ratio of MIL-101(Cr) and rGO, an electrochemical sensor was prepared using MIL-101(Cr)/rGO-20 and used for detection of metronidazole, Cd^{2+} and Pb^{2+}, respectively. The linear range for metronidazole can be divided into two parts, *i.e.* from 0.5 to 200 mM and from 200 to 900 mM and limit of detection is 0.24 mM (S/N=3, n=10). Another one is the wide linear range from 0.05 to 6.0 mM for the detection of Cd^{2+} and Pb^{2+} with detection limits found to be 5.2 and 3.0 × 10^{-6} M (S/N=3, n= 10), respectively [68]. Yu et al. displayed the carboxyl-functionalized graphene, a highly sensitive electrochemical sensor for Pb^{2+} was proposed by using 8–17 DNA enzymes as the recognition element and the reaction of H_2O_2 catalyzed by Fe-MIL-101 as the signal probe. It shows excellent sensing performance with a wide linear and LOD of 1.74 × 10^{-14} mol·L^{-1} and high selectivity to Pb^{2+} [69]. Tu et al. reported the rGO-encapsulated with Ce-MOF (rGO@Ce-MOF) composite was synthesized by using a simple and green method. Then, the obtained rGO@Ce-MOF composite was utilized as electrode material for the detection of dichlorophen (Dcp) and wide linear range from 0.02 to 10 × 10^{-6} M and a low detection limit of 0.007 × 10^{-6} M (S/N = 3) [70].

Kung et al. demonstrated the graphene nanoribbons (GNRs) are incorporated with the nanocrystals of a porphyrinic-MOF by solvothermally growing MOF-525 in a suspension of well-dispersed GNRs. The uniform thin films of MOF-525/GNR nanocomposite can be simply deposited on conducting glass substrates by drop casting and then applied for electrochemical nitrite sensors. The sensor showed wide linear range from 100 to 2500 × 10^{-3} M (R^2 = 0.998), and sensitivity of the nitrite can be estimated to be 93.8 mA mM^{-1} cm^{-2} with LOD is calculated to be 0.75 × 10^{-3} M [71]. Xie et al. reported the graphene aerogels (GAs) and ZIF-8 nanocrystals were orderly grown on the surface of PPy/GAs through the coordination interaction between Zn_{2p} and amino groups in PPy chains. This hybrid 3D architecture provided open channels for electrolyte transportation and improved electron transfer between the electrode and the catalysts, which was exploited as electrochemical sensing platform for the detection of highly toxic 2, 2methylenebis (4-chlorophenol) (dichlorophenol, Dcp) with LOD 0.1 × 10^{-9} M [72]. Zhang et al. reported the chromium terephthalate MOFs (MIL-101(Cr) with rGO was proposed to improve the electrical conductivity of MIL-101(Cr) crystals. The electrocatalytic applications of MIL-101(Cr)@rGO nanocomposites were employed for the voltammetric study of 4-nonylphenol and linear range concentration from 0.1 to 12.5 × 10^{-6} M with a LOD of 33 × 10^{-9} M [73]. Yang et al. reported the $Cu_3(btc)_2$ MOF using BTC is covalently immobilized at chitosan (CS) with electrochemically reduced graphene oxide (ERGO) hybrid film modified electrode. The modified electrode utilized for electrocatalytic application of HQ, CT and RC and linear range from 5.0 to 400 × 10^{-6} M, 2.0 to 200 × 10^{-6} M and 1.0 to 200 × 10^{-9} M with LOD 0.44, 0.41 and 0.33 × 10^{-9} M for HQ, CT and RC [74] Figure 11.

Yan et al. reported the bimetallic MOF-818@reduced graphene oxide/multi-walled carbon nanotubes (RGO/MWCNTs) composite was successfully synthesized by a facile solvothermal method. The MOF-818@RGO/MWCNTs/GCE exhibits an excellent electrocatalytic activity towards phenolic acid compounds: caffeic acid (CA), chlorogenic acid (CGA) and gallic acid (GA). The sensor shows two linear ranges from 0.2 to 7 × 10^{-6} M and 7 to 50 × 10^{-6} M with a high sensitivity of 12.89 µA/µM for the detection of CA, a low detection limit of 5.7 × 10^{-9} M and an excellent sensitivity of 12.50 µA/µM in the ranges of 0.1 to 3 × 10^{-6} M and 3 to 20 ×10^{-6} M for CGA detection, as well as a comparable electrochemical performance for GA [75]. Ling *et al.* reported the amine-functionalized MOF/reduced graphene oxide (NH_2-MIL-125/RGO) was prepared by a one-pot hydrothermal reaction. Then it was successfully hybridized with GO, and NH_2-MIL-125/RGO then modified on the GCE. NH_2-MIL-125/RGO/GCE displayed a high sensitivity and outstanding stability for the detection of BPA because of the synergistic effects of large surface area of GO and high porosity of NH_2-MIL-125. The sensor showed linear with the wide concentration of BPA from 2 to 200 × 10^{-6} M, and LOD of 0.7966 × 10^{-6} M [76].

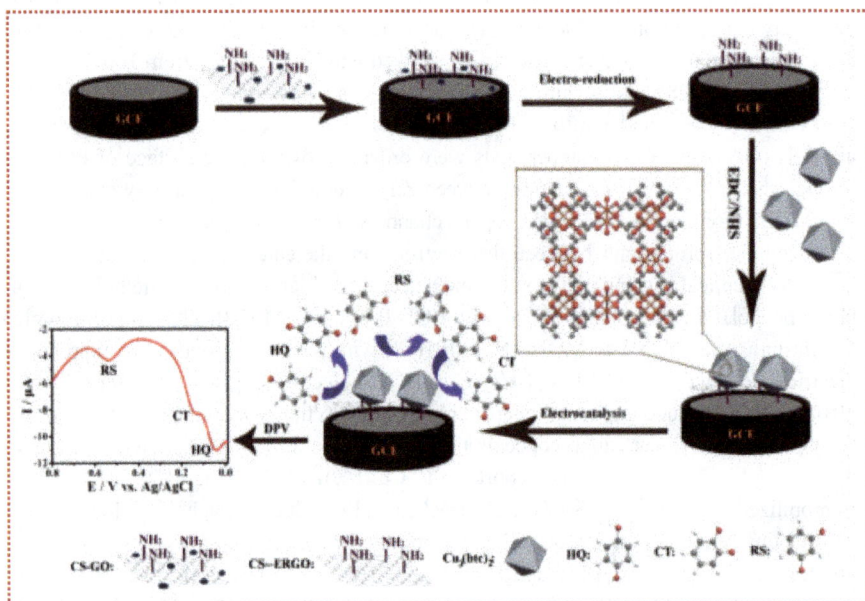

Figure 11. Scheme showing the synthesis of hybride based Cu₃(btc)₂/CS−ERGO/GCE for electrocatalytic application of HQ, CT and RS [74]. Reprinted with permission from ACS [74] (2016).

Table 3. *Comparison of performance of the various MOFs-graphene based electrochemical sensors towards the detection of toxic chemicals, environmental pollutants and hydroxyl phenols.*

Sensing material	Technique	Analytes	Linear range (M)	Detection limit (M)	Ref
Zn-MOF@rGO	AMP	Hydrazine	$0.001 - 100 \times 10^{-6}$	8.7×10^{-3}	66
MIL-Cr/rGO	DPV	Cd^{2+}	$0.5 - 200 \times 10^{-3}$	5.2×10^{-6}	68
		Pb^{2+}	$0.05 - 6.0 \times 10^{-3}$	3.0×10^{-6}	
rGO@Ce-MOF	DPV	Dcp	$0.02 - 10 \times 10^{-6}$	0.073×10^{-6}	70
GNIR-MOF-525	AMP	Nitrite	$100 - 2500 \times 10^{-3}$	0.75×10^{-3}	71
MIL-101 (Cr)/rGO	DPV	4-nonylphenol	$0.1 - 12.5 \times 10^{-6}$	33×10^{-9}	73
Cu₃(btc)₂/CS-ERGO	DPV	HQ	$5 - 400 \times 10^{-6}$	0.44×10^{-6}	74
		CT	$2 - 200 \times 10^{-6}$	0.41×10^{-6}	
		RS	$1 - 200 \times 10^{-9}$	0.33×10^{-6}	
NH₂-MIL-125/rGO	DPV	BPA	$2 - 200 \times 10^{-6}$	0.7966×10^{-6}	76
Ni-MOF-GO	DPV	p-CNB	$0.1 - 300 \times 10^{-6}$	8.0×10^{-9}	77
AuNRs@ZIF-8	DPV	DCP	$0.01 - 15 \times 10^{-6}$	3.0×10^{-9}	78
		CBZ	$0.002 - 2.5 \times 10^{-6}$	0.33×10^{-9}	
		DU	$0.001 - 20 \times 10^{-6}$	0.26×10^{-9}	
3D Co-MOF/rGO	DPV	RS	$0.1 - 800 \times 10^{-6}$	1.9×10^{-8}	79

*Bisphenol A **(BPA)**; Hydroquinone **(HQ)**; Catechol **(CT)**; Resorcinol **(RS)**, Dichlorophen **(Dcp)**; p-chloronitrobenzene **(p-CNB)**; Carbendazim **(CBZ)**.*

Gao *et al*, reported the sensitive determination of toxic organic pollutant p-chloronitrobenzene (p-CNB) using Ni based MOF (Ni-MOF) and GO. The composite sensor showed superior electrocatalytic performance towards the oxidation of p-CNB in a concentration range of 0.1 to 300×10^{-6} M, and LOD 8.0×10^{-9} M (S/N = 3) [77]. Gan *et al.* reported the AuNRs@ZIF-8 with encapsulated in GO nanosheets to enhance the chemical resistance of the multicore–shell support. The composite material was modified on electrode surface and then sensor showed wide linear range from 0.01 to 15×10^{-6} M for DCP, from 0.002 to 2.5×10^{-6} M for CBZ, and from 0.001 to 20×10^{-6} M for DU with LOD are 3.0, 0.33, and 0.26×10^{-9} M for DCP, CBZ, and DU, respectively [78]. Topcu *et al.* demonstrated the 3D Co-MOF with rGO (Co-MOF/rGO) composite paper was prepared with a facile electrochemical deposition of Co-MOF on the surface of rGO paper electrode. And then utilized for electrocatalytic application of RS with a large linear range from $0.1–800 \times 10^{-6}$ M and a LOD 1.9×10^{-8} M [79]. The performance of the MOF with graphene based composite fabricated electrochemical sensors was compared and the results are given in Table 3 [66,68,70,71,73,74,76-79].

4.   Challenges and opportunity

The above research on the MOF with graphene-based nanocomposite fabricated electrochemical sensors results showed that MOFs are becoming a very effective tool for electrochemical sensing applications [80-82]. Generally, MOF materials are less conductance and their electrical conductivity and redox activity must be improved to enhance their performance. The MOFs with a variety of graphene oxide composites for facilitate an excellent conductivity to the composite material. Finally, most of the MOFs with graphene-based nanocomposites were highly exploited towards the determination of biologically important compounds. However, their utility in the detection of toxic metal ion, environmental pollutants, nitro compounds and hazards vapor with solvents are still not elaborated in the research. Hence, their application in the electrochemical sensing of nitro compound, environmentally pollutant, toxic anion and toxic chemicals is very limited in the literature. Thus, the utilization of the MOFs with graphene based nanostructures in the electrochemical sensing of the aforementioned compounds will make a new platform with excellent sensitivity in the near future.

Conclusions

In this chapter, we discussed the various MOFs with graphene-based nanocomposites for electrochemical sensing of toxic chemicals, environmental pollutants, nitro compounds, heavy metal ions, hazards solvents and vapors. The MOFs with graphene oxide, reduced graphene oxide and other graphene-based nanocomposites have been synthesized by various synthetic approaches like solvothermal, reflux, stirring, microwave and sonochemical. Initially, electrochemical sensing application of heavy metal, toxic compounds, nitro compounds and environmentally hazards using MOF modified electrodes. Then MOFs with graphene oxide based nanocomposites utilized to enhance the electrocatalytic sensing application through synergetic effect of both components. It has been detected in many studies that the MOFs with graphene based nanocomposites can enhance the electrocatalytic activities for different species like metal ions, anions, pesticides, environmental pollutants, nitro compounds and important biologically important compounds. In recent years has been witnessed significant interest in the development of MOFs with graphene-based nanocomposite for electrochemical sensors. The electrodes modified with different functionalized MOFs with graphene-based nanocomposites introduced in this review were generally seen to have improved sensitivities than the ones modified with MOF, graphene oxide, reduced graphene oxide and bare electrodes. This may be due to the fact that the nanocomposite modified electrodes had increased electrochemical surface coverage of the MOFs and graphene based modified electrodes. Many of the MOFs with different graphene-based nanocomposites discussed in this analysis showed good sensitive and selective towards the analytes due to the synergetic effect, π–π interactions of analytes with compound and active surface area. Finally, MOFs with graphene based nanocomposites were highly exploited towards the determination of toxic chemicals, environmental pollutants, nitro compounds and carcinogenic compounds.

References

[1] S-Y. Ding, W. Wang, Covalent organic frameworks (COFs): from design to applications, Chem. Soc. Rev., 42 (2013) 548-568. https://doi.org/10.1039/C2CS35072F

[2] X. Feng, X. Ding, D. Jiang, Covalent organic frameworks, Chem. Soc. Rev., 41 (2012) 6010 - 6022. https://doi.org/10.1039/c2cs35157a

[3] Q-L. Zhu, Q. Xu, Metal–organic framework composites, Chem. Soc. Rev., 43 (2014) 5468 - 5512. https://doi.org/10.1039/C3CS60472A

[4] P.J. Waller, F. Gasndara, O.M. Yaghi, Chemistry of covalent organic frameworks, Acc. Chem. Res., 48 (2015) 3053 – 3063. https://doi.org/10.1021/acs.accounts.5b00369

[5] B.R. Pimentel, A.W. Fultz, K.V. Presnell, R.P. Lively, Synthesis of water-sensitive metal–organic frameworks within fiber sorbent modules, Ind. Eng. Chem. Res. 56 (2017) 5070-5077. https://doi.org/10.1021/acs.iecr.7b00630

[6] S. Yuan, L. Feng, K. Wang, J. Pang, M. Bosch, C. Lollar, Y. Sun, J. Qin, X. Yang, P. Zhang, Q.Wang, L. Zou, Y. Zhang, L. Zhang, Y. Fang, J. Li, H-C. Zhou, Stable

metal–organic frameworks: design, synthesis, and applications, Adv. Mater., 30 (2018) 1704303. https://doi.org/10.1002/adma.201704303

[7] M. Alhamami, H. Doan, C-H. Cheng, A review on breathing behaviors of metal-organic-frameworks (MOFs) for gas adsorption, Materials, 7 (2014) 3198 - 3250. https://doi.org/10.3390/ma7043198

[8] Y. Xue, S. Zheng, H. Xue, H. Pang, Metal–organic framework composites and their electrochemical applications, J. Mater. Chem. A, 7 (2019) 7301 - 7327. https://doi.org/10.1039/C8TA12178H

[9] R.F. Mendes, F.A. Almeida Paz, Transforming metal–organic frameworks into functional materials, Inorg. Chem. Front., 2 (2015) 495 - 509. https://doi.org/10.1039/C4QI00222A

[10] W. Huang, Y. Jiang, X. Li, X. Li, J. Wang, Q. Wu, X. Liu, Solvothermal synthesis of microporous, crystalline covalent organic framework nanofibers and their colorimetric nanohybrid structures, ACS Appl. Mater. Interfaces, 5 (2013) 8845 - 8849. https://doi.org/10.1021/am402649g

[11] L.K. Ritchie, A. Trewin, A.R. Galan, T. Hasell, A.I. Cooper, Synthesis of COF-5 using microwave irradiation and conventional solvothermal routes, Microporous and Mesoporous Mat., 132 (2010) 132 - 136. https://doi.org/10.1016/j.micromeso.2010.02.010

[12] J. Cravillon, C.A. Schroder, H. Bux, A. Rothkirch, J. Caro, M. Wiebcke, Formate modulated solvothermal synthesis of ZIF-8 investigated using time-resolved in situ X-ray diffraction and scanning electron microscopy, Cryst Eng Comm, 14 (2012) 492 - 498. https://doi.org/10.1039/C1CE06002C

[13] Y. Ban, Y. Li, X. Liu, Y. Peng, W. Yang, Solvothermal synthesis of mixed-ligand metal–organic framework ZIF-78 with controllable size and morphology, Microporous and Mesoporous Mat., , 173 (2013) 29 – 36. https://doi.org/10.1016/j.micromeso.2013.01.031

[14] L.K. Ritchie, A. Trewin, A.R. Galan, T. Hasell, A.I. Cooper, Synthesis of COF-5 using microwave irradiation and conventional solvothermal routes, Microporous and Mesoporous Mat., 132 (2010) 132 - 136. https://doi.org/10.1016/j.micromeso.2010.02.010

[15] J. Klinowski, F.A. Almeida Paz, P. Silva, J. Rocha, Microwave-assisted synthesis of metal–organic frameworks, Dalton Trans., 40 (2011) 321 - 330. https://doi.org/10.1039/C0DT00708K

[16] F. Hillman, J.M. Zimmerman, S-M. Paek, M.R.A. Hamid, W.T. Lim, H-K. Jeong, Rapid microwave-assisted synthesis of hybrid zeolitic–imidazolate frameworks with mixed metals and mixed linkers, J. Mater. Chem. A, 5 (2017) 6090 - 6099. https://doi.org/10.1039/C6TA11170J

[17] C. Vaitsis, G. Sourkouni, C. Argirusis, Metal organic frameworks (MOFs) and ultrasound: A review, Ultrason. Sonochem., 52 (2019) 106 - 119. https://doi.org/10.1016/j.ultsonch.2018.11.004

[18] S-T. Yang, J. Kim, H-Y. Cho, S. Kim, W-S. Ahn, Facile synthesis of covalent organic frameworks COF-1 and COF-5 by sonochemical method, RSC Adv., 2 (2012) 10179 - 10181. https://doi.org/10.1039/c2ra21531d

[19] H-Y. Cho, J. Kim, S-N. Kim, W-S. Ahn, High yield 1-L scale synthesis of ZIF-8 via a sonochemical route, Microporous and Mesoporous Mat., 169 (2013) 180 - 184. https://doi.org/10.1016/j.micromeso.2012.11.012

[20] W-J. Li, M. Tu, R. Cao and R.A. Fischer, Metal–organic framework thin films: electrochemical fabrication techniques and corresponding applications & perspectives, J. Mater. Chem. A, 4 (2016) 12356 - 12369. https://doi.org/10.1039/C6TA02118B

[21] W-J. Li, J. Lu, S-Y. Gao, Q-H. Li, R. Cao, Electrochemical preparation of metal–organic framework films for fast detection of nitro explosives, J. Mater. Chem. A, 2 (2014) 19473 - 19478. https://doi.org/10.1039/C4TA04203D

[22] L. Ji, J. Wang, K. Wu, N. Yang, Tunable electrochemistry of electrosynthesized copper metal–organic frameworks, Adv. Funct. Mater., 28 (2018) 1706961. https://doi.org/10.1002/adfm.201706961

[23] S.D. Worrall, H.Mann, A. Rogers, M.A. Bissett, M.P. Attfield, R.A.W. Dryfe, Electrochemical deposition of zeolitic imidazolate framework electrode coatings for supercapacitor electrodes, Electrochim. Acta, 197 (2016) 228 - 240. https://doi.org/10.1016/j.electacta.2016.02.145

[24] C. Lu, T. Ben, S. Xu, S. Qiu, Electrochemical synthesis of a microporous conductive polymer based on a metal–organic framework thin film, Angew. Chem. Int. Ed., 53 (2014) 1. https://doi.org/10.1002/anie.201402950

[25] D. Braga, S.L. Giaffreda, F. Grepioni, A. Pettersen, L. Maini, M. Curzia, M. Polito, Mechanochemical preparation of molecular and supramolecular organometallic materials and coordination networks, Dalton Trans., (2006) 1249 - 1263. https://doi.org/10.1039/b516165g

[26] S.L. James, C.J. Adams, C. Bolm, D. Braga, P. Collier, T. Friščić, F. Grepioni, K.D.M. Harris, G. Hyett, W. Jones, A. Krebs, J. Mack, L. Maini, A.G. Orpen, I.P. Parkin, W.C. Shearouse, J.W. Steed, D.C. Waddell, Mechanochemistry: opportunities for new and cleaner synthesis, Chem. Soc. Rev., 41 (2012) 413 - 447. https://doi.org/10.1039/C1CS15171A

[27] M. Klimakow, P. Klobes, A.F. Thünemann, K. Rademann, F. Emmerling, Mechanochemical synthesis of metal–organic frameworks: A fast and facile approach toward quantitative yields and high specific surface areas, Chem. Mater., 22 (2010) 5216 - 5221. https://doi.org/10.1021/cm1012119

[28] B.V. Harbuzaru, A. Corma, F. Rey, P. Atienzar, J.L. Jord, H. Garc, D. Ananias, L.D. Carlos, J. Rocha, Metal–organic nanoporous structures with anisotropic photoluminescence and magnetic properties and their use as sensors, Angew. Chem. Int. Ed. 47 (2008) 1080 –1083. https://doi.org/10.1002/anie.200704702

[29] X-W. Liu, T-J. Sun, J-L. Hu, S-D. Wang, Composites of metal–organic frameworks and carbon-based materials: preparations, functionalities and applications, J. Mater. Chem. A, 4 (2016) 3584 - 3616. https://doi.org/10.1039/C5TA09924B

[30] S. Li, F. Huo, Metal–organic framework composites: from fundamentals to applications, Nanoscale, 7 (2015) 7482 - 7501. https://doi.org/10.1039/C5NR00518C

[31] P. Wen, P. Gong, J. Sun, J. Wang, S. Yang, Design and synthesis of Ni-MOF/CNT composites and rGO/carbon nitride composites for an asymmetric supercapacitor with high energy and power density, J. Mater. Chem. A, 3 (2015) 13874 – 13883. https://doi.org/10.1039/C5TA02461G

[32] X. Fang, B. Zong, S. Mao, Metal–organic framework-based sensors for environmental contaminant sensing, Nano-Micro Lett. 10 (2018) 64. https://doi.org/10.1007/s40820-018-0218-0

[33] M. Cui, J. Li, D. Lu, Z. Shao, Development of a metal-organic framework for the sensitive determination of 2,4-Dichlorophenol, Int. J. Electrochem. Sci., 13 (2018) 3420 – 3428. https://doi.org/10.20964/2018.04.48

[34] Y. Song, M. Xu, Z. Li, L. He, M. Hu, L. He, Z. Zhang, M. Du, A bimetallic CoNi-based metal−organic framework as efficient platform for label-free impedimetric sensing toward hazardous substances, Sens. Actuators B: Chem., 311 (2020) 127927. https://doi.org/10.1016/j.snb.2020.127927

[35] Y. Li, T. Xia, J. Zhang, Y. Cui, B. Li, Y. Yang, G. Qian, A manganese-based metal-organic framework electrochemical sensor for highly sensitive cadmium ions detection, J. Solid State Chem., 275 (2019) 38–42. https://doi.org/10.1016/j.jssc.2019.03.051

[36] M. Sohail, M. Altaf, N. Baig, R. Jamil, M. Sherc and A. Fazald, A new water stable zinc metal organic framework as an electrode material for hydrazine sensing, New. J. Chem., 42 (2018) 12486 - 12491. https://doi.org/10.1039/C8NJ01507D

[37] X. Wang, Y. Qi, Y. Shen, Y. Yuan, L. Zhang, C. Zhang, Y. Sun, A ratiometric electrochemical sensor for simultaneous detection of multiple heavy metal ions based on ferrocene-functionalized metal-organic framework, Sens. Actuators B: Chem., 310 (2020) 127756. https://doi.org/10.1016/j.snb.2020.127756

[38] J. Zhang, X. Xu, L. Chen, An ultrasensitive electrochemical bisphenol A sensor based on hierarchical Ce-metal-organic framework modified with cetyltrimethylammonium bromide, Sens. Actuators B: Chem., 261 (2018) 425 - 433. https://doi.org/10.1016/j.snb.2018.01.170

[39] L. Ji, Q. Cheng, K. Wu, X. Yang, Cu-BTC frameworks-based electrochemical sensing platform for rapid and simple determination of Sunset yellow and Tartrazine, Sens. Actuators B Chem., 231 (2016) 12 -17. https://doi.org/10.1016/j.snb.2016.03.012

[40] Z. Wang, B. Ma, C. Shen, L-Z. Cheong, Direct, selective and ultra-sensitive electrochemical bio sensing of methyl parathion in vegetables using Burkholderia cepacia lipase @ MOF nanofibers based biosensor, Talanta 197 (2019) 356–362. https://doi.org/10.1016/j.talanta.2019.01.052

[41] M. Roushani, A. Valipour, Z. Saedi, Electroanalytical sensing of Cd^{2+} based on metal−organic framework modified carbon paste electrode, Sens. Actuators B Chem., 233 (2016) 419–425. https://doi.org/10.1016/j.snb.2016.04.106

[42] W. Ye, Y. Li, J. Wang, B. Li, Y. Cui, Y. Yang, G. Qian, Electrochemical detection of trace heavy metal ions using a Ln-MOF modified glass carbon electrode, J. Solid State Chem., 281 (2020) 121032. https://doi.org/10.1016/j.jssc.2019.121032

[43] Z. Zeng, X. Fang, W. Miao, Y. Liu, T. Maiyalagan, S. Mao, Electrochemically sensing of trichloroacetic acid with Iron(II) phthalocyanine and Zn-based metal organic framework nanocomposites, ACS Sens. 4 (2019) 1934 - 1941. https://doi.org/10.1021/acssensors.9b00894

[44] Z. Zhang, H. Ji, Y. Song, S. Zhang, M. Wang, C. Jia, J-Y. Tian, L. He, X. Zhang, C-S. Liu, Fe(III)-based metal–organic framework-derived core–shell nanostructure: sensitive electrochemical platform for high trace determination of heavy metal ions, Biosens. Bioelectron., 94 (2017) 358 – 364. https://doi.org/10.1016/j.bios.2017.03.014

[45] H. Guo, Z. Zheng, Y. Zhang, H. Lin, Q. Xu, Highly selective detection of Pb^{2+} by a nanoscale Ni-based metal-organic framework fabricated through one-pot hydrothermal reaction, Sens. Actuators B: Chem., 248 (2017) 430 - 436. https://doi.org/10.1016/j.snb.2017.03.147

[46] C-W. Kung, T-H. Chang, L-Y. Chou, J. T. Hupp, O.K. Farha, K-C. Ho, Porphyrin-based metal-organic framework thin films for electrochemical nitrite detection, Electrochem Commun., 58 (2015) 51 – 56. https://doi.org/10.1016/j.elecom.2015.06.003

[47] M.U.A. Prathap, S. Gunasekaran, Rapid and scalable synthesis of zeolitic imidazole framework (ZIF-8) and its use for the detection of trace levels of nitroaromatic explosives, Adv. Sustainable Syst., 2 (2018) 1800053. https://doi.org/10.1002/adsu.201800053

[48] C-H. Su, C-W. Kung, T-H. Chang, H-C. Lu, K-C. Ho, Y-C. Liao, Inkjet-printed porphyrinic metal-organic framework thin films for electrocatalysis, J. Mater. Chem. A, 4 (2016) 11094 - 11102. https://doi.org/10.1039/C6TA03547G

[49] C. Bao, Q. Niu, Z-A. Chen, X. Cao, H. Wang, W. Lu, Ultrathin nickel-metal–organic framework nanobelt based electrochemical sensor for the determination of urea in human body fluids, RSC Adv., 9 (2019) 29474 - 29481. https://doi.org/10.1039/C9RA05716A

[50] P. Arul, S.A. John, Size controlled synthesis of Ni-MOF using polyvinylpyrrolidone: New electrode material for the trace level determination of nitrobenzene, J. Electroanal. Chem., 829 (2018) 168 – 176. https://doi.org/10.1016/j.jelechem.2018.10.014

[51] C. Li, Y. Zhou, X. Zhu, B. Ye, M. Xu, Construction of a sensitive bisphenol A electrochemical sensor based on metal-organic framework/graphene composites, Int. J. Electrochem. Sci., 13 (2018) 4855 – 4867. https://doi.org/10.20964/2018.05.52

[52] M. Saraf, R. Rajak, S.M. Mobin, A fascinating multitasking Cu-MOF/rGO hybrid for high performance supercapacitors and highly sensitive and selective electrochemical nitrite sensors, J. Mater. Chem. A, 4 (2016) 16432 - 16445. https://doi.org/10.1039/C6TA06470A

[53] N. Karimian, H. Fakhri, S. Amidi, A. Hajian, F. Arduinie, H. Bagheri, A novel sensing layer based on metal–organic framework UiO-66 modified with TiO_2–graphene oxide: application to rapid, sensitive and simultaneous determination of paraoxon and chlorpyrifos, New J. Chem., 43 (2019) 2600 - 2609. https://doi.org/10.1039/C8NJ06208K

[54] S.K. Bhardwaj, G.C. Mohanta, A.L. Sharma, K-H. Kim, A. Deep, A three-phase copper MOF-graphene-polyaniline composite for effective sensing of ammonia, Analytica Chimica Acta., 1043 (2018) 89 - 97. https://doi.org/10.1016/j.aca.2018.09.003

[55] M. Lu, Y. Deng, Y. Luo, J. Lv, T. Li, J. Xu, S-W. Chen, J. Wang, Graphene aerogel−metal−organic framework-based electrochemical method for simultaneous detection of multiple heavy-metal ions, Anal. Chem., 91 (2019) 888−895. https://doi.org/10.1021/acs.analchem.8b03764

[56] X. Li, C. Li, C. Wu, K. Wu, Strategy for highly sensitive electrochemical sensing: In situ coupling of a metal−organic framework with ball-mill-exfoliated graphene, Anal. Chem. 91 (2019) 6043 − 6050. https://doi.org/10.1021/acs.analchem.9b00556

[57] J. Li, J. Xia, F. Zhang, Z. Wang, Q. Liu, An electrochemical sensor based on copper-based metal-organic frameworks-graphene composites for determination of dihydroxybenzene isomers in water, Talanta, 181 (2018) 80-86. https://doi.org/10.1016/j.talanta.2018.01.002

[58] M. Baghayeri, M.G. Motlagh, R. Tayebee, M. Fayazi, F. Narenji, Application of graphene/zinc-based metal-organic framework nanocomposite for electrochemical sensing of As(III) in water resources, Anal Chim Acta 1099 (2020) 60 - 67. https://doi.org/10.1016/j.aca.2019.11.045

[59] M. Peng, G. Guan, H. Deng, B. Han, C. Tian, J. Zhuang, Y. Xu, W. Liu, Z. Lin, PCN-224/rGO nanocomposite based photoelectrochemical sensor with intrinsic

recognition ability for efficient p-arsanilic acid detection, Environ. Sci.: Nano, 6 (2019) 207. https://doi.org/10.1039/C8EN00913A

[60] T. Gan, J. Li, H. Li, Y. Liu, Z. Xu, Synthesis of Au nanorod-embedded and graphene oxide-wrapped microporous ZIF-8 with high electrocatalytic activity for the sensing of pesticides, Nanoscale, 11 (2019) 7839 - 7849. https://doi.org/10.1039/C9NR01101C

[61] D. Ding, Q. Xue, W. Lu, Y. Xiong, J. Zhang, X. Pan, B. Tao, Chemically functionalized 3D reticular graphene oxide frameworks decorated with MOF-derived Co_3O_4: Towards highly sensitive and selective detection to acetone, Sens. Actuators B: Chem., 259 (2018) 289 – 298. https://doi.org/10.1016/j.snb.2017.12.074

[62] N.A. Travlou, K. Singh, E.R. Castellon, T.J. Bandosz, Cu–BTC MOF–graphene-based hybrid materials as low concentration ammonia sensors, J. Mater. Chem. A, 3 (2015) 11417 – 11429. https://doi.org/10.1039/C5TA01738F

[63] Q. Chen, X. Li, X. Min, D. Cheng, J. Zhou, Y. Li, Z. Xie, P. Liu, W. Cai, C. Zhang, Determination of catechol and hydroquinone with high sensitivity using MOF-graphene composites modified electrode, J. Electroanal Chem., 789 (2017) 114 – 122. https://doi.org/10.1016/j.jelechem.2017.02.033

[64] H. Wang, Q. Hu, Y. Meng, Z. Jin, Z. Fang, Q. Fu, W. Gao, L. Xu, Y. Song, F. Lu, Efficient detection of hazardous catechol and hydroquinone with MOF-rGO modified carbon paste electrode, J. Hazardous Mater., 353 (2018) 151–157. https://doi.org/10.1016/j.jhazmat.2018.02.029

[65] Y. Wang, W. Cao, L. Wang, Q. Zhuang, Y. Ni, Electrochemical determination of 2,4,6-trinitrophenol using a hybrid film composed of a copper-based metal organic framework and electroreduced graphene oxide, Microchim Acta 185 (2018) 315. https://doi.org/10.1007/s00604-018-2857-8

[66] S. Rani, S. Kapoor, B. Sharma, S. Kumar, R. Malhotra, N. Dilbaghi, Fabrication of Zn-MOF@rGO based sensitive nanosensor for the real time monitoring of hydrazine, J. Alloys and Compounds 816 (2020) 152509. https://doi.org/10.1016/j.jallcom.2019.152509

[67] T.T. Tung, M.T. Tran, J-F. Feller, M. Castro, T.V. Ngo, K. Hassan, M.J. Nine, D. Losic, Graphene and metal organic frameworks (MOFs) hybridization for tunable chemoresistive sensors for detection of volatile organic compounds (VOCs)

biomarkers, Carbon, 159 (2020) 333 - 344.
https://doi.org/10.1016/j.carbon.2019.12.010

[68] J. Gu, X. Yin, X. Bo, L. Guo, High performance electrocatalyst based on MIL-101 (Cr)/reduced graphene oxide composite: facile synthesis and electrochemical detections, ChemElectroChem., 5 (2018) 1 – 10.
https://doi.org/10.1002/celc.201800588

[69] Z. Yu, N. Li, X. Hu, Y. Dong, Y. Lin, H. Cai, Z. Xie, D. Qu, X. Li, Highly efficient electrochemical detection of lead ion using metal-organic framework and graphene as platform based on DNAzyme, Synthetic Metals 254 (2019) 164 – 171.
https://doi.org/10.1016/j.synthmet.2019.06.017

[70] X. Tu, Y. Xie, X. Ma, F. Gao, L. Gong, D. Wang, L. Lu, G. Liu, Y. Yu, X. Huang, Highly stable reduced graphene oxide-encapsulated Ce-MOF composite as sensing material for electrochemically detecting dichlorophen, J. Electroanal. Chem, 848 (2019) 113268. https://doi.org/10.1016/j.jelechem.2019.113268

[71] C-W. Kung, Y-S. Li, M-H. Lee, S-Y. Wang, W-H. Chiang, K-C. Ho, In situ growth of porphyrinic metal–organic framework nanocrystals on graphene nanoribbons for the electrocatalytic oxidation of nitrite, J. Mater. Chem. A, 4 (2016) 10673 - 10682.
https://doi.org/10.1039/C6TA02563C

[72] Y. Xie, X. Tu, X. Ma, M. Xiao, G. Liu, F. Qu, R. Dai, L. Lu, W. Wang, In-situ synthesis of hierarchically porous polypyrrole@ZIF-8/graphene aerogels for enhanced electrochemical sensing of 2, 2-methylenebis (4-chlorophenol), Electrochim. Acta 311 (2019) 114 - 122. https://doi.org/10.1016/j.electacta.2019.04.132

[73] Y. Zhang, P. Yan, Q. Wan, N. Yang, Integration of chromium terephthalate metal-organic frameworks with reduced graphene oxide for voltammetry of 4-nonylphenol, Carbon, 134 (2018) 540-547. https://doi.org/10.1016/j.carbon.2018.02.072

[74] Y. Yang, Q. Wang, W. Qiu, H. Guo, F. Gao, Covalent Immobilization of $Cu_3(btc)_2$ at chitosan−electroreduced graphene oxide hybrid film and its application for simultaneous detection of dihydroxybenzene isomers, J. Phys. Chem. C 120 (2016) 9794 − 9803. https://doi.org/10.1021/acs.jpcc.6b01574

[75] Y. Yan, X. Bo, L. Guo, MOF-818 metal-organic framework-reduced graphene oxide/multiwalled carbon nanotubes composite for electrochemical sensitive detection of phenolic acids, Talanta, 218 (2020) 121123.
https://doi.org/10.1016/j.talanta.2020.121123

[76] L.J. Ling, J.P. Xu, Y.H. Deng, Q. Peng, J.H. Chen, Y.S. He, Y.J. Nie, One-pot hydrothermal synthesis of amine functionalized metal–organic framework/ reduced graphene oxide composites for the electrochemical detection of bisphenol A, Anal. Methods, 10 (2018) 2722 - 2730. https://doi.org/10.1039/C8AY00052B

[77] J. Gao, P. He, T. Yang, X. Wang, L. Zhou, Q. He, L. Jia, H. Heng, H. Zhang, B. Jia, X. He, Short rod-like Ni-MOF anchored on graphene oxide nanosheets: A promising voltammetric platform for highly sensitive determination of p-chloronitrobenzene, J. Electroanal Chem, 861 (2020) 113954. https://doi.org/10.1016/j.jelechem.2020.113954

[78] T. Gan, J. Li, H. Li, Y. Liu, Z. Xu, Synthesis of Au nanorod-embedded and graphene oxide-wrapped microporous ZIF-8 with high electrocatalytic activity for the sensing of pesticides, Nanoscale, 11 (2019) 7839 - 7849. https://doi.org/10.1039/C9NR01101C

[79] E. Topcu, Three-dimensional, free-standing, and flexible cobalt-based metal-organic frameworks/graphene composite paper: A novel electrochemical sensor for determination of resorcinol, Mater Research Bulletin 121 (2020) 110629. https://doi.org/10.1016/j.materresbull.2019.110629

[80] P. Arul, N.S.K. Gowthaman, S.A. John, M. Tominaga, Tunable electrochemical synthesis of 3D nucleated microparticles like Cu-BTC MOF-carbon nanotubes composite: Enzyme free ultrasensitive determination of glucose in a complex biological fluid, Electrochim. Acta 254 (2020) 136673. https://doi.org/10.1016/j.electacta.2020.136673

[81] P. Arul, N.S.K. Gowthaman, S.A. John, H.N. Lim, Ultrasonic assisted synthesis of size-controlled Cu-metal–organic framework decorated graphene oxide composite: sustainable electrocatalyst for the trace-level determination of nitrite in environmental water samples, ACS Omega 5 (2020) 14242–14253. https://doi.org/10.1021/acsomega.9b03829

[82] P. Arul, S.A. John, Organic solvent free in situ growth of flower like Co-ZIF microstructures on nickel foam for glucose sensing and supercapacitor applications, Electrochim. Acta 306 (2019) 254-263. https://doi.org/10.1016/j.electacta.2019.03.117

Keyword Index

About the Editors

Dr. Alagarsamy Pandikumar is currently working as Scientist in Functional Materials Division, CSIR-Central Electrochemical Research Institute, Karaikudi, India. He obtained his Ph.D. in Chemistry (2014) from the Madurai Kamaraj University, Madurai and then successfully completed his post-doctoral fellowship tenure (2014-2016) at the University of Malaya, Malaysia under High Impact Research Grant. His current research involves development of novel materials with graphene, graphitic carbon nitride, in combination to metals, metal oxides, polymers and carbon nanotubes for energy conversion and storage and dye-sensitized solar cells applications. His results outcomes were documented in 119 peer-reviewed journals including 9 review articles and have more than 3600 citations with the h−index of 36. He served as Guest Editor for a special issue in Materials Focus journal and edited 11 books for reputed publishers.

Dr. Perumal Rameshkumar is currently working as an Assistant Professor of Chemistry at Kalasalingam Academy of Research and Education, India. He obtained his M.Sc. (chemistry) (2009) from Madurai Kamaraj University. He joined as Junior Research Fellow (2010) at the same University and subsequently was promoted to Senior Research Fellow (2012). His doctoral thesis focused on 'polymer encapsulated metal nanoparticles for sensor and energy conversion applications'. He worked as Post-Doctoral Research Fellow (2014) at University of Malaya, Malaysia in the field of 'graphene-inorganic nanocomposite materials for electrochemical sensor and energy conversion'. His current research interests include synthesis of functionalized nanomaterials, electrochemical sensors, energy-related electrocatalysis and photoelectrocatalysis. His research findings were documented in 34 peer reviewed journals including 01 review article. For his credit, he edited 03 books under Elsevier publications.

www.ingramcontent.com/pod-product-compliance
Lightning Source LLC
Chambersburg PA
CBHW071329210326
41597CB00015B/1380